# GRAVITATION AND COGRAVITATION

Oleg D. Jefimenko
WEST VIRGINIA UNIVERSITY

# GRAVITATION AND COGRAVITATION

DEVELOPING NEWTON'S THEORY OF GRAVITATION TO ITS PHYSICAL AND MATHEMATICAL CONCLUSION

Electret Scientific Company
Star City

Library of Congress Control Number: 2006909417

ISBN 0-917406-00-1 (hardcover)
ISBN 0-917406-15-X (paperback)

Printed in the United States of America

# PREFACE

Newton's theory of gravitation is the basic working theory of astronomers and of all the scientists dealing with space exploration and celestial mechanics. On the basis of Newton's theory of gravitation we determine the motion of planets and their satellites, predict the existence and celestial coordinates of planets not previously observed, launch artificial satellites and space ships. Nevertheless, Newton's theory of gravitation has serious defects. As far as its practical applications are concerned, it is incapable of accounting for certain fine details of planetary motion. As far as its conceptual content is concerned, it is a theory of "gravitational state" rather than a theory of "gravitational process," since it does not provide any information on the temporal aspect of gravitation[1]. Furthermore, Newton's theory of gravitation cannot be reconciled with the principle of causality and with the law of conservation of momentum when it is applied to time-dependent gravitational systems.

And yet, the fundamental validity of Newton's theory of gravitation is indisputable and its essential reliability has been established beyond any doubt. It is plausible therefore that Newton's theory of gravitation is merely incomplete and requires a further development. The purpose of this book is to extend and to generalize Newton's theory of gravitation so as to make it free from the above defects and to make it fully applicable to all possible gravitational systems and interactions.

The starting point of the generalization of Newton's theory of gravitation presented in this book is the idea that gravitational interactions in time-dependent gravitational systems are mediated

---

[1]I am indebted to Professor Yu. G. Kosarev for an illuminating discussion of this property of Newton's theory.

by two force fields: the gravitational field proper created by all masses and acting upon all masses, and by the "cogravitational" field created by moving masses only and acting upon moving masses only. In accordance with the principle of causality, the two fields are represented by *retarded* field integrals, which, for static or slowly-varying gravitational systems, reduce to the ordinary Newtonian gravitational field integral. As the readers will see, the generalized Newtonian theory of gravitation developed on this basis yields extremely significant and far-reaching results.

An immediate consequence of the generalized Newtonian theory of gravitation is that gravitational interactions normally involve not just one single force of gravitational attraction, but at least four additional forces associated with velocities, accelerations and rotations of interacting bodies.

Another direct consequence of the generalized theory of gravitation is an astonishing complexity of gravitational interactions and a surprising variety of effects of gravitational interactions. Here are some examples: a fast-moving point mass passing a spherically-symmetric body causes the latter to rotate; a mass moving with rapidly-decreasing velocity exerts both an attractive and a repulsive force on neighboring bodies; a fast-moving mass passing a stationary mass exerts an explosion-like force on the latter; a rotating mass that is suddenly stopped causes neighboring bodies to rotate; the period of revolution of a planet or satellite is affected by the rotation of the central body.

The generalized theory of gravitation provides a large variety of methods for calculating gravitational interactions between bodies of all shapes and sizes. Among these methods are: calculations using gravitational-cogravitational force equations, calculations based on the gravitational-cogravitational field energy, calculations based on gravitational-cogravitational Maxwell's stress integral, direct calculations in terms of scalar and vector potentials without using gravitational or cogravitational fields.

The generalized theory of gravitation is fully compatible with the laws of conservation of energy and momentum. A very important result of this compatibility is the definitive explanation provided by the generalized theory of gravitation for the process of conversion of potential energy (field energy) into the kinetic energy of bodies falling under the action of a gravitational field.

The generalized theory of gravitation provides explanations for certain peculiarities of the motion of celestial bodies, and for the differential rotation of the Sun in particular.

The generalized theory of gravitation is compatible with the special theory of relativity. As a result, gravitational equations for stationary gravitational systems can be easily converted into the corresponding equations for moving gravitational systems, thus providing an additional method for analyzing and computing gravitational effects associated with moving bodies.

The generalized theory of gravitation predicts the existence of gravitation-cogravitational waves and explains how such waves can be generated.

The generalized theory of gravitation indicates the existence of a link between gravitation and electromagnetism by showing that beams of electromagnetic radiation (light beams) are deflected and bent by gravitational fields. This means that a gravitational field can be regarded as a medium whose index of refraction is larger than that of a pure vacuum in the absence of a gravitational field. Since the index of refraction is associated with the permittivity and permeability of the medium, and since electromagnetic forces are affected by permittivity and permeability, electromagnetic forces become weaker in the presence of gravitational fields, and electromagnetic processes (such as the rate of electromagnetic clocks, for example) become slower when taking place in gravitational fields.

The generalized theory of gravitation also indicates the existence of antigravitational (repulsive) fields and mass formations. A cosmological consequence of such fields and mass formations is a periodic expansion and contraction of the Universe. Another consequence is that the actual mass of the Universe may be much larger than the mass revealed by analyzing gravitational attraction in the galaxies, since antigravitational mass formations do not attract other masses.

# Part II APPLICATIONS AND PREDICTIONS OF THE GENERALIZED THEORY OF GRAVITATION

## 13 SOME ELEMENTARY APPLICATIONS OF THE GENERALIZED THEORY OF GRAVITATION



## 14 TORQUE EXERTED BY A MOVING MASS ON A STATIONARY MASS



## 15 MORE ABOUT ORBITAL MOTION AND ROTATION

# I

# NEWTON'S GRAVITATIONAL THEORY GENERALIZED TO TIME-DEPENDENT SYSTEMS

# 1

# NEWTON'S THEORY OF GRAVITATION AND THE NEED FOR ITS FURTHER DEVELOPMENT

In this chapter we shall summarize Newton's theory of gravitation and shall analyze it from the viewpoint of causality and from the viewpoint of the law of conservation of momentum. We shall find that it has two major defects: when applied to moving or time-dependent systems, it violates the principle of causality and violates the conservation of momentum law. We shall then discuss the means for correcting these defects.

## 1-1. Newton's Theory of Gravitation

Newton's theory of gravitation is based on the gravitational force law

$$F = G\frac{mM}{r^2}, \tag{1-1.1}$$

where $F$ is the force with which two point masses $m$ and $M$ attract each other, $G$ is the universal constant of gravitation, and $r$ is the distance between the two masses. In vector notation, the gravitational force law, Eq. (1-1.1), can be written as

$$\mathbf{F}/m = -G\frac{M}{r^2}\mathbf{r}_u, \tag{1-1.2}$$

where $\mathbf{F}$ is the force exerted on the point mass $m$ by the point mass $M$, $G$ and $r$ are as before, and $\mathbf{r}_u$ is the unit vector directed from $M$ to $m$; the minus sign indicates that the force is directed toward the mass exerting the force.

In modern presentations, Newton's theory of gravitation is based not on Eq. (1-1.1) or (1-1.2) directly, but on two equations that formulate his theory as a force-field theory in terms of the gravitational field vector $\mathbf{g}$. These equations are

$$\nabla \times \mathbf{g} = 0 \tag{1-1.3}$$

and

$$\nabla \cdot \mathbf{g} = -4\pi G\rho. \tag{1-1.4}$$

The gravitational field vector $\mathbf{g}$ is defined as

$$\mathbf{g} = \mathbf{F}/m_t, \tag{1-1.5}$$

where $\mathbf{F}$ is the force exerted by the gravitational field on a test mass $m_t$ which is at rest in an inertial reference frame ("laboratory"). In Eq. (1-1.4), $\rho$ is the mass density defined as

$$\rho = dm/dV', \tag{1-1.6}$$

where $dm$ is a mass element contained in the volume element $dV'$.

With the help of Eqs. (1-1.5) and (1-1.6), Eq. (1-1.2) can be reformulated into a more general equation

$$\mathbf{g} = -G\int \frac{\rho}{r^3}\mathbf{r}dV', \tag{1-1.7}$$

where $\mathbf{g}$ is the gravitational field created by the mass $m$ distributed in space with density $\rho$, $r = [(x - x')^2 + (y - y')^2 + (z - z')^2]^{1/2}$ is the distance from the *source point* $(x', y', z')$, where the volume element of integration $dV'$ is located, to the *field point* $(x, y, z)$, where $\mathbf{g}$ is being observed or computed, $\mathbf{r}$ is the radius vector directed from $dV'$ to the field point. The integral is extended over the region of space occupied by the mass $m$; the minus in front of the integral indicates that the field **g** is directed toward the mass that creates this field.

The force **F** acting on a mass of density $\rho$ located in the gravitational field **g** is then found from the equation

$$\mathbf{F} = \int \rho \mathbf{g} dV, \tag{1-1.8}$$

where the integration is over the space occupied by the mass experiencing the force.

For practical applications of Newton's theory, and in celestial mechanics in particular, the gravitational field vector **g** is seldom computed directly. Instead, one usually computes the gravitational potential $\varphi$, connected with the field vector **g** by the equation

$$\mathbf{g} = -\nabla\varphi, \tag{1-1.9}$$

and connected with the mass density $\rho$ by the equation

$$\nabla^2\varphi = 4\pi G\rho \tag{1-1.10}$$

obtained by substituting Eq. (1-1.9) into Eq. (1-1.4). Integrating Eq. (1-1.10), one obtains the equation

$$\varphi = -G\int \frac{\rho}{r} dV', \tag{1-1.11}$$

from which $\varphi$ can be found directly in terms of $\rho$. For a point mass $m$, Eq. (1-1.11) reduces to

$$\varphi = -G\frac{m}{r}. \tag{1-1.12}$$

## 1-2. Newton's Gravitational Theory and Causality

One of the most important tasks of physics is to establish causal relations between physical phenomena. No physical theory can be complete unless it provides a clear statement and description of causal links involved in the phenomena encompassed by that theory. In establishing and describing causal relations it is important not to confuse equations which we call "basic laws" with "causal equations." A "basic law" is an equation (or a system of equations) from which we can derive most (hopefully all) possible correlations between the various quantities involved in a particular group of phenomena subject to the "basic law." A "causal equation," on the other hand, is an equation that unambiguously relates a quantity representing an effect to one or more quantities representing the cause of this effect. Clearly, a "basic law" need not constitute a causal relation, and an equation depicting a causal relation may not necessarily be among the "basic laws" in the above sense.

Causal relations between phenomena are governed by the *principle of causality*. According to this principle, all present phenomena are exclusively determined by past events. Therefore equations depicting causal relations between physical phenomena must, in general, be equations where a present-time quantity (the effect) relates to one or more quantities (causes) that existed at some previous time. An exception to this rule are equations constituting causal relations by definition; for example, if force is defined as the cause of acceleration, then the equation $\mathbf{F} = m\mathbf{a}$, where $\mathbf{F}$ is the force and $\mathbf{a}$ is the acceleration, is a causal equation by definition.

In general, then, according to the principle of causality, an equation between two or more quantities simultaneous in time cannot represent a causal relation between these quantities because, according to this principle, the cause *must precede* its effect. Therefore the only kind of equations representing causal relations between physical quantities, other than equations representing cause and effect by definition, must be equations involving "retarded" (previous-time) quantities.

Let us apply these considerations to Newton's law of gravitation. Since neither of the Eqs. (1-1.1)-(1-1.12) is defined to be a causal relation, and since all these equations connect quantities simultaneous in time, neither of these equations represents a causal relation. In particular, Newton's gravitational law Eq. (1-1.1), even though it is a basic law, does not represent a cause-and-effect relation between the quantities involved. It is clear therefore that Newton's gravitational theory has a serious flaw.

## 1-3. Newton's Gravitational Theory and Conservation of Momentum

One of the most fundamental laws of mechanics is Newton's law of action and reaction. It is typically stated as follows: "Whenever a body exerts a force (action) on a second body, the second body exerts an equal and opposite force (reaction) on the first." However, in gravitational systems the law does not always hold. Consider the following example.

Suppose that a stationary mass is located in the gravitational field created by another, distant, stationary mass. The two masses exert upon each other equal and opposite forces, as required by the law of action and reaction. Suppose now that the first mass is allowed to move under the action of the field of the second mass and arrives at a new position. But the second mass, being far away from the first, does not yet "know" that the first mass has moved (because, by the principle of causality, a gravitational field cannot propagate instantaneously) and continues to experience the same force as before. The forces are now unequal in magnitude and direction, and the action and reaction law no longer holds!

In two-body systems, the law of action and reaction is equivalent to the law of the conservation of mechanical momentum. Therefore, if the law of action and reaction in a gravitational system involving a moving mass does not hold, then the mechanical momentum of the system is not conserved. Hence, Newton's gravitational law conflicts with the conservation of momentum law, which is one of the most fundamental laws of nature. This conflict constitutes another very serious flaw of Newton's theory of gravitation.

## 1-4. Is Newton's Gravitational Theory Wrong or Incomplete?

Newton's theory of gravitation is the basic working theory of astronomers and of all the scientists dealing with space exploration and celestial mechanics. The reliability of Newton's gravitational theory is indisputable. On the basis of this theory we determine the motion of planets and of their satellites, predict the existence and celestial coordinates of planets not previously observed, launch artificial satellites and space ships. Thus, even though Newton's gravitational theory does not agree with the principle of causality and conflicts with the conservation of momentum law, the essential validity of the theory has been established beyond any doubt. It is plausible, therefore, that Newton's theory of gravitation is merely incomplete and requires further development, but does not need to be replaced by another theory of gravitation. In particular, it must be refined and reformulated so as to satisfy the principle of causality and to comply with the momentum conservation law, without destroying the fundamental relations represented by Eqs. (1-1.1)-(1-1.12).

In order to reformulate Newton's gravitational theory in accordance with the principle of causality, we must establish causal gravitational equations that agree with Eqs. (1-1.1)-(1-1.12). What should be the form of such causal gravitational equations? Since an effect can be a combined or cumulative result of several causes, it is plausible that in causal equations a physical quantity representing an effect should be expressed in terms of *integrals* involving physical quantities representing the various causes of that effect. And since, by the principle of causality, the cause must precede its effect, the integrals in causal equations must be *retarded*, that is, the integrands in these integrals must involve quantities as they existed at a time prior to the time for which the quantity representing the effect is being computed. Thus, causal gravitational equations must involve retarded integrals.

However, as was explained above, the reformulated Newton's gravitational theory must also satisfy the law of conservation of momentum. It is well known that there exists a strong similarity between equations of Newton's gravitational theory and equations of electrostatics. It is also well known that in Maxwellian electromagnetic theory, the momentum conservation law is satisfied because time-dependent electromagnetic interactions involve not only the electric field but also the magnetic field. We may assume therefore that time-dependent gravitational interactions, just like electromagnetic interactions, involve not only the gravitational field, but also a second force field, not taken into account in Newton's theory. In fact, such a field was proposed in 1893 by Oliver Heaviside.[1] According to Heaviside, this second force field is created by moving masses only and acts exclusively on moving masses. We shall call it the "cogravitational," or "Heaviside's," field and shall denote it by the letter **K**.

As we shall presently see, by accepting the existence of the cogravitational field and by expressing the gravitational and cogravitational fields in terms of retarded integrals, it is possible to develop and reformulate Newton's single field theory of gravitation so that it becomes a special case of the reformulated theory of gravitation, and so that the reformulated theory satisfies both the principle of causality and the momentum conservation law. The reformulated theory basically generalizes Newton's original theory to gravitational systems involving moving and time-dependent masses. Accordingly, we shall call it the "generalized Newtonian theory of gravitation," or, simply, the "generalized theory of gravitation."

## References and Remarks for Chapter 1

**1.** Oliver Heaviside, "A Gravitational and Electromagnetic Analogy," The Electrician, **31**, 281-282 and 359 (1893). This article is reproduced in Oleg D. Jefimenko, *Causality, Electromagnetic Induction and Gravitation,* 2nd ed., (Electret Scientific, Star City, 2000) pp. 189-202.

# 2

# PHYSICAL AND MATHEMATICAL BASIS OF THE GENERALIZED THEORY OF GRAVITATION

In this chapter the conceptual content of the generalized theory of gravitation is described, the mathematical apparatus used for the formulations of the theory is explained, and the fundamental equations of the theory are presented. The main difference between Newton's original theory of gravitation and the generalized theory of gravitation is elucidated.

## 2-1. Conceptual Content of the Generalized Theory of Gravitation

The generalized theory of gravitation assumes that gravitational interactions are mediated by gravitational and cogravitational force fields.

A gravitational field is a region of space where a mass experiences a gravitational force. Quantitatively, a gravitational field is defined in terms of the gravitational field vector **g** by the same equation by which it is defined in Newton's theory:

$$\mathbf{g} = \mathbf{F}/m_t, \qquad (1\text{-}1.5)$$

where **F** is the force exerted by the gravitational field on a stationary test mass $m_t$.

A cogravitational field is a region of space where a mass experiences a cogravitational force. Quantitatively, a cogravitational field is defined in terms of the field vector **K** by the equation

$$\mathbf{F} = m_t(\mathbf{v} \times \mathbf{K}), \tag{2-1.1}$$

where **F** is the force exerted by the cogravitational field on a test mass $m_t$ moving with velocity **v**. As noted in Chapter 1, cogravitational fields are created by moving masses only and act upon moving masses only.[1,2]

It is assumed that both gravitational and cogravitationl fields propagate in space with finite velocity. This velocity is not yet known, but is believed to be equal to the velocity of light. However, the generalized theory of gravitation is compatible with a propagation velocity of gravitation different than the velocity of light and is not affected by the actual speed with which gravitation propagates.[3]

The generalized theory of gravitation agrees with the principle of causality because, as we shall presently see, in this theory the gravitational and cogravitational fields are expressed in terms of retarded integrals whose integrands are the causative sources of the fields.

The generalized theory of gravitation agrees also with the law of conservation of momentum because, according to this theory, gravitational-cogravitational fields are repositories of gravitational-cogravitational field momentum, and because mechanical momentum of a body moving in a gravitational-cogravitational field can be converted into the field momentum and the field momentum can be converted into the mechanical momentum of the body. As a result of this conversion, the sum of the mechanical and field momentum of the combined field-body system is always the same, and the total momentum of the system is thus conserved (see Chapter 8 for a general proof of momentum conservation in such systems).

According to the generalized theory of gravitation, gravitational-cogravitational fields are also repositories of field energy. Kinetic energy of a body moving in a gravitational-cogravitational field can be converted into the energy of the field, and the energy of the field can be converted into the kinetic energy of the body. As a result of this conversion, the sum of the mechanical and field energy of the combined field-body system is always the same, and the total energy of the system is thus conserved (see Chapter 8 for a general proof of energy conservation in such systems).

## 2-2. Fundamental Equations of the Generalized Theory of Gravitation

The two principal equations of the generalized theory of gravitation are the equations for the gravitational field **g** and the cogravitational field **K**:

$$\mathbf{g} = -G\int\left\{\frac{[\rho]}{r^3} + \frac{1}{r^2c}\left[\frac{\partial\rho}{\partial t}\right]\right\}\mathbf{r}\,dV' + \frac{G}{c^2}\int\frac{1}{r}\left[\frac{\partial(\rho\mathbf{v})}{\partial t}\right]dV' \quad (2\text{-}2.1)$$

and

$$\mathbf{K} = -\frac{G}{c^2}\int\left\{\frac{[\rho\mathbf{v}]}{r^3} + \frac{1}{r^2c}\frac{\partial[\rho\mathbf{v}]}{\partial t}\right\}\times\mathbf{r}\,dV', \quad (2\text{-}2.2)$$

where $G$, $\rho$, $r$, $\mathbf{r}$, and $dV'$ are the same as in Eq. (1-1.7), $\mathbf{v}$ is the velocity with which the mass distribution $\rho$ moves (the product $\rho\mathbf{v}$ constitutes the "mass-current density"), and $c$ is the velocity of the propagation of gravitation (usually assumed to be the same as the velocity of light). The square brackets in these equations are the retardation symbol indicating that the quantities between the brackets are to be evaluated for the "retarded" time, $t' = t - r/c$, where $t$ is the time for which $\mathbf{g}$ and $\mathbf{K}$ are evaluated. The integration in the integrals of Eqs. (2-2.1) and (2-2.2) is over all space (unless stated otherwise, all integrals in this book are over all space).

According to Eqs. (2-2.1) and (2-2.2), the gravitational field has three causative sources: the mass density $\rho$, the time derivative of $\rho$, and the time derivative of the mass-current density $\rho\mathbf{v}$; the cogravitational field has two causative sources: the mass-current density $\rho\mathbf{v}$ and the time derivative of $\rho\mathbf{v}$.

Let us note that for time-independent stationary masses Eq. (2-2.2) disappears and Eq. (2-2.1) becomes

$$\mathbf{g} = -G\int \frac{\rho}{r^3}\mathbf{r}dV', \tag{1-1.7}$$

that is, Eqs. (2-2.1) and (2-2.2) reduce to the gravitational field equation of Newton's theory of gravitation. Therefore, in the light of the generalized theory of gravitation, Newton's gravitational theory is an approximate theory in which the dependence of the gravitational interactions on the motion and temporal variations of interacting masses is not taken into account.

Equations (2-2.1) and (2-2.2) should be preferably considered as postulates. Therefore it is not necessary to discuss the original considerations that led to their formulation (these considerations can be found elsewhere[2]). The proof of their validity lies not in the considerations that led to their formulation, but rather in the agreement of all the known consequences of these equations with experimental and observational data within the limits of experimental errors imposed upon these data by the available techniques of measurements and observations.

It is important to note that although in Eqs. (2-2.1) and (2-2.2) the mass density, the mass current, and their derivatives are retarded, retardation can frequently be neglected, in which case these equations can be used with ordinary (unretarded) mass density, mass current, and their derivatives. Let us define the "characteristic time" of a gravitational-cogravitational system as the time $T$ during which the mass density, the mass current, or their temporal derivatives experience a significant change. For example, in the case of periodic variation of mass and mass current, $T$ may be assumed to be the period of the variation; in the case of planetary motion, $T$ may be assumed to be the period of revolution; and in the case of monotonously changing masses and mass currents, $T$ may be assumed to be the time during which the mass density, the mass current, or their temporal derivatives change by a factor of two. Let us now assume that the largest linear dimensions of the system under consideration is $L$. If $T$ and $L$ satisfy the relation

$$T \gg L/c, \tag{2-2.3}$$

then no significant change occurs in the system during the time that the gravitational or cogravitational "field signal" moves across the system, and therefore the retardation in the propagation of the gravitational or cogravitational fields within the system is negligible.

In addition to Eqs. (2-2.1) and (2-2.2) for the gravitational and cogravitational fields, the following equations constitute the mathematical foundation of the generalized theory of gravitation.

*The mass conservation equation ("continuity law")*:

$$\nabla \cdot (\rho \mathbf{v}) = -\frac{\partial \rho}{\partial t}, \tag{2-2.4}$$

or, in the integral form,

$$\oint \rho \mathbf{v} \cdot d\mathbf{S} = -\frac{\partial}{\partial t}\int \rho \, dV. \tag{2-2.5}$$

According to these equations, whenever a mass contained in a region of space diminishes or increases, there is an outflow or inflow of mass from or into this region.

*Force acting on a mass distribution of density ρ:*

$$\mathbf{F} = \int \rho(\mathbf{g} + \mathbf{v} \times \mathbf{K})\,dV, \tag{2-2.6}$$

where **v** is the velocity of $\rho$ and the integral is extended over the region of space containing the mass under consideration [for a stationary point mass this equation reduces to Eq. (1-1.8)].

*Density of the field energy contained in the gravitational-cogravitational field:*[4]

$$U_v = -\frac{1}{8\pi G}(\mathbf{g}^2 + c^2\mathbf{K}^2). \tag{2-2.7}$$

*Field energy contained in a region of the gravitational-cogravitational field:*[4]

$$U = -\frac{1}{8\pi G}\int(\mathbf{g}^2 + c^2\mathbf{K}^2)\,dV, \tag{2-2.8}$$

where the integration is extended over the region under consideration.

*Energy flow vector in the gravitational and cogravitational field ("gravitational Poynting vector"):*

$$\mathbf{P} = \frac{c^2}{4\pi G}\mathbf{K} \times \mathbf{g}. \tag{2-2.9}$$

This vector represents the direction and rate of gravitational-cogravitational energy flow per unit area at a point of space under consideration. Equation (2-2.9) together with Eqs. (2-2.1), (2-2.2), (2-2.4), and (2-2.7) ensures the conservation of energy in gravitational-cogravitational interactions.

*Density of the field momentum contained in the gravitational-cogravitational field:*

$$\mathbf{G}_{vf} = \frac{1}{4\pi G}\mathbf{K} \times \mathbf{g}. \tag{2-2.10}$$

*Field momentum contained in a region of the gravitational-cogravitational field:*

$$\mathbf{G}_f = \frac{1}{4\pi G}\int \mathbf{K}\times\mathbf{g}\,dV, \tag{2-2.11}$$

where the integration is extended over the region under consideration.

*Correlations between the mechanical momentum,* $\mathbf{G}_M$, *and the gravitational-cogravitational field momentum*:

$$\frac{d\mathbf{G}_M}{dt} = -\frac{1}{4\pi G}\int \frac{\partial}{\partial t}(\mathbf{K}\times\mathbf{g})\,dV + \frac{1}{4\pi G}\left[\frac{1}{2}\oint(\mathbf{g}^2 + c^2\mathbf{K}^2)d\mathbf{S} - \oint \mathbf{g}(\mathbf{g}\cdot d\mathbf{S}) - c^2\oint \mathbf{K}(\mathbf{K}\cdot d\mathbf{S})\right], \tag{2-2.12}$$

where **g** and **K** are the gravitational and cogravitational fields in the system under consideration. In this equation, the derivative on the left represents the rate of change of the momentum of a body located in a gravitational-cogravitationl field, the volume integral represents the rate of change of the field momentum in the region of the field where the body is located, and the surface integrals represent the flux of the field momentum through the surface enclosing the region under consideration. Together with Eqs. (2-2.1), (2-2.2), (2-2.4), (2-2.6), and (2-2.10) this equation ensures the conservation of momentum in gravitational-cogravitational interactions.

Equations (2-2.4), (2-2.6), (2-2.7), (2-2.9), (2-2.10) and (2-2.12) should preferably be considered as postulates, although Eq. (2-2.9) can be derived from Eqs (2-2.1), (2-2.2), (2-2.4), and (2-2.7) if conservation of energy is assumed to hold for gravitational-cogravitational interactions. Likewise, Eq. (2-2.12) can be derived from Eqs. (2-2.1), (2-2.2), (2-2.4), (2-2.7) and (2-2.11), if conservation of momentum is assumed to hold for gravitational-cogravitational interactions.

***Note on the mathematical apparatus and techniques used in the generalized theory of gravitation***. The mathematical apparatus used in the generalized theory of gravitation is mainly vector analysis, as specifically developed for field-theoretical applications in the author's book *Electricity and Magnetism*[5] and, for operation with retarded quantities, in the author's book *Electromagnetic Retardation and Theory of Relativity*[6]. Most of the mathematical operations used in the generalized theory of gravitation are simply transformations of vector-analytical expressions by means of vector identities. The vector identities used in this book are listed in the Appendix and are identified by the prefix "V".

Mathematical formulation of the generalized theory of gravitation is very similar to that of Maxwellian electrodynamics. Because of this similarity, many electromagnetic equations have their counterparts in the generalized theory of gravitation. As a result, it is possible to convert many electromagnetic equations to gravitational and cogravitational equations by a mere substitution of symbols. A table of corresponding electromagnetic and gravitational-cogravitational symbols for the substitution is presented in Chapter 7, and some particularly useful gravitational and cogravitational equations obtained by the substitution are also shown there.

▼

**Example 2-2.1** A thin, heavy circular ring of radius $a$ and cross-sectional area $s$ has a uniformly distributed mass $m$. At $t = 0$ the ring starts to rotate with constant angular acceleration $\alpha$ about its symmetry axis which is also the $x$ axis of rectangular coordinates (Fig. 2.1). Find the gravitational and cogravitational fields at a point $x$ on the axis for $t > 0$.

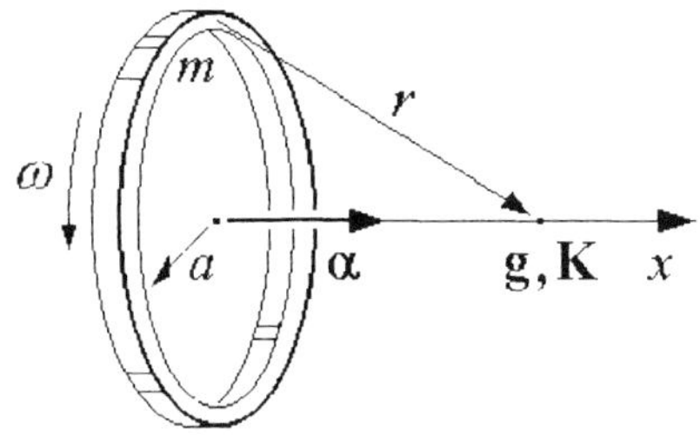


*Fig. 2.1 Calculation of the gravitational and cogravitational fields on the axis of a heavy ring rotating with angular acceleration* $\boldsymbol{\alpha}$.

For brevity, let us designate the mass-current density of the ring by the symbol **J**. We then have for the rotating ring $\mathbf{J} = \rho\mathbf{v} = \rho\omega a\boldsymbol{\theta}_u = \rho\alpha t a\boldsymbol{\theta}_u$, where $\rho$ is the mass density in the ring, $\omega$ is the angular velocity of the ring, and $\boldsymbol{\theta}_u$ is a unit vector in the circular direction (right-handed with respect to $x$). The time derivative of **J** is $\partial\mathbf{J}/\partial t = \rho\alpha a\boldsymbol{\theta}_u$. In terms of $m$, the mass-current density and the derivative are $\mathbf{J} = (m\alpha t/2\pi s)\boldsymbol{\theta}_u$ and $\partial\mathbf{J}/\partial t = (m\alpha/2\pi s)\boldsymbol{\theta}_u$.

To find the gravitational field of the ring, we use Eq. (2-2.1). Since $\partial\mathbf{J}/\partial t$ is in the circular direction, and since $r$ is the same for all points of the ring, the second integral in Eq. (2-2.1) makes no contribution to the gravitational field on the axis (the contributions of any two volume elements on the opposite ends of a diameter cancel each other). Since the mass density does not depend on time, the contribution of the first integral is

$$\mathbf{g} = -G\int\frac{\rho}{r^3}\mathbf{r}\,dV', \qquad (2\text{-}2.13)$$

which is identical with the expression for the gravitational field produced by a stationary mass of density $\rho$. Integrating Eq. (2-2.13), we obtain[7]

$$\mathbf{g} = -G\frac{mx}{(a^2+x^2)^{3/2}}\mathbf{i}. \qquad (2\text{-}2.14)$$

To find the cogravitational field, we use Eq. (2-2.2). Expressing [**J**] and [$\partial\mathbf{J}/\partial t$] in Eq. (2-2.2) in terms of $m$, $\alpha$, s, and

$\theta_u$, we have

$$\mathbf{K} = -\frac{G}{c^2}\int\left\{\frac{m\alpha(t - r/c)}{2\pi s r^3}\theta_u + \frac{m\alpha}{r^2 c 2\pi s}\theta_u\right\}\times \mathbf{r}\,dV'$$

$$= -\frac{G}{c^2}\int\left\{\frac{m\alpha t}{2\pi s r^3}\theta_u - \frac{m\alpha}{r^2 c 2\pi s}\theta_u + \frac{m\alpha}{r^2 c 2\pi s}\theta_u\right\}\times \mathbf{r}\,dV' \quad (2\text{-}2.15)$$

$$= -\frac{G}{c^2}\int\left\{\frac{m\alpha t}{2\pi s r^3}\theta_u\right\}\times \mathbf{r}\,dV' .$$

The mass current formed by the ring is filamentary. Its magnitude is $I = Js = m\alpha t/2\pi$. Since the mass current is filamentary, the volume element $dV'$ in Eq. (2-2.15) can be written as $sdl'$, where $dl'$ is a length element along the circumference of the ring. Furthermore, we can combine $\theta_u$ and $dl'$ into the vector $d\mathbf{l}' = dl'\boldsymbol{\theta}_u$. Transposing $\boldsymbol{\theta}_u$ and $\mathbf{r}$, we then have from Eq. (2-2.15)

$$\mathbf{K} = \frac{G}{c^2}\oint\frac{I}{r^3}\mathbf{r}\times d\mathbf{l}' , \quad (2\text{-}2.16)$$

Integrating Eq. (2-2.16), we obtain

$$\mathbf{K} = -\frac{2\pi G}{c^2}\frac{Ia^2}{(a^2 + x^2)^{3/2}}\mathbf{i} , \quad (2\text{-}2.17)$$

or, substituting $I = m\alpha t/2\pi$,

$$\mathbf{K} = -G\frac{m\alpha t a^2}{c^2(a^2 + x^2)^{3/2}}\mathbf{i} . \quad (2\text{-}2.18)$$

Note that the cogravitational field is left-handed relative to the mass current that produces it.

The surprising result of this example is that, once the fields have reached the point of observation, neither the gravitational nor the cogravitational field on the axis of the rotating ring is affected by retardation.

▲

## 2-3. Gravitational and Cogravitational Forces According to the Generalized Theory of Gravitation

One of the most important differences between Newton's original theory of gravitation and the generalized theory of gravitation is in the interpretation of the mechanism of gravitational interactions. Whereas in Newton's original theory of gravitation gravitational interaction between two bodies involves one single force of gravitational attraction, in the generalized theory of gravitation gravitational interaction between two bodies involves an intricate juxtaposition of several different forces. Mathematically, these forces result from Eqs. (2-2.1), (2-2.2) and (2-2.6). When Eqs. (2-2.1) and (2-2.2) are written as five separate integrals, they become, using **J** for $\rho\mathbf{v}$,

$$\mathbf{g} = -G\int\frac{[\rho]}{r^3}\mathbf{r}\,dV' - G\int\frac{1}{r^2c}\left[\frac{\partial\rho}{\partial t}\right]\mathbf{r}\,dV' + \frac{G}{c^2}\int\frac{1}{r}\left[\frac{\partial\mathbf{J}}{\partial t}\right]dV' \qquad (2\text{-}3.1)$$

and

$$\mathbf{K} = -\frac{G}{c^2}\int\frac{[\mathbf{J}]}{r^3}\times\mathbf{r}\,dV' - \frac{G}{c^2}\int\frac{1}{r^2c}\frac{\partial[\mathbf{J}]}{\partial t}\times\mathbf{r}\,dV'. \qquad (2\text{-}3.2)$$

Each of these integrals represents a force field. Therefore, according to the generalized theory of gravitation, gravitational interactions between two bodies involve at least five different forces. Let us consider the physical sources of these forces.

First let us consider Eq. (2-3.1). The field represented by the first integral of this equation is the ordinary Newtonian gravitational field created by the mass distribution $\rho$ corrected for the finite speed of the propagation of the field, as indicated by the square brackets (the retardation symbol) in the numerator. The field represented by the second integral is created by a mass whose density varies with time. Like the ordinary Newtonian gravitational field, these two fields are directed toward the masses which create them. The field represented by the last integral in

Eq. (2-3.1) is created by a mass current whose magnitude and/or direction varies with time. The direction of this field is parallel to the direction along which the mass current increases. All three fields in Eq. (2-3.1) act on stationary as well as on moving masses.

Consider now Eq. (2-3.2). The first integral in this equation represents the cogravitational field created by the mass current. The direction of this field is normal to the mass current vector. The second integral represents the field created by a time-variable mass current. The direction of this field is normal to the direction along which the mass current increases. By Eq. (2-2.6), both fields in Eq. (2-3.2) act on moving masses only.

If the mass under consideration does not move and does not change with time, then there is no retardation and no mass current. In this case both integrals in Eq. (2-3.2) vanish and only the first integral remains in Eq. (2-3.1). As a result, one simply obtains the integral representing the ordinary Newtonian gravitational field. Thus, the ordinary Newtonian gravitational theory is a special case of the generalized theory, as it should be.

As far as the gravitational interaction between two masses is concerned, the meaning of the five integrals discussed above can be explained with the help of Fig. 2.2. The upper part of Fig. 2.2 shows the force which the mass $m_1$ experiences under the action of the mass $m_2$ according to the ordinary Newtonian theory. The lower part of Fig. 2.2 shows five forces which the same mass $m_1$ experiences under the action of the mass $m_2$ according to the generalized theory. The time for which the positions of the two masses and the force experienced by $m_1$ are observed is indicated by the letter $t$. Let us note first of all that, according to the ordinary Newtonian theory, the mass $m_1$ is subjected to one single force directed to the mass $m_2$ at its present location, that is, to its location at the time $t$. However, according to the generalized theory, all forces acting on the mass $m_1$ are associated not with the position of the mass $m_2$ at the time of observation, but with

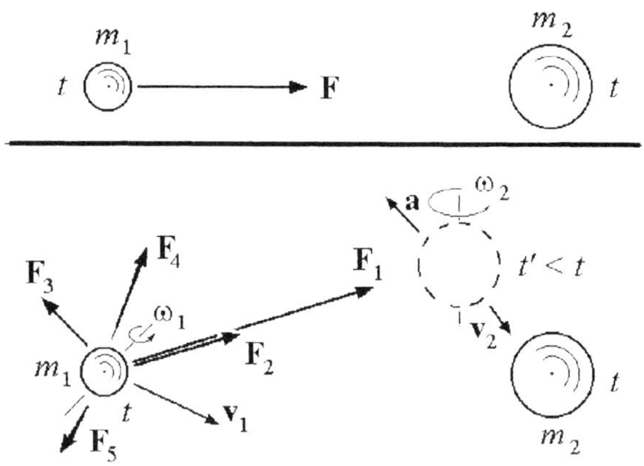


*Fig. 2.2 The upper part of this figure shows the force that the mass $m_1$ experiences under the action of the mass $m_2$ according to the ordinary Newtonian theory. The lower part shows five forces which the same mass $m_1$ experiences under the action of the mass $m_2$ according to the generalized Newtonian theory.*

the position of $m_2$ at an earlier time $t' < t$. Therefore, the magnitude of the mass $m_2$, its position and its state of motion at the present time $t$ have no effect at all on the mass $m_1$.

The subscripts identifying the five forces shown in the lower part of Fig. 2.2 correspond to the five integrals in the Eqs. (2-3.1) and (2-3.2). The force $\mathbf{F}_1$ is associated simply with the mass $m_2$ and differs from the ordinary Newtonian gravitational force only insofar as it is directed not to the mass $m_2$ at its present position, but to the place where $m_2$ was located at the past time $t'$. The force $\mathbf{F}_2$ is associated with the variation of the density of the mass $m_2$ with time; the direction of this force is the same as that of $\mathbf{F}_1$. The force $\mathbf{F}_3$ is associated with the time variation of the mass current produced by $m_2$; this force is directed along the acceleration vector $\mathbf{a}$ (or along the velocity vector $\mathbf{v}_2$) which the mass $m_2$ had at the time $t'$. The three forces are produced by the gravitational field **g** (if $m_2$ is a point mass moving at constant velocity, **g** and the resultant of the three forces are directed toward the *present position* of $m_2$; see Chapter 5).

The forces $\mathbf{F}_4$ and $\mathbf{F}_5$ are due to the cogravitational field **K**. The force $\mathbf{F}_4$ is associated with the mass current created by the mass $m_2$ and with the velocity of the mass $m_1$. Its direction is normal to the velocity vector $\mathbf{v}_2$ which the mass $m_2$ had at the time $t'$ and normal to the velocity vector $\mathbf{v}_1$ which the mass $m_1$ has at the present time $t$. The force $\mathbf{F}_5$ is associated with the velocity of the mass $m_1$ and with the variation of the mass current of the mass $m_2$ with time; the direction of this force is normal to the acceleration vector (or to the velocity vector) that the mass $m_2$ had at the time $t'$ and normal to the velocity vector that the mass $m_1$ has at the present time $t$. Although not shown in Fig. 2.2, additional forces associated with the rotation of $m_2$ and $m_1$ (angular velocities $\omega_2$ and $\omega_1$) are generally involved in the interaction between the two masses (see Chapters 14 and 15).

The forces $\mathbf{F}_2$, $\mathbf{F}_3$, $\mathbf{F}_4$, and $\mathbf{F}_5$ are usually much weaker than the force $\mathbf{F}_1$ because of the presence of the speed of gravitation $c$ (usually assumed to be the same as the speed of light) in the denominators of the integrals representing the fields responsible for these four forces. This means that only when the translational or rotational velocity of $m_2$ or $m_1$ is close to $c$, are the forces $\mathbf{F}_2$, $\mathbf{F}_3$, $\mathbf{F}_4$, and $\mathbf{F}_5$ dominant. Of course, the cumulative effect of these forces in long-lasting gravitational systems (such as the Solar system, for example) may be significant regardless of the velocities of the interacting masses.

## References and Remarks for Chapter 2

**1.** It should be noted that the cogravitational field **K** has not yet been actually observed. However, it is very likely that it will be revealed by the *Gravity Probe B* launched in 2004 by NASA in a polar orbit around the Earth.

**2.** For the various theoretical considerations demanding the existence of the cogravitational field see Oleg D. Jefimenko, *Causality, Electromagnetic Induction and Gravitation,* 2nd ed., (Electret Scientific, Star City, 2000), pp. 80-100.

**3.** Although we say that gravitational and cogravitational fields "propagate," it is not entirely clear what physical entity actually propagates, since by definition gravitational and cogravitational fields are "regions of space." It is conceivable that what actually propagates is some particles that somehow create the gravitational and cogravitational fields. It is possible that these particles have already been described [see M. R. Edwards, Ed., *Pushing Gravity* (Apeiron, Montreal, 2002)], and it is possible that some of their effects have already been observed [see I. A. Eganova, *The Nature of Space-Time* (Publishing House of SB RAS, Novosibirsk, 2005), pp. 137-223]. Yet, there is not enough information about these particles for making any definite statement about their existence, nature, composition, or properties.

**4.** It is important to note that the gravitational-cogravitational field energy is negative. This means that no energy can be extracted from the gravitational-cogravitational field by destroying the field. On the contrary, energy must be delivered to the field in order to destroy the field.

**5.** Oleg D. Jefim*enko, Electricity and Magnetism*, 2nd ed., (Electret Scientific, Star City, 1989), pp. 18-62.

**6.** Oleg D. Jefimenko, *Electromagnetic Retardation and Theory of Relativity,* 2nd ed., (Electret Scientific, Star City, 2004), pp. 6-14.

**7.** Here and throughout this book we use the standard notation **i**, **j**, **k** for the unit vectors along the *x*, *y* and *z* axis, respectively, of rectangular system of coordinates.

# 3

# ALTERNATIVE FORMS OF THE PRINCIPAL FIELD EQUATIONS OF THE GENERALIZED THEORY OF GRAVITATION

The principal field equations of the generalized theory of gravitation can be converted into several different equations that may be more useful for practical applications than the original equations themselves. In this chapter we will present several such equations, will show their derivations, and will demonstrate some of their applications.

## 3-1. Alternative Expressions for the Principal Field Equations in Terms of Volume Integrals

As will be shown below, the principal equations of the generalized theory of gravitation, Eqs. (2-2.1) and (2-2.2), can be converted into

$$\mathbf{g} = G\int \frac{[\nabla'\rho]}{r}dV' + \frac{G}{c^2}\int \frac{1}{r}\left[\frac{\partial \mathbf{J}}{\partial t}\right]dV' \qquad (3\text{-}1.1)$$

and

$$\mathbf{K} = -\frac{G}{c^2}\int \frac{[\nabla' \times \mathbf{J}]}{r}dV', \qquad (3\text{-}1.2)$$

where, for brevity, the letter **J** is used for the mass-current density $\rho\mathbf{v}$, and where the primed operator $\nabla'$ operates on the source-point coordinates only. The integration, as usual, is over all space.

Furthermore, equations (2-2.1) and (2-2.4) can be combined into the single equation

$$\mathbf{g} = -G\int \frac{[\rho]}{r^3}\mathbf{r}\,dV' + \frac{G}{c}\int\left\{\frac{[\mathbf{J}]}{r^2} - 2\mathbf{r}\frac{[\mathbf{J}]\cdot\mathbf{r}}{r^4} - \frac{\mathbf{r}}{r^3 c}\left[\frac{\partial \mathbf{J}}{\partial t}\right]\cdot\mathbf{r} + \frac{1}{rc}\left[\frac{\partial \mathbf{J}}{\partial t}\right]\right\}dV'. \tag{3-1.3}$$

**Derivation of Eq. (3-1.1).** We start with Eq. (2-2.1)

$$\mathbf{g} = -G\int\left\{\frac{[\rho]}{r^3} + \frac{1}{r^2 c}\left[\frac{\partial\rho}{\partial t}\right]\right\}\mathbf{r}\,dV' + \frac{G}{c^2}\int\frac{1}{r}\left[\frac{\partial(\rho\mathbf{v})}{\partial t}\right]dV'. \tag{2-2.1}$$

Using vector identity (V-35) with $\mathbf{r}_u = \mathbf{r}/r$ for replacing the two terms in the integrand of the first integral of Eq. (2-2.1) by a single term, we obtain (note that we now use **J** in place of $\rho\mathbf{v}$)

$$\mathbf{g} = G\int\nabla\frac{[\rho]}{r}dV' + \frac{G}{c^2}\int\frac{1}{r}\left[\frac{\partial\mathbf{J}}{\partial t}\right]dV'. \tag{3-1.4}$$

Transforming the integrand in the first integral of Eq. (3-1.4) by means of vector identity (V-34), we obtain (note that the ordinary operator $\nabla$ operates upon the field-point coordinates, whereas the primed operator $\nabla'$ operates upon the source-point coordinates)

$$\mathbf{g} = G\int\frac{[\nabla'\rho]}{r}dV' - G\int\nabla'\frac{[\rho]}{r}dV' + \frac{G}{c^2}\int\frac{1}{r}\left[\frac{\partial\mathbf{J}}{\partial t}\right]dV'. \tag{3-1.5}$$

The second integral in the last equation can be transformed into a surface integral by means of vector identity (V-20). But this surface integral vanishes, because $\rho$ is confined to a finite region of space, while the surface of integration is at infinity. We thus obtain

$$\mathbf{g} = G\int \frac{[\nabla'\rho]}{r} dV' + \frac{G}{c^2}\int \frac{1}{r}\left[\frac{\partial \mathbf{J}}{\partial t}\right] dV'. \tag{3-1.1}$$

**Derivation of Eq. (3-1.2).** We start with Eq. (2-2.2)

$$\mathbf{K} = -\frac{G}{c^2}\int\left\{\frac{[\rho\mathbf{v}]}{r^3} + \frac{1}{r^2 c}\frac{\partial[\rho\mathbf{v}]}{\partial t}\right\}\times \mathbf{r}\, dV'. \tag{2-2.2}$$

Applying vector identity (V-35) to Eq. (2-2.2) and noting that $\mathbf{r}\times = -\times\mathbf{r}$, we obtain (note that we now use $\mathbf{J}$ in place of $\rho\mathbf{v}$)

$$\mathbf{K} = -\frac{G}{c^2}\int \nabla \times \frac{[\mathbf{J}]}{r} dV'. \tag{3-1.6}$$

Transforming Eq. (3-1.6) by means of vector identity (V-34) and eliminating $\nabla' \times ([\mathbf{J}]/r)$ by means of vector identity (V-21) [see the explanation below Eq. (3-1.5); note that $\mathbf{J}$ is confined to a finite region of space], we obtain for the cogravitational field

$$\mathbf{K} = -\frac{G}{c^2}\int \frac{[\nabla' \times \mathbf{J}]}{r} dV'. \tag{3-1.2}$$

**Derivation of Eq. (3-1.3).** We start again with Eq. (2-2.1)

$$\mathbf{g} = -G\int\left\{\frac{[\rho]}{r^3} + \frac{1}{r^2 c}\left[\frac{\partial \rho}{\partial t}\right]\right\}\mathbf{r}\, dV' + \frac{G}{c^2}\int \frac{1}{r}\left[\frac{\partial(\rho\mathbf{v})}{\partial t}\right] dV'. \tag{2-2.1}$$

By Eq. (2-2.4), the contribution that $\partial\rho/\partial t$ makes to the first integral in Eq. (2-2.1) can be expressed as

$$\int \frac{1}{r^2 c}\left[\frac{\partial \rho}{\partial t}\right]\mathbf{r}\,dV' = -\int \frac{[\nabla' \cdot \mathbf{J}]}{r^2 c}\mathbf{r}\,dV'. \tag{3-1.7}$$

Transforming the last integral by using vector identities (V-31) and (V-36) with $\mathbf{r}_u = \mathbf{r}/r$, and using vector identity (V-8), we obtain

$$\begin{aligned}\int \frac{[\nabla' \cdot \mathbf{J}]}{r^2 c}\mathbf{r}\,dV' &= \int\left(\frac{\nabla' \cdot [\mathbf{J}]}{r^2 c}\mathbf{r} - \frac{\mathbf{r}\cdot[\partial \mathbf{J}/\partial t]}{r^3 c^2}\mathbf{r}\right)dV' \\ &= \int\left(\frac{\mathbf{r}}{c}\nabla' \cdot \frac{[\mathbf{J}]}{r^2} - \frac{\mathbf{r}}{c}[\mathbf{J}]\cdot\nabla'\frac{1}{r^2} - \frac{\mathbf{r}\cdot[\partial \mathbf{J}/\partial t]}{r^3 c^2}\mathbf{r}\right)dV'.\end{aligned} \tag{3-1.8}$$

Next, using vector identity (V-23), we transform the first term in the integrand of the last integral of Eq. (3-1.8), obtaining

$$\int \frac{\mathbf{r}}{c}\nabla' \cdot \frac{[\mathbf{J}]}{r^2}dV' = \oint \frac{\mathbf{r}}{c}\left(\frac{[\mathbf{J}]}{r^2}\cdot d\mathbf{S}'\right) - \int\left(\frac{[\mathbf{J}]}{r^2}\cdot\nabla'\right)\frac{\mathbf{r}}{c}dV'. \tag{3-1.9}$$

Since the integration is over all space, and since there is no mass current at infinity, the surface integral in Eq. (3-1.9) vanishes. Applying vector identity (V-4) to the integrand of the remaining integral on the right of Eq. (3-1.9) and noting that a $\nabla'$ operation upon $\mathbf{r}$ is the negative of the same $\nabla$ operation, we then have

$$\int \frac{\mathbf{r}}{c}\nabla' \cdot \frac{[\mathbf{J}]}{r^2}dV' = \int \frac{[\mathbf{J}]}{cr^2}dV'. \tag{3-1.10}$$

From Eqs. (3-1.7), (3-1.8), (3-1.9), and (3-1.10), we obtain therefore

$$\int \frac{1}{r^2 c}\left[\frac{\partial \rho}{\partial t}\right]\mathbf{r}\,dV' = -\int\left(\frac{[\mathbf{J}]}{r^2 c} - \frac{\mathbf{r}}{c}[\mathbf{J}]\cdot\nabla'\frac{1}{r^2} - \frac{\mathbf{r}\cdot[\partial \mathbf{J}/\partial t]}{r^3 c^2}\mathbf{r}\right)dV'. \tag{3-1.11}$$

Substituting Eq. (3-1.11) into Eq. (2-2.1) and taking into account that $\nabla'(1/r^2) = 2\mathbf{r}/r^4$, we finally obtain

$$\mathbf{g} = -G\int \frac{[\rho]}{r^3}\mathbf{r}\,dV' + \frac{G}{c}\int\left\{\frac{[\mathbf{J}]}{r^2} - 2\mathbf{r}\frac{[\mathbf{J}]\cdot\mathbf{r}}{r^4} - \frac{\mathbf{r}}{r^3 c}\left[\frac{\partial\mathbf{J}}{\partial t}\right]\cdot\mathbf{r} + \frac{1}{rc}\left[\frac{\partial\mathbf{J}}{\partial t}\right]\right\}dV'. \tag{3-1.3}$$

## 3-2. Expressing the Principal Field Equations in Terms of Surface Integrals

A remarkable feature of Eqs. (3-1.1) and (3-1.2) is that they correlate the gravitational field with the *gradient* of the mass distribution and correlate the cogravitational field with the *curl* of the mass-current distribution rather than with the mass and mass-current distribution as such. Hence, the equations may be interpreted as indicating that the gravitational and cogravitational fields are associated not with masses and mass currents, but rather with the *inhomogeneities* in the distribution of masses and mass currents (a homogeneous, or uniform, mass distribution has zero gradient, and a homogeneous, or uniform, mass-current distribution has zero curl).

Particularly interesting in this connection is a mass or mass-current distribution in which the mass or mass-current changes abruptly from a finite value in the interior of the distribution to zero outside the distribution. For this type of mass and mass-current distribution, Eqs. (3-1.1) and (3-1.2) can be transformed into special forms that are more convenient to use than Eqs. (3-1.1) and (3-1.2) themselves.

As is shown below, for the gravitational field the following equation can be used:

$$\mathbf{g} = -G\oint_{Boundary}\frac{[\rho]}{r}d\mathbf{S}' + G\int_{Int}\frac{[\nabla'\rho]}{r}dV' + \frac{G}{c^2}\int\frac{1}{r}\left[\frac{\partial\mathbf{J}}{\partial t}\right]dV', \tag{3-2.1}$$

where the surface integral is extended over the boundary layer of the mass distribution, and the first volume integral is extended over the interior of the mass distribution. This equation becomes especially simple in the case of a constant (uniform) mass distribution surrounded by a free space. In this case $\nabla\rho$ in the interior of the distribution is zero, and Eq. (3-2.1) simplifies to

$$\mathbf{g} = -G\oint_{Boundary}\frac{[\rho]}{r}d\mathbf{S}' + \frac{G}{c^2}\int\frac{1}{r}\left[\frac{\partial\mathbf{J}}{\partial t}\right]dV' \qquad (3\text{-}2.2)$$

For the cogravitational field, the following equations can be used:

$$\mathbf{K} = -\frac{G}{c^2}\oint_{Boundary}\frac{[\mathbf{J}]}{r}\times d\mathbf{S}' - \frac{G}{c^2}\int_{Interior}\frac{[\nabla'\times\mathbf{J}]}{r}dV'. \qquad (3\text{-}2.3)$$

and, for the special case of $\nabla' \times \mathbf{J} = 0$ in the interior of the mass-current distribution,

$$\mathbf{K} = -\frac{G}{c^2}\oint_{Boundary}\frac{[\mathbf{J}]}{r}\times d\mathbf{S}'. \qquad (3\text{-}2.4)$$

**Derivation of Eq. (3-2.1).** We start with Eq. (3-1.1). In this equation the integral involving $\nabla'\rho$ can be separated into two integrals: the integral over the boundary layer of the mass distribution under consideration and the integral over the interior of the mass distribution:

$$G\int\frac{[\nabla'\rho]}{r}dV' = G\int_{B.layer}\frac{[\nabla'\rho]}{r}dV' + G\int_{Int}\frac{[\nabla'\rho]}{r}dV'. \qquad (3\text{-}2.5)$$

The first integral on the right of Eq. (3-2.5) can be transformed by using vector identity (V-34):

$$G\int_{B.\,layer}\frac{[\nabla'\rho]}{r}dV' = G\int_{B.\,layer}\nabla\frac{[\rho]}{r}dV' + G\int_{B.\,layer}\nabla'\frac{[\rho]}{r}dV'. \quad (3\text{-}2.6)$$

In Eq. (3-2.6), the operator $\nabla$ in the first integral on the right operates upon the field point coordinates only. Therefore it can be factored out from under the integral sign. The integrand in this integral will then be $[\rho]/r$. Since both $[\rho]$ and $r$ are finite, while the integration is over the volume of the boundary layer whose thickness, and therefore volume, can be assumed to be as small as we please, the integral vanishes. The second integral on the right of Eq. (3-2.6) can be transformed into a surface integral by using vector identity (V-20). Equation (3-2.6) can be written therefore as

$$G\int_{B.\,layer}\frac{[\nabla'\rho]}{r}dV' = G\oint_{B.\,layer}\frac{[\rho]}{r}d\mathbf{S}', \quad (3\text{-}2.7)$$

where the surface integral is extended over *both* surfaces (exterior and interior) of the boundary layer.

In Eq. (3-2.7), the surface element vector $d\mathbf{S}'$ of the exterior surface is directed into the space outside the mass distribution, while $d\mathbf{S}'$ of the interior surface is directed into the mass distribution. However, since there is no mass outside the mass distribution, the integral over the exterior surface vanishes. Since the boundary layer can be made as thin as we please, we can make the interior surface of the boundary layer coincide with the surface of the mass distribution. Reversing the sign in front of the surface integral, we can write then Eq. (3-2.7) as

$$G\int_{B.\,layer}\frac{[\nabla'\rho]}{r}dV' = -G\oint_{Boundary}\frac{[\rho]}{r}d\mathbf{S}', \quad (3\text{-}2.8)$$

where the integration is now over the surface of the mass distribution, and where the surface element vector $d\mathbf{S}'$ is directed, as usual, from the mass distribution into the surrounding space.

From Eqs. (3-1.1), (3-2.5) and (3-2.8) we obtain

$$\mathbf{g} = -G\oint_{Boundary}\frac{[\rho]}{r}d\mathbf{S}' + G\int_{Int}\frac{[\nabla'\rho]}{r}dV' + \frac{G}{c^2}\int\frac{1}{r}\left[\frac{\partial \mathbf{J}}{\partial t}\right]dV'. \quad (3\text{-}2.1)$$

**Derivation of Eq. (3-2.3).** We start with Eq. (3-1.2). Just as in the case of Eq. (3-1.1), we can separate the integral in Eq. (3-1.2) into an integral over the boundary layer of the mass-current distribution and an integral over the interior of the distribution. By the same reasoning as that used to simplify Eq. (3-2.5), we find that the integral over the boundary layer can be written as

$$\frac{G}{c^2}\int_{B.\,layer}\frac{[\nabla'\times\mathbf{J}]}{r}dV' = \frac{G}{c^2}\int_{B.\,layer}\nabla'\times\frac{[\mathbf{J}]}{r}dV'. \quad (3\text{-}2.9)$$

Transforming the integral on the right of Eq. (3-2.9) into a surface integral by means of vector identity (V-21), and taking into account that there is no mass current in the space outside the mass-current distribution, we obtain, just as we obtained Eq. (3-2.8),

$$\frac{G}{c^2}\int_{B.\,layer}\frac{[\nabla'\times\mathbf{J}]}{r}dV' = \frac{G}{c^2}\oint_{Boundary}\frac{[\mathbf{J}]}{r}\times d\mathbf{S}', \quad (3\text{-}2.10)$$

where the integration is over the surface of the mass-current distribution, and the surface element vector $d\mathbf{S}'$ is directed from the mass-current distribution into the surrounding space.

Equation (3-1.2) can be written therefore as

$$\mathbf{K} = -\frac{G}{c^2}\oint_{Boundary}\frac{[\mathbf{J}]}{r}\times d\mathbf{S}' - \frac{G}{c^2}\int_{Interior}\frac{[\nabla'\times\mathbf{J}]}{r}dV'. \quad (3\text{-}2.3)$$

▼

**Example 3-2.1** A thin heavy disk of uniform mass density $\rho$, radius $a$, and thickness $b$ rotates with constant angular acceleration $\alpha$ about its symmetry axis, which is also the $x$ axis of rectangular coordinates. The midplane of the disk coincides with the $yz$ plane of the coordinates, and the rotation of the disk is right-handed relative to the $x$ axis (Fig. 3.1). Using Eqs. (3-2.2) and (3-2.3), find the gravitational and cogravitational fields produced by the disk at a point of the $x$ axis, if at $t = 0$ the angular velocity of the disk is $\omega = 0$.

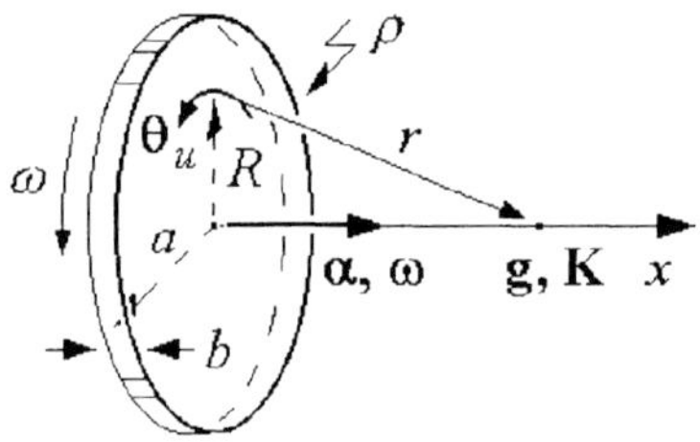


*Fig. 3.1 Calculation of the gravitational and cogravitational fields on the axis of a heavy disk rotating with constant angular acceleration* $\alpha$.

The disk creates a convection mass-current $\mathbf{J} = \rho\mathbf{v} = \rho\omega R\boldsymbol{\theta}_u = \rho\alpha t R\boldsymbol{\theta}_u$, where $R$ is the distance from the center of the disk, and $\boldsymbol{\theta}_u$ is a unit vector in the circular direction (right-handed with respect to $\boldsymbol{\alpha}$). The time derivative of $\mathbf{J}$ is $\partial\mathbf{J}/\partial t = \rho\alpha R\boldsymbol{\theta}_u$. To find $\boldsymbol{\nabla}' \times \mathbf{J}$, we use the relation $\mathbf{v} = \boldsymbol{\omega} \times \mathbf{R}$ and vector identity (V-12). Taking into account that $\boldsymbol{\omega}$ is not a function of coordinates, we then obtain

$$\boldsymbol{\nabla}' \times \mathbf{J} = \boldsymbol{\nabla}' \times (\rho\,\boldsymbol{\omega} \times \mathbf{R}) = \rho\,[\boldsymbol{\omega}\,(\boldsymbol{\nabla}' \cdot \mathbf{R}) - (\boldsymbol{\omega} \cdot \boldsymbol{\nabla}')\mathbf{R}], \qquad (3\text{-}2.11)$$

and since $\mathbf{R} = y'\mathbf{j} + z'\mathbf{k}$, while $\boldsymbol{\omega} \cdot \boldsymbol{\nabla}' = \omega\partial/\partial x'$, we have

$$\boldsymbol{\nabla}' \times \mathbf{J} = 2\rho\,\boldsymbol{\omega} = 2\rho\,\boldsymbol{\alpha} t = 2\rho\,\alpha t\,\mathbf{i}. \qquad (3\text{-}2.12)$$

Examining now Eq. (3-2.2) and taking into account that $\partial\mathbf{J}/\partial t$ is in the circular direction, we recognize that the second integral in

Eq. (3-2.2) vanishes by symmetry (see Example 2-2.1). And since $\rho$ does not depend on time, we see from Eq. (3-2.2) that the gravitational field of the disk is the ordinary Newtonian field given by

$$\mathbf{g} = -G\oint_{Boundary}\frac{\rho}{r}d\mathbf{S}' = -G\rho\oint_{Boundary}\frac{d\mathbf{S}'}{r}. \quad (3\text{-}2.13)$$

Let us now evaluate the last surface integral in Eq. (3-2.13). By the symmetry of the system, only the two flat surfaces of the disk contribute to the field on the axis. The back surface is located at $x' = -b/2$, the front surface is located at $x' = +b/2$. The direction of the surface element vector $d\mathbf{S}'$ is $-\mathbf{i}$ for the back surface and $+\mathbf{i}$ for the front surface. We have therefore

$$\mathbf{g} = G\rho\mathbf{i}\int_0^a\frac{2\pi RdR}{[R^2+(x+b/2)^2]^{1/2}} - G\rho\mathbf{i}\int_0^a\frac{2\pi RdR}{[R^2+(x-b/2)^2]^{1/2}} \quad (3\text{-}2.14)$$
$$= 2\pi G\rho\mathbf{i}\{[a^2+(x+b/2)^2]^{1/2}-(x+b/2)-[a^2+(x-b/2)^2]^{1/2}+(x-b/2)\}.$$

Since $b \ll x$, we can use the relation

$$[a^2+(x\pm b/2)^2]^{1/2} = [a^2+x^2\pm xb]^{1/2} = (a^2+x^2)^{1/2}[1\pm xb/2(a^2+x^2)]. \quad (3\text{-}2.15)$$

Substituting Eq. (3-2.15) into Eq. (3-2.14), we obtain after elementary simplifications

$$\mathbf{g} = -2\pi G\rho b\left[1-\frac{x}{(a^2+x^2)^{1/2}}\right]\mathbf{i}. \quad (3\text{-}2.16)$$

To find the cogravitational field, we use Eq. (3-2.3). Substituting $[\mathbf{J}] = \rho\alpha R(t-r/c)\boldsymbol{\theta}_u$ and $[\boldsymbol{\nabla}'\times\mathbf{J}] = 2\rho\alpha(t-r/c)\mathbf{i}$ into Eq. (3-2.3), we have

$$\mathbf{K} = -\frac{G}{c^2}\oint_{Boundary}\frac{\rho\alpha R(t-r/c)}{r}\boldsymbol{\theta}_u \times d\mathbf{S}' - \mathbf{i}\frac{G}{c^2}\int_{Int}\frac{2\rho\alpha(t-r/c)}{r}dV'. \tag{3-2.17}$$

By the symmetry of the system, only the curved surface of the disk contributes to the first integral. At this surface $R = a$, $r = (a^2 + x^2)^{1/2}$, $\boldsymbol{\theta}_u \times d\mathbf{S}' = -\,\mathbf{i}\,dS'$, and the surface itself is $S' = 2\pi ab$. In the second integral $r$ is $r = (R^2 + x^2)^{1/2}$ and the volume element is $dV' = b2\pi RdR$. The cogravitational field is therefore

$$\mathbf{K} = \mathbf{i}\frac{G\rho\alpha a[t-(a^2+x^2)^{1/2}/c]2\pi ab}{c^2(a^2+x^2)^{1/2}} - \frac{\mathbf{i}2G\rho\alpha}{c^2}\int_0^a\frac{t-(R^2+x^2)^{1/2}/c}{(R^2+x^2)^{1/2}}2\pi bRdR$$
$$= \mathbf{i}2\pi G\left(\frac{\rho\alpha t a^2 b}{c^2(a^2+x^2)^{1/2}} - \frac{\rho\alpha a^2 b}{c^3} - \frac{2\rho\alpha t b(a^2+x^2)^{1/2}}{c^2} + \frac{\rho\alpha a^2 b}{c^3}\right), \tag{3-2.18}$$

or

$$\mathbf{K} = -\,\mathbf{i}4\pi G\frac{\rho\alpha b t(a^2+x^2)^{1/2}}{c^2}\left[1 - \frac{a^2}{2(a^2+x^2)}\right]. \tag{3-2.19}$$

It is interesting to note that neither the gravitational nor the cogravitational field of the rotating disk is retarded, just as was the case with the fields of the rotating ring discussed in Example 2-2.1.

▲

## 3-3. Expressing Gravitational and Cogravitational Fields in Terms of Potentials

The calculation of time-dependent gravitational and cogravitational fields can sometimes be simplified by using retarded gravitational and cogravitational potentials.

As we shall presently see, the cogravitational field can be obtained by using the equation

$$\mathbf{K} = \nabla \times \mathbf{A}. \tag{3-3.1}$$

where **A** is the retarded cogravitational vector potential defined as

$$\mathbf{A} = -\frac{G}{c^2}\int \frac{[\mathbf{J}]}{r} dV'. \tag{3-3.2}$$

By vector identity (V-17), Eq. (3-3.1) can be written also as

$$\oint \mathbf{A} \cdot d\mathbf{l} = \int \mathbf{K} \cdot d\mathbf{S}. \tag{3-3.3}$$

The gravitational field can be obtained by using the equation

$$\mathbf{g} = -\nabla\varphi - \frac{\partial \mathbf{A}}{\partial t}, \tag{3-3.4}$$

where $\varphi$ is the retarded scalar potential defined by the equation

$$\varphi = -G\int \frac{[\rho]}{r} dV'. \tag{3-3.5}$$

For time-independent gravitational and cogravitational systems, the retarded vector potential reduces to

$$\mathbf{A} = -\frac{G}{c^2}\int \frac{\mathbf{J}}{r} dV' \tag{3-3.6}$$

and the retarded scalar potential reduces to

$$\varphi = -G\int \frac{\rho}{r} dV'. \tag{3-3.7}$$

Thus the scalar potential becomes the familiar scalar potential of the Newtonian theory of gravitation, once again indicating that the Newtonian theory is an incomplete theory and constitutes a special case of the generalized theory of gravitation.

In a mass-free region of space, the gravitational field can also be expressed in terms of the *gravitational vector potential* according to the equation

$$\mathbf{g} = \nabla \times \mathbf{A}_g, \tag{3-3.8}$$

which, by vector identity (V-17), can be written also as

$$\oint \mathbf{A}_g \cdot d\mathbf{l} = \int \mathbf{g} \cdot d\mathbf{S}. \tag{3-3.9}$$

Likewise, in a mass-free region of space, the cogravitational field can also be expressed in terms of the *cogravitational scalar potential*, according to the equation

$$\mathbf{K} = -\nabla\varphi_c + \frac{1}{c^2}\frac{\partial \mathbf{A}_g}{\partial t}. \tag{3-3.10}$$

The validity of Eqs. (3-3.8) and (3-3.10) follows from the fact that these equations are in accord with Eqs. (7-1.1)-(7-1.4) (see Chapter 7), which can be easily verified by direct substitution.[2]

**Derivation of Eqs. (3-3.1)-(3-3.5).** Factoring out the operator $\nabla$ from under the first integral of Eq. (3-1.4), we immediately obtain the relation for the retarded gravitational scalar potential $\varphi$

$$\mathbf{g} = -\nabla\varphi + \frac{G}{c^2}\int \frac{1}{r}\left[\frac{\partial \mathbf{J}}{\partial t}\right] dV', \tag{3-3.11}$$

where

$$\varphi = -G\int \frac{[\rho]}{r} dV'. \tag{3-3.5}$$

Likewise, factoring out the operator $\nabla$ from under the integral of Eq. (3-1.6), we immediately obtain the relation for the retarded cogravitational vector potential $\mathbf{A}$

$$\mathbf{K} = \nabla \times \mathbf{A}, \tag{3-3.1}$$

where

$$\mathbf{A} = -\frac{G}{c^2}\int \frac{[\mathbf{J}]}{r}dV'. \tag{3-3.2}$$

Next, using vector identity (V-36), factoring out the time derivative from under the integral sign in Eq. (3-3.11), and eliminating the integral by means of Eq. (3-3.2), we obtain

$$\mathbf{g} = -\nabla\varphi - \frac{\partial \mathbf{A}}{\partial t}. \tag{3-3.4}$$

▼

**Example 3-3.1** Show that the retarded potentials $\varphi$ and $\mathbf{A}$ satisfy the relation

$$\nabla \cdot \mathbf{A} = -\frac{1}{c^2}\frac{\partial\varphi}{\partial t}. \tag{3-3.12}$$

From Eqs. (3-3.5) and vector identity (V-36) we have

$$-\frac{1}{c^2}\frac{\partial\varphi}{\partial t} = \frac{G}{c^2}\int \frac{\partial}{\partial t}\frac{[\rho]}{r}dV' = \frac{G}{c^2}\int \frac{1}{r}\left[\frac{\partial\rho}{\partial t}\right]dV'. \tag{3-3.13}$$

But according to the continuity law, Eq. (2-2.4), remembering that we now use $\mathbf{J}$ for $\rho\mathbf{v}$, and noting that $\rho$ in Eq. (3-3.13) is a function of primed coordinates ,

$$-\left[\frac{\partial\rho}{\partial t}\right] = [\nabla' \cdot \mathbf{J}], \tag{3-3.14}$$

so that

$$-\frac{1}{c^2}\frac{\partial\varphi}{\partial t} = -\frac{G}{c^2}\int \frac{[\nabla' \cdot \mathbf{J}]}{r}dV'. \tag{3-3.15}$$

Transforming the integral in Eq. (3-3.15) by means of vector identity (V-34), we have

$$-\frac{1}{c^2}\frac{\partial\varphi}{\partial t} = -\frac{G}{c^2}\int \nabla \cdot \frac{[\mathbf{J}]}{r}dV' - \frac{G}{c^2}\int \nabla' \cdot \frac{[\mathbf{J}]}{r}dV'. \tag{3-3.16}$$

The last integral in Eq. (3-3.16) can be transformed into a surface integral by means of the vector identity (V-19), and since there is no mass current at infinity, the surface integral is zero, and so is the last integral. In the first integral, $\nabla$ can be factored out from under the integral sign. Therefore we obtain

$$-\frac{1}{c^2}\frac{\partial\varphi}{\partial t} = -\nabla\cdot\frac{G}{c^2}\int\frac{[\mathbf{J}]}{r}dV'. \qquad (3\text{-}3.17)$$

Eliminating the last integral in Eq. (3-3.17) by means of Eq. (3-3.2), we obtain Eq. (3-3.12). Note that the analogous equation in the electromagnetic theory is known as *Lorenz's condition.*

**Example 3-3.2** Using gravitational and cogravitational potentials, find gravitational and cogravitational fields at all points of space far from the rotating ring described in Example 2-2.1 (Fig. 3.2).

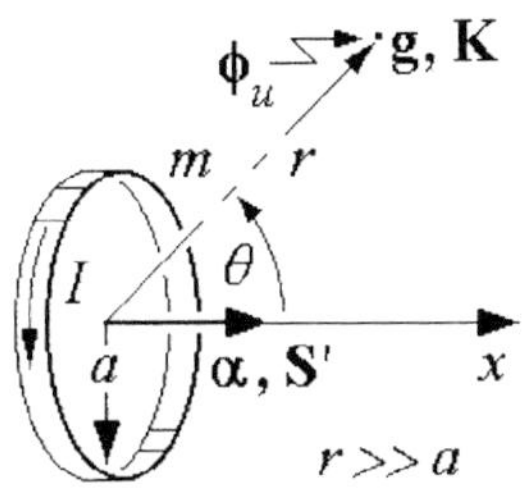


*Fig. 3.2 Calculation of the gravitational and cogravitational fields far from the heavy ring rotating with constant angular acceleration. (The unit vector $\phi_u$ is directed into the page.)*

At large distances from the ring, the ring constitutes a point mass $m$, which does not depend on time. Therefore the gravitational potential of the ring is the ordinary Newtonian potential

$$\varphi = -G\frac{m}{r}. \qquad (3\text{-}3.18)$$

Since the ring constitutes a filamentary convection mass-current $I = m\alpha t/2\pi$, the cogravitational vector potential of the ring is, by Eq. (3-3.2) with $\mathbf{J}dV'$ replaced by $Id\mathbf{l}' = (m\alpha t/2\pi)d\mathbf{l}'$ and the volume integral replaced by a line integral,

$$\mathbf{A} = -\frac{G}{c^2}\oint \frac{m\alpha(t-r/c)/2\pi}{r}\,d\mathbf{l}' = -\frac{Gm\alpha t}{2\pi c^2}\oint \frac{d\mathbf{l}'}{r} + \frac{Gm\alpha}{2\pi c^3}\oint d\mathbf{l}' . \tag{3-3.19}$$

The last integral in Eq. (3-3.19) (closed line integral over vector length elements) is zero. The remaining integral can be transformed into a surface integral by means of vector identity (V-18). We then obtain

$$\mathbf{A} = -\frac{Gm\alpha t}{2\pi c^2}\oint \frac{d\mathbf{l}'}{r} = -\frac{Gm\alpha t}{2\pi c^2}\int \frac{\mathbf{r}'_u}{r^2} \times d\mathbf{S}', \tag{3-3.20}$$

where $\mathbf{r}'_u$ is a unit vector directed from the point of observation toward the surface element $d\mathbf{S}'$.

Now, since the point of observation is far from the ring, the integral can be replaced by the (vector) product of the integrand and the surface area $\mathbf{S}'$ of the ring, so that the vector potential is

$$\mathbf{A} = -\frac{Gm\alpha t}{2\pi c^2 r^2}\mathbf{r}'_u \times \mathbf{S}' = \frac{Gm\alpha t}{2\pi c^2 r^2}\mathbf{r}_u \times \mathbf{S}', \tag{3-3.21}$$

where $\mathbf{r}_u$ is a unit vector directed from the ring toward the point of observation. The magnitude of the vector $\mathbf{S}'$ is $\pi a^2$, and the direction is along the $x$ axis. Designating the angle between $\mathbf{r}_u$ and $\mathbf{S}'$ as $\theta$, we then have for the vector potential

$$\mathbf{A} = \frac{Ga^2\alpha t}{2c^2 r^2}\sin\theta\,\boldsymbol{\phi}_u, \tag{3-3.22}$$

where $\boldsymbol{\phi}_u$ is a unit vector in the circular direction left-handed relative to the $x$ axis.

By Eq. (3-3.1), the cogravitational field associated with this vector potential is

$$\mathbf{K} = \boldsymbol{\nabla} \times \mathbf{A} = -\frac{Gma^2\alpha t}{2c^2 r^3}(2\cos\theta\,\mathbf{r}_u + \sin\theta\,\boldsymbol{\theta}_u) \tag{3-3.23}$$

(we do not reproduce the actual calculation of $\nabla \times \mathbf{A}$, since it is not important for the purpose of the present example; the calculation is done by using the expressions for the curl of a vector in spherical coordinates[1]). It is interesting to note that this field is an ordinary (unretarded) "dipole field." On the $x$ axis ($\theta = 0$) it reduces to the field found in Example 2-2.1 (for $x \gg a$).

Let us now find the gravitational field of the ring. By Eqs. (3-3.4), (3-3.18), and (3-3.22), we have

$$\mathbf{g} = -G\frac{m}{r^2}\mathbf{r}_u - \frac{Gma^2\alpha}{2c^2r^2}\sin\theta\,\boldsymbol{\phi}_u \tag{3-3.24}$$

It is interesting to note that although the gravitational field of the ring does not depend on $t$, the presence of the $\boldsymbol{\phi}_u$ term makes the field different from the ordinary Newtonian field of the ring. This term is due to $[\partial\mathbf{J}/\partial t]$ in Eq. (3-1.1) and represents the "gravikinetic field" (see Chapter 11). In the case under consideration, the gravikinetic field is circular and is in the same direction as the mass current in the ring.

On the $x$ axis, the gravitational field of the ring reduces to the field found in Example 2-2.1 (for $x \gg a$).

▲

## References and Remarks for Chapter 3

**1.** See, for example, Oleg D. Jefimenko, *Electricity and Magnetism*, 2nd ed., (Electret Scientific, Star City, 1989) p. 55.
**2.** See also Sections 10.2 and 10.4.

# 4

# RETARDED INTEGRALS FOR GRAVITATIONAL AND COGRAVITATIONAL FIELDS AND POTENTIALS OF MOVING MASSES

In this chapter we shall learn how retarded integrals for gravitational and cogravitational fields and potentials can be used for finding gravitational and cogravitational fields and potentials of moving mass distributions. We shall also discover important relations between the gravitational and cogravitational fields for two special cases of moving mass distributions: an arbitrary mass distribution moving with constant velocity and a point mass in arbitrary motion.

## 4-1. Using Retarded Integrals for Finding Fields and Potentials of Moving Mass Distributions

A time-variable mass distribution always involves a movement of masses. For example, if the density of a mass distribution changes with time, then some masses change their location within the mass distribution or move to or from the mass distribution. Conversely, a moving mass distribution is inevitably a time-variable mass distribution because it increases mass density in regions of space which it enters and decreases mass density from the regions of space which it leaves. Consequently, the gravitational and cogravitational fields of a moving mass distribution can be determined from retarded field (or potential) integrals presented in Chapters 2 and 3 for the general case of time-dependent mass and mass-current distributions.

To use retarded field integrals for finding gravitational and cogravitational fields of moving mass distributions, we need to express the time derivatives $\partial\rho/\partial t$ and $\partial\mathbf{J}/\partial t$ in terms of the velocity of the mass distribution under consideration. This can be done as follows. Consider a stationary mass distribution (hereafter called "mass") of density $\rho$ as a function of $x'$, $y'$, $z'$,

$$\rho = \rho(x',y',z'). \tag{4-1.1}$$

If this mass moves with velocity $\mathbf{v}$ without changing its density, the total time derivative of $\rho$ is

$$\frac{d\rho}{dt} = \frac{\partial\rho}{\partial t} + \frac{\partial\rho}{\partial x'}\frac{dx'}{dt} + \frac{\partial\rho}{\partial y'}\frac{dy'}{dt} + \frac{\partial\rho}{\partial z'}\frac{dz'}{dt} = \frac{\partial\rho}{\partial t} + \mathbf{v}\cdot\nabla'\rho. \tag{4-1.2}$$

Since $\rho$ remains the same as the mass moves, $d\rho/dt = 0$, so that

$$\frac{\partial\rho}{\partial t} = -\,\mathbf{v}\cdot\nabla'\rho. \tag{4-1.3}$$

A moving mass constitutes a mass-current whose density is $\mathbf{J} = \rho\mathbf{v}$. Therefore, differentiating by parts,

$$\frac{\partial\mathbf{J}}{\partial t} = \frac{\partial(\rho\,\mathbf{v})}{\partial t} = -(\mathbf{v}\cdot\nabla'\rho)\,\mathbf{v} + \rho\frac{\partial\mathbf{v}}{\partial t} = -(\mathbf{v}\cdot\nabla'\rho)\,\mathbf{v} + \rho\,\dot{\mathbf{v}}. \tag{4-1.4}$$

Observe that in the retarded field integrals presented in Chapters 2 and 3, the denominator $r$ representing the distance between the volume element $dV'$ and the point of observation is not a function of time. Therefore it is not a function of time also in the case of moving masses. A moving mass must be considered as moving past different volume elements of space associated with different but fixed $r$'s. The question arises, if $dV'$ is a volume element of space, rather than a volume element of a moving mass, how does one introduce the volume of the mass into the field integrals? To answer this question, let us examine how the gravitational and cogravitational fields of a moving mass are created.

The phenomenon of retardation indicates that time-dependent masses send out gravitational (and cogravitational) field "signals" that propagate in all directions with a finite speed $c$ (as stated in Chapter 2, it is usually assumed that gravitation propagates with the speed of light). The gravitational or cogravitational field created by a time-variable mass at the point of observation is the result of the signals sent out by all the elementary masses within the mass and simultaneously "received" at the point of observation at the instant $t$. But different mass elements within the mass are at different distances from the point of observation, and the times needed for the signals originating from the different mass elements to arrive at the point of observation are different. Therefore the signals that are received at the point of observation simultaneously at the instant $t$ are sent out from the different mass elements within the mass at different retarded times $t' = t - r/c$. For a moving mass these times are different not only because different mass elements within the mass are located at different distances from the point of observation, but also because the location of these mass elements changes as the mass moves. As a result, the region of space from which the field signals responsible for the field at the point of observation are sent is not equal to the region of space, or volume, occupied by the mass when it is at rest.

Consider a mass of length $l$ moving against the $x$ axis with a constant velocity $\mathbf{v}$. The gravitational field $\mathbf{g}$ of the mass is observed at the point $O$ (Fig. 4.1). A field signal is sent from the trailing end of the mass when this end is at the distance $r_1$ from the point of observation. A field signal is sent from the leading end, when this end is at the distance $r_2$ from the point of observation. Since the leading end is closer to the point of observation than the trailing end, the field signal from the leading end must be sent at a later time if it is to arrive at the point of observation simultaneously with the signal sent from the trailing end. The difference in the times needed for the two signals to arrive at the point of observation is $r_1/c - r_2/c$. During this time the mass moves a distance $(r_1/c - r_2/c)v$. Hence the distance $l^*$ between the two points from which the two signals are sent is

$$l^* = (r_1 - r_2)v/c + l. \tag{4-1.5}$$

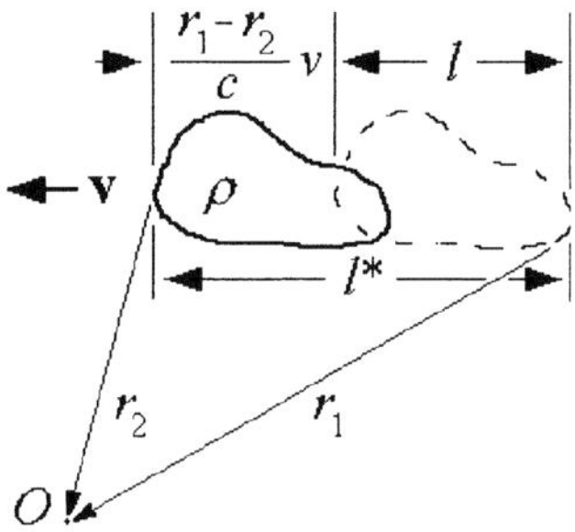


*Fig. 4.1 For the two field signals to arrive simultaneously at O, the field signal originating from the leading end of the moving mass must be sent later than the field signal originating from the trailing end of the mass.*

In this chapter we shall be mainly concerned with the special case of masses for which $r_1, r_2 \gg l^*$. In this case (see Fig. 4.2), $r_1 - r_2 = l^* \cos\phi = l^*(\mathbf{r}\cdot\mathbf{v})/rv$, where $r$ is the distance between the midpoint of $l^*$ and the point of observation, and $\phi$ is the angle between $\mathbf{r}$ and $\mathbf{v}$. Substituting this expression for $r_1 - r_2$ in Eq. (4-1.5), we have

$$l^* = l^*(\mathbf{r}\cdot\mathbf{v})/rc + l, \tag{4-1.6}$$

or

$$l^* = \frac{l}{1 - (\mathbf{r}\cdot\mathbf{v})/rc}. \tag{4-1.7}$$

Therefore, as already mentioned, the region of space from which

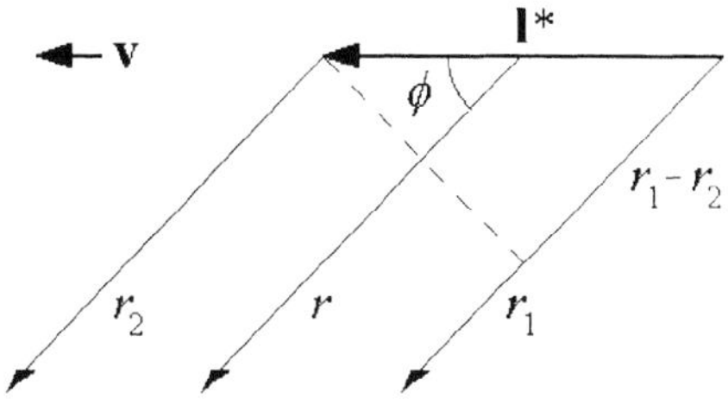


*Fig. 4.2 Geometrical relations between $r$, $\phi$, and $l^*$ when $r_1, r_2 \gg l^*$. The significance of the vector* $\mathbf{l}^*$ *will be explained later.*

the moving mass sends out the field signals resulting in the gravitational and cogravitational fields created at the point of observation is not equal to the region of space (volume) actually occupied by the mass. In the case of a mass whose linear dimensions are small compared with the distance from the mass to the point of observation, this region of space, usually called the *effective volume*, or the *retarded volume*, $\Delta V'_{ret}$ is

$$\Delta V'_{ret} = \frac{\Delta V'}{1-(\mathbf{r}\cdot\mathbf{v})/rc}, \tag{4-1.8}$$

where $\Delta V'$ is the actual volume of the mass [this equation is obtained from Eq. (4-1.7) by noting that the volume dimensions perpendicular to the direction of motion are not affected by retardation, and that the dimensions along the direction of motion change in accordance with Eq. (4-1.7)].

Although the distance $l^*$ given by Eq. (4-1.5) or Eq. (4-1.7) is a distance between two points in space rather than a length of an object, it is usually called the *retarded length*. In fact, it is actually the "visual" length of a rapidly moving body, as the length of the body would appear to a stationary observer. As follows from Eq. (4-1.7), the retarded length of a body moving toward the observer is longer, and the retarded length of a body moving away from the observer is shorter, than the actual length of the body.[1] It should be emphasized that Eqs. (4-1.6)-(4-1.8) hold only for masses or bodies observed from a distance much greater than the linear dimensions of the mass or body. For a general case, the retarded length or volume of a body cannot be expressed by a simple formula, but can be calculated in terms of the actual length of the body once the position of the body at the time of observation is given (see Section 5-3).

Another effect of retardation that needs to be taken into account when applying retarded field equations to moving masses is an apparent distortion of the shape of a moving mass. The mass appears to change its shape because the retarded times for different points within the mass are different.

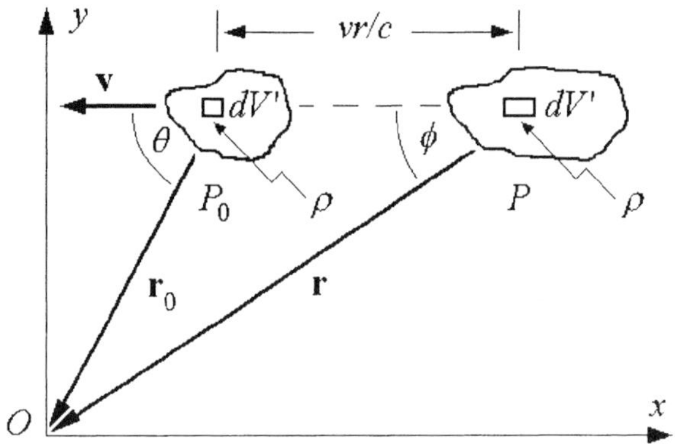


*Fig. 4.3 Geometrical relations between the "present position vector"* $\mathbf{r}_0$ *and the "retarded position vector"* $\mathbf{r}$ *for a mass moving with velocity* $\mathbf{v}$ *in the negative x direction.*

Consider a mass moving against the $x$ axis with a velocity $\mathbf{v}$ and observed from a point $O$ (Fig. 4.3). The retarded volume element $dV'$ of the mass is at the point $P$ and is represented by the vector $\mathbf{r}$. The present position of the same volume element is at the point $P_0$ and is represented by the vector $\mathbf{r}_0$. The distance $\Delta x'$ from $P$ to $P_0$ is the distance that the mass travels during the time that it takes the field signal to propagate from $P$ to $O$, that is, $\Delta x'$

$= v(r/c)$. We shall now show that, within the mass, any line parallel to the $y$ axis when the mass is at rest or at its present position appears to be slanted when the mass is moving and is at a retarded position (Fig. 4.4).

First, let us note that, according to Fig. 4.4, the relation between the $x$ and $y$ components of the retarded position vector $\mathbf{r}$ and the $x$ component of the present position vector $\mathbf{r}_0$ is (as usual, we use primes to indicate source-point coordinates)

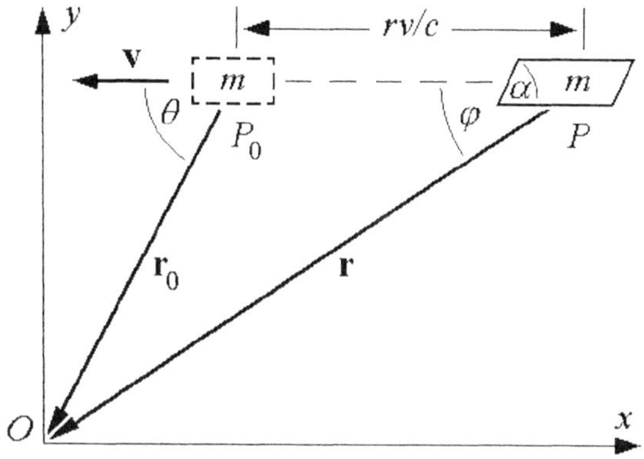


*Fig. 4.4 A mass at its retarded position appears to be elongated and its vertical lines appear to be slanted.*

$$x' = x_0' + vr/c, \tag{4-1.9}$$

or

$$x' = x_0' + (x'^2 + y'^2)^{1/2} v/c. \tag{4-1.10}$$

Differentiating Eq. (4-1.10) while keeping $x_0'$ constant, we have

$$\frac{dx'}{dy'} = \frac{y'(v/c)}{r[1 - (v/c)(x'/r)]}, \tag{4-1.11}$$

which can be written as

$$\frac{dx'}{dy'} = \frac{y'v/c}{r[1-(v/c)\cos\varphi]} = \frac{y'v/c}{r[1-(\mathbf{r}\cdot\mathbf{v})/rc]} = \frac{(v/c)\sin\varphi}{1-(\mathbf{r}\cdot\mathbf{v})/rc}. \tag{4-1.12}$$

Thus, according to Eq. (4-1.12), a vertical line ($x_0'$ = constant, $dx_0'/dy_0' = 0$) within the mass at the present position appears to be slanted when the mass is viewed at its retarded position, and the angle $\alpha$ of the slant is given by

$$\cot\alpha = \frac{y'v/c}{r[1-(\mathbf{r}\cdot\mathbf{v})/rc]}. \tag{4-1.13}$$

In the derivations presented later in Chapter 5, we shall consider a moving mass in the shape of a rectangular prism of length $l$ and thickness $a$. For determining the gravitational and cogravitational field of such a mass we shall make use of two special vectors shown in Fig. 4.5: the vector $\mathbf{l}^*$ representing the retarded length of the mass, given by

$$\mathbf{l}^* = -\frac{l}{1-(\mathbf{r}\cdot\mathbf{v})/rc}\mathbf{i}, \tag{4-1.14}$$

and the vector $\mathbf{a}^*$ representing the "slanted" thickness of the mass, given by (note that $\mathbf{r}\cdot\mathbf{v} = x'v$)

$$\mathbf{a}^* = -\frac{ay'v/c}{r[1-(\mathbf{r}\cdot\mathbf{v})/rc]}\mathbf{i} - a\mathbf{j} = -\frac{ay'v/c}{r[1-(\mathbf{r}\cdot\mathbf{v})/rc]}\mathbf{i} - \frac{a(r-x'v/c)}{r[1-(\mathbf{r}\cdot\mathbf{v})/rc]}\mathbf{j}. \tag{4-1.15}$$

We shall also use the following relation derived in Example 4-1.1 for a mass moving with acceleration $\dot{\mathbf{v}} = \partial\mathbf{v}/\partial t'$

$$\nabla' \frac{1}{[r-(\mathbf{r}\cdot\mathbf{v})/c]} = \frac{\mathbf{r} - r\mathbf{v}/c + (\mathbf{r}\cdot\dot{\mathbf{v}})\mathbf{r}/c^2}{r^3[1-(\mathbf{r}\cdot\mathbf{v})/rc]^2}. \tag{4-1.16}$$

Note that if $\dot{\mathbf{v}} = 0$ (motion with constant velocity), Eq. (4-1.16) becomes

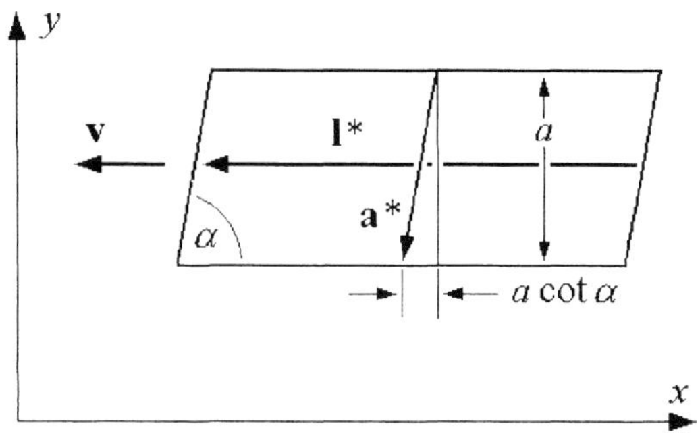


*Fig. 4.5 Explanation of the vectors* **l*** *and* **a***. *The vector* **l*** *represents the retarded length of the moving mass, the vector* **a*** *represents the "slanted" thickness of the mass.*

$$\nabla' \frac{1}{[r-(\mathbf{r}\cdot\mathbf{v})/c]} = \frac{\mathbf{r}-r\mathbf{v}/c}{r^3[1-(\mathbf{r}\cdot\mathbf{v})/rc]^2}. \tag{4-1.17}$$

In dealing with retarded integrals for moving masses, we shall frequently use the expression

$$r-(\mathbf{r}\cdot\mathbf{v})/c, \tag{4-1.18}$$

where $\mathbf{r}$ is the retarded position vector joining a retarded volume element $dV'$ of a moving mass with the point of observation. If the mass moves with a constant velocity $\mathbf{v}$, this expression can be converted to the *present position* of the mass, that is, to the position occupied by the volume element $dV'$ of the mass at the instant for which the gravitational and cogravitational fields are being determined. This can be done as follows.

First, assuming that the mass moves in the negative $x$ direction and assuming that $dV'$ is in the $xy$ plane, we see from Fig. 4.3 that the present position vector $\mathbf{r}_0$ of $dV'$ can be expressed in terms of the retarded position vector $\mathbf{r}$ as

$$\mathbf{r}_0 = \mathbf{r} - r\mathbf{v}/c. \tag{4-1.19}$$

Next, we write Eq. (4-1.18) as

$$\begin{aligned}[r-(\mathbf{r}\cdot\mathbf{v})/c] &= [r-x'v/c] \\ &= [(r-x'v/c)^2]^{1/2} = [r^2-2rx'v/c+x'^2v^2/c^2]^{1/2}.\end{aligned} \tag{4-1.20}$$

Adding and subtracting $x'^2$ and $r^2v^2/c^2$ to the right side of Eq. (4-1.20), we then have

$$\begin{aligned}&[r-(\mathbf{r}\cdot\mathbf{v})/c] \\ &= [r^2-2rx'v/c+x'^2v^2/c^2+x'^2-x'^2+r^2v^2/c^2-r^2v^2/c^2]^{1/2}.\end{aligned} \tag{4-1.21}$$

Let us now collect the terms on the right of Eq. (4-1.21) into three groups:

$$x'^2 - 2rx'v/c + r^2v^2/c^2, \tag{4-1.22}$$

$$r^2 - x'^2, \tag{4-1.23}$$

and

$$x'^2v^2/c^2 - r^2v^2/c^2. \tag{4-1.24}$$

By Eq. (4-1.9), the first group represents $x_0'^2$, where $x_0'$ is the distance between the $yz$ plane and the volume element $dV'$ of the moving mass at its present position. The second group is simply $y'^2$, where $y'$ is the (constant) $y$ coordinate of the volume element $dV'$. And the third group is $-y'^2v^2/c^2$. We can write therefore

$$\begin{aligned}[r-(\mathbf{r}\cdot\mathbf{v})/c] &= (x_0'^2+y'^2-y'^2v^2/c^2)^{1/2} \\ &= (x_0'^2+y'^2)^{1/2}\{1-(v^2/c^2)y'^2/(x_0'^2+y'^2)\}^{1/2}.\end{aligned} \tag{4-1.25}$$

But, as can be seen from Fig. 4.3, $x_0'^2 + y'^2 = r_0^2$, and $y'^2/(x_0^2 + y'^2) = \sin^2\theta$, where $\theta$ is the angle between $\mathbf{r}_0$ and the velocity vector $\mathbf{v}$. Therefore

$$[r-(\mathbf{r}\cdot\mathbf{v})/c] = r[1-(\mathbf{r}\cdot\mathbf{v})/rc] = r_0\{1-(v^2/c^2)\sin^2\theta\}^{1/2}, \tag{4-1.26}$$

where all the quantities in the last expression are present time quantities. In obtaining Eqs. (4-1.25) and (4-1.26) we assumed that the volume element $dV'$ of the moving mass was located in the $xy$ plane. Clearly, however, the two equations are valid even if $dV'$ is not in that plane, provided that we replace in these equations $y'^2$ by $y'^2 + z'^2$.

Expressions involving the retarded position vector $\mathbf{r}$ and its magnitude $r$ have a very peculiar and important property which should be kept in mind when dealing with moving masses and mass-currents. As already mentioned, a moving mass is assumed to move through different but *fixed* points of space. Therefore neither the retarded position vector $\mathbf{r}$ nor its magnitude $r$ explicitly appearing in retarded integrals is a function of time. On the other hand, in the case of moving masses and currents, the distance $r$ appearing in the retarded time $t' = t - r/c$ is variable and therefore is a function of time. The same applies to Eqs. (4-1.7) - (4-1.17) presented above and to all similar expressions.

▼

**Example 4-1.1** Derive Eq. (4-1.16).

Let us arrange a rectangular system of coordinates so that the acceleration vector of the moving mass is in the $xy$ plane and the velocity vector is in the negative $x$ direction. Let the point of observation be at the origin. The position vector of the mass is then $\mathbf{r} = -x'\mathbf{i} - y'\mathbf{j}$. Using vector identity (V-7), we have

$$\nabla' \frac{1}{[r - (\mathbf{r} \cdot \mathbf{v})/c]} = -\frac{\nabla'[r - (\mathbf{r} \cdot \mathbf{v})/c]}{[r - (\mathbf{r} \cdot \mathbf{v})/c]^2}. \qquad (4\text{-}1.27)$$

In differentiating the numerator in Eq. (4-1.27), we should remember that the numerator is retarded. However, as explained in Section 4-1, neither the position vector $\mathbf{r}$ nor its magnitude $r$ appearing in retarded integrals is a function of time and therefore neither is affected by retardation (the mass moves through different but *fixed* points of space). The only quantity in the numerator affected by retardation is the velocity $\mathbf{v}$ which is a function of the retarded time $t - r/c$ and does change as the mass moves. Hence we can write, making use of vector identity (V-5),

$$\nabla' \frac{1}{[r-(\mathbf{r}\cdot\mathbf{v})/c]} = -\frac{\nabla' r - \nabla'[(\mathbf{r}\cdot\mathbf{v})/c]}{[r-(\mathbf{r}\cdot\mathbf{v})/c]^2} = -\frac{-\mathbf{r}_u - (1/c)\nabla'[\mathbf{r}\cdot\mathbf{v}]}{[r-(\mathbf{r}\cdot\mathbf{v})/c]^2}. \tag{4-1.28}$$

To evaluate $\nabla'[\mathbf{r}\cdot\mathbf{v}]$, we first use vector identity (V-31), obtaining

$$\nabla'[\mathbf{r}\cdot\mathbf{v}] = [\nabla'(\mathbf{r}\cdot\mathbf{v})] + \frac{\mathbf{r}_u}{c}\left[\frac{\partial(\mathbf{r}\cdot\mathbf{v})}{\partial t}\right]. \tag{4-1.29}$$

The first expression on the right can be evaluated with the help of vector identity (V-6). Note that in this expression $\nabla'$ operates upon unretarded quantities. Therefore we have

$$\nabla'(\mathbf{r}\cdot\mathbf{v}) = (\mathbf{r}\cdot\nabla')\mathbf{v} + \mathbf{r}\times(\nabla'\times\mathbf{v}) + (\mathbf{v}\cdot\nabla')\mathbf{r} + \mathbf{v}\times(\nabla'\times\mathbf{r}). \tag{4-1.30}$$

Since all the quantities in this equation are unretarded, and since the unretarded $\mathbf{v}$ does not depend on spatial coordinates, the first two terms on the right of this equation vanish. Since $\nabla'\times\mathbf{r} = 0$, the last term vanishes also. By vector identity (V-4), the remaining term is simply $-\mathbf{v}$. We thus obtain

$$\nabla'(\mathbf{r}\cdot\mathbf{v}) = -\mathbf{v}. \tag{4-1.31}$$

Taking into account that $\mathbf{r}$ in the last term of Eq. (4-1.29) is not a function of time, we have

$$\frac{\mathbf{r}_u}{c}\left[\frac{\partial(\mathbf{r}\cdot\mathbf{v})}{\partial t}\right] = \frac{\mathbf{r}_u}{c}\left[\mathbf{r}\cdot\frac{\partial\mathbf{v}}{\partial t}\right] = \frac{\mathbf{r}_u}{c}[\mathbf{r}\cdot\dot{\mathbf{v}}]. \tag{4-1.32}$$

Combining Eqs. (4-1.28), (4-1.29), (4-1.31), and (4-1.32), factoring out $r$ in the denominator, and multiplying the numerator and the denominator by $r$, we finally obtain

$$\nabla' \frac{1}{[r-(\mathbf{r}\cdot\mathbf{v})/c]} = \frac{\mathbf{r} - r\mathbf{v}/c + (\mathbf{r}\cdot\dot{\mathbf{v}})\mathbf{r}/c^2}{r^3[1-(\mathbf{r}\cdot\mathbf{v})/rc]^2}. \tag{4-1.16}$$

Although all quantities in Eq. (4-1.16) refer to the retarded position of the mass, to avoid an exceedingly cumbersome notation we do not place them between the retardation brackets.

▲

## 4-2. Correlation Between the Gravitational and the Cogravitational Field of a Moving Mass Distribution

There are two special cases of moving mass distributions for which there exist simple correlations between the gravitational and the cogravitational field produced by the distributions. The first case is that of an arbitrary mass distribution moving with constant velocity. The second case is that of a point mass moving with acceleration.

Consider first a mass distribution of arbitrary size and shape moving with constant velocity **v**. Let us form the vector product of **v** and Eq. (3-1.1) with the two integrals in Eq. (3-1.1) combined into a single integral. Since **v** is a constant vector, we can place it under the integral sign, so that

$$\mathbf{v} \times \mathbf{g} = G\int \frac{\mathbf{v} \times \left[\nabla'\rho + \dfrac{1}{c^2}\dfrac{\partial \mathbf{J}}{\partial t}\right]}{r}\, dV' . \qquad (4\text{-}2.1)$$

If a mass distribution moves with constant velocity **v**, then by Eq. (4-1.4) the derivative $\partial\mathbf{J}/\partial t$ is parallel to **v**. Therefore the product $\mathbf{v} \times [\partial\mathbf{J}/\partial t]$ vanishes, and since **v** is not affected by retardation, Eq. (4-2.1) simplifies to

$$\mathbf{v} \times \mathbf{g} = G\int \frac{[\mathbf{v} \times \nabla'\rho]}{r}\, dV' . \qquad (4\text{-}2.2)$$

Using now vector identity (V-11), taking into account that $\mathbf{v} \times \nabla'\rho = -\nabla'\rho \times \mathbf{v}$, and taking into account that $\nabla' \times \mathbf{v} = 0$ and that $\mathbf{v}\rho = \mathbf{J}$, we obtain from Eq. (4-2.2)

$$\mathbf{v}\times\mathbf{g} = -G\int\frac{[\nabla'\times\mathbf{J}]}{r}dV', \tag{4-2.3}$$

which, by Eq. (3-1.2), is equivalent to

$$\mathbf{K} = (\mathbf{v}\times\mathbf{g})/c^2. \tag{4-2.4}$$

Note that **K** is perpendicular to **v** and **g**, and that **g** in Eq. (4-2.4) is the gravitational field produced by a *moving* mass distribution.

It is interesting to note that since, in the present case, the term $\partial\mathbf{J}/\partial t$ in Eq. (4-2.1) makes no contribution to $\mathbf{v}\times\mathbf{g}$, we can write Eq. (4-2.4), using Eq. (3-1.1), as

$$\mathbf{K} = \mathbf{v}\times\frac{G}{c^2}\int\frac{[\nabla'\rho]}{r}dV', \tag{4-2.5}$$

and, assuming that the velocity is along the $x$ axis, so that $\mathbf{v}\times\mathbf{i} = 0$, as

$$\mathbf{K} = \mathbf{v}\times\frac{G}{c^2}\int\frac{[(\nabla_y'+\nabla_z')\rho]}{r}dV', \tag{4-2.6}$$

where only the components of $\nabla'$ perpendicular to **v** occur. Furthermore, using Eq. (2-2.1) and taking into account that $\partial\mathbf{J}/\partial t$ makes no contribution to $\mathbf{v}\times\mathbf{g}$ and that $\mathbf{v}\times\mathbf{i} = 0$, we can write Eq. (4-2.4) as

$$\mathbf{K} = -\mathbf{v}\times\frac{G}{c^2}\int\left\{\frac{[\rho]}{r^3}+\frac{1}{r^2c}\left[\frac{\partial\rho}{\partial t}\right]\right\}(y\mathbf{j}+z\mathbf{k})\,dV'. \tag{4-2.7}$$

As it follows from Eqs. (4-1.7) and (4-1.8), for slowly moving mass distributions the retardation can be neglected, in which case Eq. (4-2.4) reduces to

$$\mathbf{K} = (\mathbf{v} \times \mathbf{g})/c^2, \tag{4-2.8}$$

where $\mathbf{g}$ is the ordinary Newtonian gravitational field of the mass distribution under consideration. Likewise, Eqs. (4-2.5)-(4-2.7) reduce to the corresponding equations involving unretarded mass densities.

Consider now a point mass moving with acceleration. Let us assume that the retarded position of the point mass is given by the vector $\mathbf{r}$, and let us form the cross product of $\mathbf{r}/cr$ and Eq. (3-1.3). Assuming that $\mathbf{r}$ for a moving point mass can be considered the same throughout the entire volume occupied by the mass, we can place $\mathbf{r}/r$ under the integral signs.[2] Noting that $\mathbf{r} \times \mathbf{r} = 0$, we then obtain, transposing $\mathbf{r}$ in the integrand,

$$\frac{\mathbf{r} \times \mathbf{g}}{cr} = -\frac{G}{c^2}\int\left\{\frac{[\mathbf{J}]}{r^3} + \frac{1}{r^2 c}\left[\frac{\partial \mathbf{J}}{\partial t}\right]\right\} \times \mathbf{r}\, dV'. \tag{4-2.9}$$

Using now Eq. (2-2.2) with $\mathbf{J} = \rho\mathbf{v}$, and using vector identity (V-36), we immediately obtain

$$\mathbf{K} = \frac{\mathbf{r} \times \mathbf{g}}{cr}, \tag{4-2.10}$$

where $\mathbf{r}$ is the *retarded* position vector connecting the moving point mass with the point of observation. Equation (4-2.10) shows that the cogravitational field of a moving point mass is perpendicular to the gravitational field produced by the mass and to the radius vector joining the retarded position of the mass with the point of observation.[3]

It is interesting to note that for a point mass moving with constant velocity, Eq. (4-2.4) as well as Eq. (4-2.10) hold, because Eqs. (4-2.10) is true for any acceleration, including zero acceleration. However, it is important to remember that Eq. (4-2.10) involves the retarded position vector $\mathbf{r}$. If the acceleration is zero, Eq. (4-2.10) reduces to Eq. (4-2.4), as is shown in Example 5-1.1.

## References and Remarks for Chapter 4

**1.** The retarded length should not be confused with the relativistic "Lorentz-contracted length;" see Oleg D. Jefimenko, *Electromagnetic Retardation and Theory of Relativity,* 2nd ed., (Electret Scientific, Star City, 2004), pp. 207-209.

**2.** This procedure is generally applicable to stationary point masses only. For moving point masses its applicability depends on certain parameters of the system under consideration. See Section 5-7 [in particular Eqs. (5-7.1) and (5-7.2)] for details.

**3.** It is important to stress that Eq. (4-2.10) is only approximately correct. See Section 5-7 for details.

# 5

# GRAVITATIONAL AND COGRAVITATIONAL FIELDS AND POTENTIALS OF MOVING POINT AND LINE MASSES

The finite propagation speed of gravitational and cogravitational "signals" has a profound effect on the gravitational and cogravitational fields and potentials of moving mass distributions. In this chapter, starting with retarded field integrals, we shall compute and analyze gravitational and cogravitational fields and potentials of the two simplest types of moving mass distributions: a moving point mass and a moving line mass.

## 5-1. The Gravitational Field of a Uniformly Moving Point Mass

Any stationary mass distribution viewed from a sufficiently large distance constitutes a "point mass."[1] Consider a mass distribution of total mass $m$ and density $\rho$ confined to a small rectangular prism (Fig. 5.1) whose center is located at the point $x'$, $y'$ in the $xy$ plane of a rectangular system of coordinates, and whose sides $l$, $a$, and $b$ are parallel to the $x$, $y$, and $z$ axis, respectively. Let the point of observation be at the origin of the

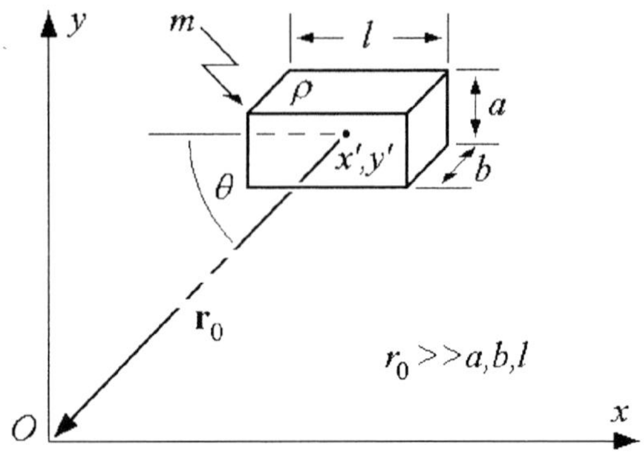


*Fig. 5.1 A mass of uniform density ρ is confined to a small rectangular prism. The mass of the prism is m. The mass constitutes a point mass when viewed from a distance large compared to the linear dimensions of the prism.*

coordinates, and let the distance between the center of the prism and the origin be $r_0 \gg a,\ b,\ l$. Viewed from the origin, this mass distribution constitutes a point mass. Let the mass move with uniform velocity $\mathbf{v} = -v\mathbf{i}$. We want to find the gravitational and cogravitational fields of this mass at the point of observation.[2]

To find the gravitational field of this mass, we use Eq. (3-1.1). First we eliminate from Eq. (3-1.1) the term with the mass-current density **J**. We can do so with the help of Eq. (4-1.4). Since the velocity of our mass is $\mathbf{v} = v_x\mathbf{i} = -v\mathbf{i}$, and since the mass moves without acceleration so that $\dot{\mathbf{v}} = 0$, Eq. (4-1.4) gives

$$\frac{\partial \mathbf{J}}{\partial t} = -\left(v_x \frac{\partial \rho}{\partial x'}\right)\mathbf{v} = -v^2 \frac{\partial \rho}{\partial x'}\mathbf{i}. \tag{5-1.1}$$

Substituting Eq. (5-1.1) into Eq. (3-1.1), we then have for the gravitational field of the mass

$$\mathbf{g} = G\int \frac{\left[\nabla'\rho - \dfrac{v^2}{c^2}\dfrac{\partial\rho}{\partial x'}\mathbf{i}\right]}{r}\,dV'. \tag{5-1.2}$$

Observe that in this equation $\nabla'$ and $\partial/\partial x'$ operate on the *unretarded* $\rho$, so that in computing $\nabla'\rho$ and $\partial\rho/\partial x'$ we must use the ordinary, unretarded, shape and size of the prism. Since $\rho$ is constant within the prism, $\nabla'\rho = 0$ within it, and the only contribution to $\nabla'\rho$ comes from the surface layer of the prism, where $\rho$ changes from $\rho$ (inside the prism) to 0 (outside the prism). Let the thickness of the surface layer be $w$. Taking into account that $\nabla'\rho$ represents the rate of change of $\rho$ in the positive direction of the greatest rate of change, we then have $\nabla'\rho = (\rho/w)\mathbf{n}_{in}$, where $\mathbf{n}_{in}$ is a unit vector normal to the surface layer and pointing *into* the prism. Hence $\nabla'\rho$ for the right, left, top, bottom, front, and back surfaces of the mass (prism) are $-(\rho/w)\mathbf{i}$, $(\rho/w)\mathbf{i}$, $-(\rho/w)\mathbf{j}$, $(\rho/w)\mathbf{j}$, $-(\rho/w)\mathbf{k}$, and $(\rho/w)\mathbf{k}$, respectively. Likewise, $\partial\rho/\partial x'$ is zero in the interior of the mass and is different from zero only in the left and in the right surface layers of the mass, where $\partial\rho/\partial x' = \rho/w$ in the left surface layer and $\partial\rho/\partial x' = -\rho/w$ in the right surface layer.

The volume integral of Eq. (5-1.2) can be split therefore into six integrals, one over each of the six surface layers corresponding to the six surfaces of the mass (prism). However, since the center of the mass is in the $xy$ plane ($z' = 0$), the integrals over the two surface layers parallel to the $xy$ plane cancel each other, because $\nabla'\rho$ for one of the layers is opposite to that for the other layer, while $r$ is the same for both layers. Thus only the four integrals over the layers parallel to the $xz$ and $yz$ planes remain. Let us designate the retarded distances from these layers to the point of observation as $r_1$, $r_2$, $r_3$, and $r_4$ (see Figs. 5.2 and 5.3). Since the linear dimensions of the mass are much smaller than $r_1$, $r_2$, $r_3$, and $r_4$, we can replace each integral over a surface

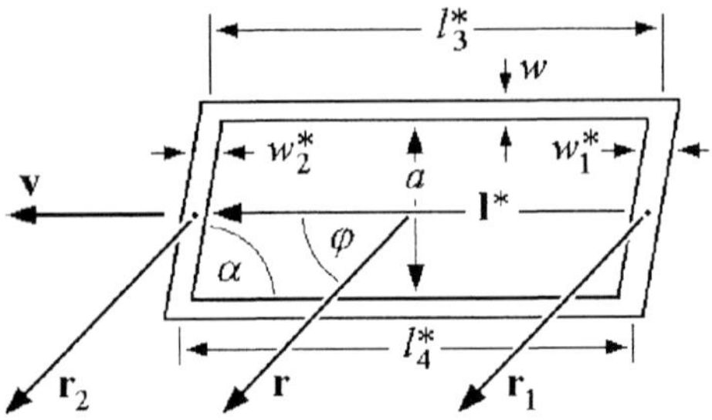


*Fig. 5.2 When the mass shown in Fig. 5.1 is moving and is at a retarded position, its apparent length, shape, and thickness of its front and back surface layers are no longer the same as for the stationary mass. (All* **r***'s meet at the origin).*

layer by the product of the integrand and the volume of the corresponding layer. However, the integration in Eq. (5-1.2) is over the *effective* (retarded) volume of the mass, and therefore we must use not the true volume of the surface layers, but their effective volume. The effective volume of the surface layers is not the same as their actual volume, because, in accordance with Eq. (4-1.7), the length $l$ of the two layers parallel to the $xz$ plane must be replaced by

$$l^* = \frac{l}{1-(\mathbf{r}\cdot\mathbf{v})/rc}, \tag{5-1.3}$$

and because, also in accordance with Eq. (4-1.7), the thickness $w$ of the two layers parallel to the $yz$ plane must be replaced by

$$w^* = \frac{w}{1-(\mathbf{r}\cdot\mathbf{v})/rc}. \tag{5-1.4}$$

Equation (5-1.2) becomes therefore

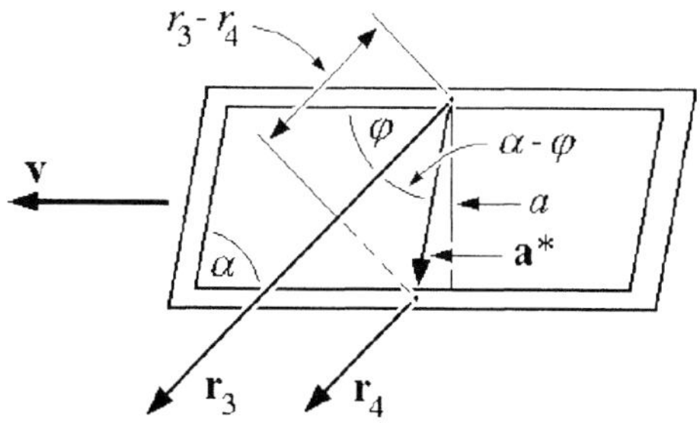


*Fig. 5.3 The relations between* $\mathbf{r}_3$, $\mathbf{r}_4$, *and* $\mathbf{a}^*$ *for the moving mass at a retarded position. (The two* $\mathbf{r}$*'s meet at the origin.)*

$$\mathbf{g} = G\left[\frac{\rho/w}{r_1}abw_1^*(-\mathbf{i}) + \frac{\rho/w}{r_2}abw_2^*\mathbf{i} + \frac{\rho/w}{r_3}bl_3^*w(-\mathbf{j}) + \frac{\rho/w}{r_4}bl_4^*w\mathbf{j} + \left(\frac{v^2}{c^2}\right)\left(\frac{\rho/w}{r_1}abw_1^*\mathbf{i} + \frac{\rho/w}{r_2}abw_2^*(-\mathbf{i})\right)\right] \tag{5-1.5}$$

or, substituting $l^*$ and $w^*$ from Eqs. (5-1.3) and (5-1.4),

$$\begin{aligned}\mathbf{g} = G\bigg[&\frac{\rho/w}{r_1 - \mathbf{r}_1\cdot\mathbf{v}/c}abw(-\mathbf{i}) + \frac{\rho/w}{r_2 - \mathbf{r}_2\cdot\mathbf{v}/c}abw\mathbf{i} \\ &+ \frac{\rho/w}{r_3 - \mathbf{r}_3\cdot\mathbf{v}/c}blw(-\mathbf{j}) + \frac{\rho/w}{r_4 - \mathbf{r}_4\cdot\mathbf{v}/c}blw\mathbf{j} \\ &+ \left(\frac{v^2}{c^2}\right)\left(\frac{\rho/w}{r_1 - \mathbf{r}_1\cdot\mathbf{v}/c}abw\mathbf{i} + \frac{\rho/w}{r_2 - \mathbf{r}_2\cdot\mathbf{v}/c}abw(-\mathbf{i})\right)\bigg],\end{aligned} \tag{5-1.6}$$

which simplifies to

$$\begin{aligned}\mathbf{g} = G\rho b\bigg[&\left(1 - \frac{v^2}{c^2}\right)\left(\frac{1}{r_2 - \mathbf{r}_2\cdot\mathbf{v}/c} - \frac{1}{r_1 - \mathbf{r}_1\cdot\mathbf{v}/c}\right)a\mathbf{i} \\ &+ \left(\frac{1}{r_4 - \mathbf{r}_4\cdot\mathbf{v}/c} - \frac{1}{r_3 - \mathbf{r}_3\cdot\mathbf{v}/c}\right)l\mathbf{j}\bigg].\end{aligned} \tag{5-1.7}$$

As can be seen from Figs. 5.2 and 5.3, the differences of the fractions in these equations are simply the increments of the function $1/(r - \mathbf{r} \cdot \mathbf{v}/c)$ associated with the displacement of the tail of $\mathbf{r}$ over the distances represented by the vector $\mathbf{l}^*$ [in the $\mathbf{i}$ component of Eq. (5-1.7)] and by the vector $\mathbf{a}^*$ [in the $\mathbf{j}$ component of Eq. (5-1.7)]. Therefore we can write Eq. (5-1.7) as[3]

$$\mathbf{g} = G\rho b\left\{\left(1 - \frac{v^2}{c^2}\right)\left[\left(\nabla' \frac{1}{r - \mathbf{r} \cdot \mathbf{v}/c}\right) \cdot \mathbf{l}^*\right]a\mathbf{i} + \left[\left(\nabla' \frac{1}{r - \mathbf{r} \cdot \mathbf{v}/c}\right) \cdot \mathbf{a}^*\right]l\mathbf{j}\right\}. \tag{5-1.8}$$

Substituting the gradient from Eq. (4-1.17) (remembering that $\dot{\mathbf{v}} = 0$) and substituting $\mathbf{l}^*$ and $\mathbf{a}^*$ from Eqs. (4-1.14) and (4-1.15), we have

$$\begin{aligned}\mathbf{g} = -G\rho b\Bigg[&\left(1 - \frac{v^2}{c^2}\right)\left(\frac{\mathbf{r} - r\mathbf{v}/c}{r^3(1 - \mathbf{r} \cdot \mathbf{v}/rc)^2} \cdot \mathbf{i}\right)\frac{la}{1 - \mathbf{r} \cdot \mathbf{v}/rc}\mathbf{i} \\ &+ \left(\frac{\mathbf{r} - r\mathbf{v}/c}{r^3(1 - \mathbf{r} \cdot \mathbf{v}/rc)^2} \cdot \mathbf{i}\right)\frac{y'v/c}{r(1 - \mathbf{r} \cdot \mathbf{v}/rc)}al\mathbf{j} \\ &+ \left(\frac{\mathbf{r} - r\mathbf{v}/c}{r^3(1 - \mathbf{r} \cdot \mathbf{v}/rc)^2} \cdot \mathbf{j}\right)\frac{r - x'v/c}{r(1 - \mathbf{r} \cdot \mathbf{v}/rc)}al\mathbf{j}\Bigg].\end{aligned} \tag{5-1.9}$$

Simplifying and taking into account that $\mathbf{r} \cdot \mathbf{i} = -x'$, $\mathbf{r} \cdot \mathbf{j} = -y'$, $\mathbf{v} \cdot \mathbf{i} = -v$, $\mathbf{v} \cdot \mathbf{j} = 0$, and $\mathbf{r} \cdot \mathbf{v} = x'v$, we obtain

$$\begin{aligned}\mathbf{g} &= -\frac{G\rho abl}{r^3[1 - \mathbf{r} \cdot \mathbf{v}/rc]^3}\Bigg[\left(1 - \frac{v^2}{c^2}\right)(-x' + rv/c)\,\mathbf{i} \\ &\quad + (-x' + rv/c)\frac{y'v/c}{r}\mathbf{j} + (-y')\frac{r - x'v/c}{r}\mathbf{j}\Bigg] \\ &= -\frac{G\rho abl}{r^3[1 - \mathbf{r} \cdot \mathbf{v}/rc]^3}\left[\left(1 - \frac{v^2}{c^2}\right)(-x'\mathbf{i} - r\mathbf{v}/c) + \left(1 - \frac{v^2}{c^2}\right)(-y)\mathbf{j}\right],\end{aligned} \tag{5-1.10}$$

and finally, noting that $\mathbf{r} = -x'\mathbf{i} - y'\mathbf{j}$, and that $\rho abl = m$,

$$\mathbf{g} = -G\frac{m(1 - v^2/c^2)}{r^3[1 - \mathbf{r}\cdot\mathbf{v}/rc]^3}\left[\mathbf{r} - \frac{r\mathbf{v}}{c}\right]. \qquad (5\text{-}1.11)$$

Equation (5-1.11) expresses $\mathbf{g}$ in terms of the *retarded* position of the mass specified by the retarded position vector $\mathbf{r}$ (see Fig. 4.4). Usually it is desirable to express $\mathbf{g}$ in terms of the *present* position of the mass specified by the present position vector $\mathbf{r}_0$ (see Fig. 4.4). We can convert Eq. (5-1.11) from $\mathbf{r}$ to $\mathbf{r}_0$ by using Eqs. (4-1.19) and (4-1.26). According to Eq. (4-1.19),

$$\mathbf{r} - r\mathbf{v}/c = \mathbf{r}_0, \qquad (5\text{-}1.12)$$

so that the last factor in Eq. (5-1.11) is simply the present position vector $\mathbf{r}_0$. Substituting Eq. (5-1.12) and Eq. (4-1.26) into Eq. (5-1.11), we obtain the desired equation for the gravitational field of a uniformly moving point mass expressed in terms of the present position of the mass (thus without gravitational aberration)

$$\mathbf{g} = -G\frac{m(1 - v^2/c^2)}{r_0^3\{1 - (v^2/c^2)\sin^2\theta\}^{3/2}}\mathbf{r}_0. \qquad (5\text{-}1.13)$$

This equation (in a different notation) was first derived by Oliver Heaviside in 1893 on the basis of the analogy between gravitation and electromagnetism.[4]

There are two interesting properties of Eq. (5-1.13). First, as was noted by Heaviside, with increasing velocity of the mass the gravitational field of the mass concentrates itself more and more about the equatorial plane, $\theta = \pi/2$, and decreases along the line of motion, $\theta = 0$. This effect is shown in Fig. 5.4a. Second, the gravitational field *appears to originate* at the mass in its present position. This, of course, is merely an illusion, because by supposition the distance between the mass and the point of observation is much greater than the linear dimensions of the mass, so that neither Eq. (5-1.11) nor Eq. (5-1.13) gives us any information concerning the structure of the field close to the mass.

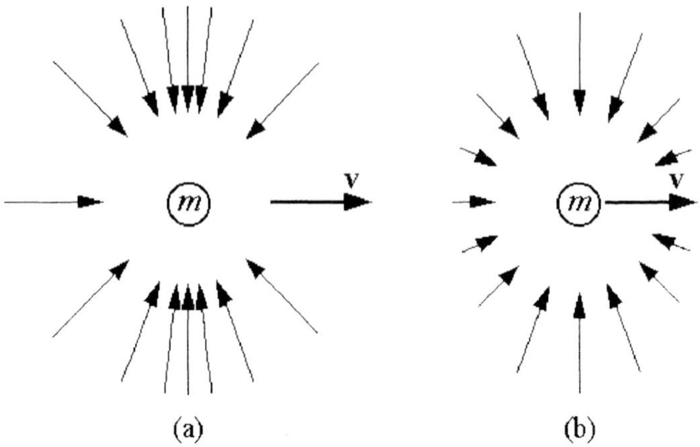


*Fig. 5.4* (a) *As first noticed by Heaviside, the gravitational field of a moving point mass concentrates itself in the direction perpendicular to the direction of motion of the mass and decreases along the line of motion.* (b) *A more accurate way to show the gravitational field of a moving point mass is to use uniformly spaced field vectors of different lengths* (*see also Section 13-3*).

Note also that because of the finite speed of the propagation of the field signals and light signals one can never *observe* the mass at its present position. In fact, the mass could have stopped after sending the field signal from its retarded position, and even then Eq. (5-1.13) would remain valid, although in this case Eq. (5-1.13) would apply to the "projected," or "anticipated," present position of the mass.

▼

**Example 5-1.1** Show that for a point mass moving without acceleration Eq. (4-2.10) reduces to (4-2.4).

According to Eq. (5-1.12), the retarded position vector of the mass can be expressed in terms of the present position as

$$\mathbf{r} = \mathbf{r}_0 + r\mathbf{v}/c. \tag{5-1.14}$$

Substituting Eq. (5-1.14) into Eq. (4-2.10), we have

$$\mathbf{K} = \frac{\mathbf{r}\times\mathbf{g}}{cr} = \frac{(\mathbf{r}_0 + r\mathbf{v}/c)\times\mathbf{g}}{cr} = \frac{\mathbf{r}_0\times\mathbf{g}}{cr} + \frac{(r\mathbf{v}/c)\times\mathbf{g}}{cr}. \tag{5-1.15}$$

Since, by Eq. (5-1.13), $\mathbf{g}$ is directed along $\mathbf{r}_0$, $\mathbf{r}_0 \times \mathbf{g} = 0$, and we obtain [compare with Eq. (4-2.4)]

$$\mathbf{K} = (\mathbf{v} \times \mathbf{g})/c^2. \tag{5-1.16}$$

**Example 5-1.2** Equation (5-1.13) represents a "snapshot" of the gravitational field of a moving point mass, since it does not express the field as a function of time. Modify Eq. (5-1.13) so that it shows how the field changes as the mass moves.

Let the "snapshot" be for $t = 0$. If the mass moves in the $-x$ direction, the functional dependence of $\mathbf{g}$ on the $x$ coordinate will be preserved for $t \neq 0$ if we express Eq. (5-1.13) in terms of $x_0'$ and replace $x_0'$ by $x_0' - vt$. From Eqs. (4-1.26) and (4-1.25), we have

$$\begin{aligned} r_0\{1-(v^2/c^2)\sin^2\theta\}^{1/2} &= (x_0'^2 + y'^2 - y'^2v^2/c^2)^{1/2} \\ &= [x_0'^2 + (1-v^2/c^2)y'^2]^{1/2}. \end{aligned} \tag{5-1.17}$$

Replacing in Eq. (5-1.17) $x_0'$ by $x_0' - vt$, we obtain

$$r_0\{1-(v^2/c^2)\sin^2\theta\}^{1/2} = [(x_0' - vt)^2 + (1-v^2/c^2)y'^2]^{1/2}, \tag{5-1.18}$$

where $x_0'$ is now the $x$ coordinate of the point mass at $t = 0$. Expressing $\mathbf{r}_0$ in terms of its components and replacing $x_0'$ by $x_0' - vt$, we similarly have $\mathbf{r}_0 = -(x_0' - vt)\mathbf{i} - y'\mathbf{j}$. Therefore Eq. (5-1.13) can be written as

$$\mathbf{g} = G\frac{n(1 - v^2/c^2)\{(x_0' - vt)\mathbf{i} + y'\mathbf{j}\}}{\{(x_0' - vt)^2 + (1 - v^2/c^2)y'^2\}^{3/2}}, \tag{5-1.19}$$

where the dependence of **g** on $t$ is shown explicitly. This equation holds for the mass moving parallel to the $x$ axis in the $xy$ plane. If it moves parallel to the $x$ axis anywhere in space, $y'^2$ in this equation should be replaced by $(y'^2 + z'^2)$.

▲

## 5-2. The Cogravitational Field of a Uniformly Moving Point Mass

Although by using Eq. (3-1.2) or Eq. (2-2.2), we can find the cogravitational field of a uniformly moving point mass in the same manner as we found the gravitational field in Section 5-1 (see Example 5-2.1), it is much easier to find it from the known gravitational field by using Eq. (4-2.4).

Applying Eq. (4-2.4) to Eq. (5-1.11), we obtain for the cogravitational field in terms of the retarded position of the mass

$$\mathbf{K} = -G\frac{m[1 - v^2/c^2]}{r^3c^2[1 - \mathbf{r}\cdot\mathbf{v}/rc]^3}[\mathbf{v}\times\mathbf{r}]. \tag{5-2.1}$$

Applying Eq. (4-2.4) to Eq. (5-1.13), we obtain for the cogravitational field in terms of the present position of the mass

$$\mathbf{K} = -G\frac{m(1 - v^2/c^2)}{r_0^3c^2\{1 - (v^2/c^2)\sin^2\theta\}^{3/2}}(\mathbf{v}\times\mathbf{r}_0). \tag{5-2.2}$$

▼

**Example 5-2.1** Find the cogravitational field of a uniformly moving point mass shown in Fig. 5.1 by using Eq. (3-1.2),

$$\mathbf{K} = -\frac{G}{c^2}\int\frac{[\nabla'\times\mathbf{J}]}{r}dV'. \tag{3-1.2}$$

To use Eq. (3-1.2), we need to know $\nabla' \times \mathbf{J}$ associated with the mass under consideration. The moving mass constitutes a mass-current density $\mathbf{J} = \rho\mathbf{v}$. Since $\mathbf{v}$ is not a function of $x'$, $y'$, $z'$, we have $\nabla' \times \mathbf{J} = \nabla'\rho \times \mathbf{v}$. But $\rho$ is constant within the mass, and therefore the only contribution to $\nabla' \times \mathbf{J}$ comes from the surface layer of the mass, where $\rho$ changes from $\rho$ (inside the mass) to 0 (outside the mass). Using the values for $\nabla'\rho$ obtained in Section 5-1, we then have for $\nabla' \times \mathbf{J}$ of the top, bottom, front, and back surface layers of the mass (prism) $-\rho v/w\mathbf{k}$, $\rho v/w\mathbf{k}$, $\rho v/w\mathbf{j}$, and $-\rho/w\mathbf{j}$, respectively; the left and right surface layers make no contribution to $\nabla' \times \mathbf{J}$, because $\mathbf{v}$ and $\nabla'\rho$ are parallel (or antiparallel) there. Furthermore, since $\nabla' \times \mathbf{J}$ in the front surface layer is opposite to $\nabla' \times \mathbf{J}$ in the back surface layer, while both surface layers are at the same distance $r$ from the point of observation, the contributions of these two layers to the integral in Eq. (3-1.2) cancel each other, so that only the top and the bottom surface layers contribute to the cogravitational field of the mass.

Since the linear dimensions of the mass are much smaller than $r_3$ and $r_4$ (see Figs. 5.1 and 5.3), we can replace the integrals over the two surface layers by the product of the integrand and the volumes of the corresponding layers. Using Eq. (3-1.2) and taking into account the effective volume of the boundary layers (see Sections 4-1 and 5-1), we have, as in Eqs. (5-1.5)-(5-1.7),

$$\begin{aligned}\mathbf{K} &= \frac{G}{c^2}\left[\frac{\rho v/w}{r_3 - \mathbf{r}_3\cdot\mathbf{v}/c}wbl\mathbf{k} + \frac{\rho v/w}{r_4 - \mathbf{r}_4\cdot\mathbf{v}/c}wbl\mathbf{k}\right]\\ &= G\frac{\rho vbl}{c^2}\left[\frac{1}{r_3 - \mathbf{r}_3\cdot\mathbf{v}/c} - \frac{1}{r_4 - \mathbf{r}_4\cdot\mathbf{v}/c}\right]\mathbf{k}.\end{aligned} \tag{5-2.3}$$

The difference of the two fractions in the last expression is simply the increment of the function $1/(r - \mathbf{r}\cdot\mathbf{v}/c)$ associated with the displacement of the tail of **r** over the distance represented by the vector **a*** (see Fig 5.3). Therefore, using Eqs. (4-1.17) and (4-1.15), we can write Eq. (5-2.3) as

$$\mathbf{K} = G\frac{\rho bvl}{c^2}\left[\left(\frac{\mathbf{r} - r\mathbf{v}/c}{r^3(1 - \mathbf{r}\cdot\mathbf{v}/rc)^2}\cdot\mathbf{i}\right)\frac{y'v/c}{r(1-\mathbf{r}\cdot\mathbf{v}/rc)}a + \left(\frac{\mathbf{r} - r\mathbf{v}/c}{r^3(1 - \mathbf{r}\cdot\mathbf{v}/rc)^2}\cdot\mathbf{j}\right)\frac{r - x'v/c}{r(1-\mathbf{r}\cdot\mathbf{v}/rc)}a\right]\mathbf{k}. \tag{5-2.4}$$

Simplifying and taking into account that $\mathbf{r}\cdot\mathbf{i} = -x'$, $\mathbf{r}\cdot\mathbf{j} = -y'$, $\mathbf{v}\cdot\mathbf{i} = -v$, $\mathbf{v}\cdot\mathbf{j} = 0$, and $\mathbf{r}\cdot\mathbf{v} = x'v$, we obtain

$$\mathbf{K} = G\frac{mv}{r^3c^2[1-\mathbf{r}\cdot\mathbf{v}/rc]^3}[(-x' + rv/c)y'v/rc + (-y')(1 - x'v/rc)]\mathbf{k}$$

$$= -G\frac{mv[1 - v^2/c^2]y'}{r^3c^2[1-\mathbf{r}\cdot\mathbf{v}/rc]^3}\mathbf{k}, \tag{5-2.5}$$

which, noting that $vy'\mathbf{k} = \mathbf{v}\times\mathbf{r}$, is the same as Eq. (5-2.1).

▲

## 5-3. Gravitational and Cogravitational Fields of a Linear Mass Uniformly Moving Along its Length

Consider a linear mass of finite length $L$, cross-sectional area $S$, mass density $\rho$, and linear mass density $\lambda = \rho S$ moving with constant velocity **v** parallel to the $x$ axis of a rectangular system of coordinates in the negative direction of the axis and at a distance $R$ above the axis (Fig. 5.5). Let the point of observation $O$ be at the origin. What is the gravitational field at $O$ at the time $t$ when the leading end of the mass is at a distance $L_2$ from the $y$ axis?

We can find the gravitational field of the moving mass by using Eq. (3-1.1) or Eq. (2-2.1) if we know its retarded position corresponding to the time for which the field is computed. We can determine this position as follows.

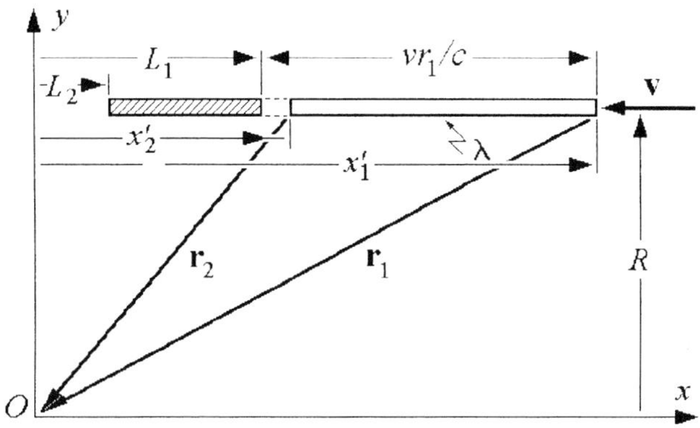


*Fig. 5.5 A line mass of linear density λ is moving with constant velocity* **v**. *The retarded positions of the trailing and leading ends of the mass are* $x_1'$ and $x_2'$, *respectively. The present positions of the two ends are* $L_1$ *and* $L_2$, *respectively. The distance between the trajectory of the mass and the x axis is R. The point of observation O is at the origin. The "retarded, " or "effective, " length of the mass is longer than its true length.*

First, let us determine the retarded position $x_2'$ of the leading end of the mass corresponding to the time $t$, that is, the position from which the leading end sends out its field signal which arrives at $O$ at the time $t$. If the retarded distance between $O$ and the leading end is $r_2$, then the time it takes for the signal to travel from the leading end to $O$ is $r_2/c$. During this time the mass travels a distance $v(r_2/c)$. Therefore at the moment when the leading end sends out its field signal, the position of the leading end is

$$x_2' = L_2 + r_2 v/c. \tag{5-3.1}$$

Next, let us find the retarded position $x_1'$ of the trailing end of the mass corresponding to the time $t$. If the retarded distance between $O$ and the trailing end is $r_1$, then the time it takes for the signal to travel from the trailing end to $O$ is $r_1/c$. During this time the mass travels a distance $v(r_1/c)$. Hence, at the moment when the trailing end sends out its signal, the position of the trailing end is

$$x_1' = L_1 + r_1 v/c. \tag{5-3.2}$$

***The x component of the gravitational field.*** We are now ready to find the gravitational field of the mass by using Eq. (3-1.1) or Eq. (2-2.1). The easiest way to find the $x$ component of the gravitational field of the mass under consideration is to use Eq. (3-1.1). According to this equation, the $x$ component of the field is due to the $x$ components of $[\nabla'\rho]$ and $[\partial\mathbf{J}/\partial t]$ of the moving mass. For the linear mass under consideration, these components exist only at the leading and trailing ends of the mass and are the same as for the moving prism discussed in the preceding sections of this chapter: $[\nabla'\rho]_x = (\rho/w)\mathbf{i}$ for the leading end, and $[\nabla'\rho]_x = -(\rho/w)\mathbf{i}$ for the trailing end, $[\partial\mathbf{J}/\partial t]_x = -(v^2\rho/w)\mathbf{i}$ for the leading end, and $[\partial\mathbf{J}/\partial t]_x = (v^2\rho/w)\mathbf{i}$ for the trailing end, where $w$ is the thickness of the surface layer of the mass (this is the actual thickness, not the retarded one). Since the surface layer of the mass may be assumed as thin as one wishes, the retarded volume integrals in Eq. (3-1.1), as far as the $x$ component of the field is concerned, reduces to the product of the integrands and the volume of the surface layers of the leading and trailing ends of the mass at their retarded positions. By Eq. (5-1.4), for the leading end this volume is, using the asterisk to indicate values evaluated at retarded positions,

$$w_2^* S = \frac{wS}{1 - (\mathbf{r}_2 \cdot \mathbf{v})/r_2 c}, \tag{5-3.3}$$

and for the trailing end it is

$$w_1^* S = \frac{wS}{1-(\mathbf{r}_1 \cdot \mathbf{v})/r_1 c} . \tag{5-3.4}$$

The $x$ component of the gravitational field is therefore

$$g_x = G\rho S(1-v^2/c^2)\left(\frac{1}{r_2[1-(\mathbf{r}_2 \cdot \mathbf{v})/r_2 c]} - \frac{1}{r_1[1-(\mathbf{r}_1 \cdot \mathbf{v})/r_1 c]}\right), \tag{5-3.5}$$

or

$$g_x = G\lambda(1-v^2/c^2)\left(\frac{1}{r_2 - x_2' v/c} - \frac{1}{r_1 - x_1' v/c}\right). \tag{5-3.6}$$

Equation (5-3.6) gives the gravitational field in terms of the retarded position of the mass. We shall now convert it to the present position of the mass (that is, the actual position of the mass at the time $t$). The calculations are similar to those used for deriving Eqs. (4-1.20)-(4-1.26). First, we note that, by Eq. (5-3.1),

$$L_2^2 = x_2'^2 - 2x_2 r_2 v/c + r_2^2 v^2/c^2. \tag{5-3.7}$$

Next, we write the denominator of the first fraction inside the parentheses of Eq. (5-3.6) as

$$r_2 - x_2' v/c = [(r_2 - x_2' v/c)^2]^{1/2} = (r_2^2 - 2r_2 x_2' v/c + x_2'^2 v^2/c^2)^{1/2}. \tag{5-3.8}$$

Adding and subtracting $x'^2$ and $r_2^2 v^2/c^2$ to the right side of Eq. (5-3.8), we then have

$$r_2 - x_2' v/c \tag{5-3.9}$$
$$= (r_2^2 - 2r_2 x_2' v/c + x_2'^2 v^2/c^2 + x_2'^2 - x_2'^2 + r_2^2 v^2/c^2 - r_2^2 v^2/c^2)^{1/2}.$$

Let us now collect the terms on the right of Eq. (5-3.9) into three groups:

$$x_2'^2 - 2r_2 x_2' v/c + r_2^2 v^2/c^2 , \tag{5-3.10}$$

$$r_2^2 - x_2'^2 , \tag{5-3.11}$$

and

$$x_2'^2 v^2/c^2 - r_2^2 v^2/c^2 . \tag{5-3.12}$$

By Eq. (5-3.7), the first group represents $L_2^2$. The second group is simply $R^2$ (see Fig. 5.5). And the third group is $-R^2v^2/c^2$.

Similar relations hold for the denominator of the second fraction inside the parentheses of Eq. (5-3.6). Therefore Eq. (5-3.6) transforms to

$$g_x = G\frac{\lambda(1-v^2/c^2)}{R}\left[\frac{1}{(L_2^2/R^2+1-v^2/c^2)^{1/2}} - \frac{1}{(L_1^2/R^2+1-v^2/c^2)^{1/2}}\right], \tag{5-3.13}$$

where only the present time quantities appear.

***The y component of the gravitational field.*** The easiest way to find the $y$ component of the gravitational field of the mass under consideration is to use Eq. (2-2.1). Only the first integral of Eq. (2-2.1) makes a contribution to the $y$ component of the field, because $\partial\mathbf{J}/\partial t$ has no $y$ component. Separating this integral into two integrals, we then have

$$g_y = G\int\frac{[\rho]}{r^3}RdV' + G\int\frac{1}{r^2c}\left[\frac{\partial\rho}{\partial t}\right]RdV' . \tag{5-3.14}$$

The first integral in Eq. (5-3.14) is the same as for a stationary mass, except that the integration must be extended over the retarded (effective) length of the mass. Designating the contribution of the first integral as $g_{1y}$ and noting that $r = (x'^2 + R^2)^{1/2}$, we obtain

$$g_{1y} = G\int\frac{\rho}{r^3}RdV' = G\rho S\int_{x_2}^{x_1}\frac{R}{(x'^2+R^2)^{3/2}}dx', \tag{5-3.15}$$

or

$$g_{1y} = G\frac{\lambda}{R}\left[\frac{x_1'}{(x_1'^2 + R^2)^{1/2}} - \frac{x_2'}{(x_2'^2 + R^2)^{1/2}}\right] = G\frac{\lambda}{R}\left(\frac{x_1'}{r_1} - \frac{x_2'}{r_2}\right). \quad (5\text{-}3.16)$$

In order to evaluate the contribution of $g_{2y}$ of the second integral of Eq. (5-3.14) to the total field, we must determine the value of the derivative $[\partial\rho/\partial t]$. According to the notation convention for retarded quantities, this derivative is the ordinary derivative $\partial\rho/\partial t$ used at the retarded position of the moving mass. By Eq. (4-1.3), taking into account that for our mass $\mathbf{v} = -v\mathbf{i}$, $[\partial\rho/\partial t]$ is then simply $v\partial\rho/\partial x'$. Since $\rho$ is constant within the line mass, only the leading and the trailing ends of the mass contribute to this expression, and the contributions are $v\rho/w$ and $-v\rho/w$, respectively. The gravitational field $g_{2y}$ is therefore

$$g_{2y} = G\frac{R}{c}\int\frac{v\rho/w}{r_2^2}dV_2' - G\frac{R}{c}\int\frac{v\rho/w}{r_1^2}dV_1', \quad (5\text{-}3.17)$$

where the integration is over the surface layers of the leading and trailing ends of the mass at the retarded positions of the mass. Since the thickness of the surface layers is much smaller than $r_1$ and $r_2$, we can replace the integrals, as before for $g_x$, by the products of the integrands and the volumes of integration (the volumes of the respective surface layers). Using the relations $dV_2' = w_2{}^*S$, $dV_1' = w_1{}^*S$, and using Eqs. (5-3.3) and (5-3.4) , we then have

$$\begin{aligned} g_{2y} &= G\frac{R}{c}\left[\frac{v\rho/w}{r_2^2 - r_2(\mathbf{r}_2\cdot\mathbf{v})/c}wS + \frac{v\rho/w}{r_1^2 - r_1(\mathbf{r}_1\cdot\mathbf{v})/c}wS\right] \\ &= G\frac{\lambda vR}{c}\left[\frac{1}{r_2(r_2 - x_2'v/c)} - \frac{1}{r_1(r_1 - x_1'v/c)}\right]. \end{aligned} \quad (5\text{-}3.18)$$

Adding Eqs. (5-3.16) and (5-3.18), we obtain for the $y$ component of the field

$$g_y = G\frac{\lambda}{R}\left[\frac{x_1'}{r_1} - \frac{R^2 v/c}{r_1(r_1 - x_1' v/c)} - \frac{x_2'}{r_2} + \frac{R^2 v/c}{r_2(r_2 - x_2' v/c)}\right]$$
$$= G\frac{\lambda}{R}\left[\frac{x_1'(r_1 - x_1' v/c) - R^2 v/c}{r_1(r_1 - x_1' v/c)} - \frac{x_2'(r_2 - x_2' v/c) - R^2 v/c}{r_2(r_2 - x_2' v/c)}\right], \tag{5-3.19}$$

or

$$g_y = G\frac{\lambda}{R}\left[\frac{x_1' r_1 - x_1'^2 v/c - R^2 v/c}{r_1(r_1 - x_1' v/c)} - \frac{x_2' r_2 - x_2'^2 v/c - R^2 v/c}{r_2(r_2 - x_2' v/c)}\right]. \tag{5-3.20}$$

But $x_1'^2 v/c + R^2 v/c = r_1^2 v/c$ and $x_2'^2 v/c + R^2 v/c = r_2^2 v/c$. Therefore

$$g_y = G\frac{\lambda}{R}\left(\frac{x_1' - r_1 v/c}{r_1 - x_1' v/c} - \frac{x_2' - r_2 v/c}{r_2 - x_2' v/c}\right). \tag{5-3.21}$$

Now, by Eq. (5-3.1), $x_2' - r_2 v/c = L_2$, and by Eq. (5-3.2), $x_1' - r_1 v/c = L_1$. Substituting $L_2$ and $L_1$ into Eq. (5-3.21) and transforming the denominators to the present position quantities by means of Eqs. (5-3.7)-(5-3.12), just as we did in Eq. (5-3.6), we finally obtain

$$g_y = G\frac{\lambda}{R^2}\left[\frac{L_1}{(L_1^2/R^2 + 1 - v^2/c^2)^{1/2}} - \frac{L_2}{(L_2^2/R^2 + 1 - v^2/c^2)^{1/2}}\right]. \tag{5-3.22}$$

***The cogravitational field.*** Although we could find the cogravitational field of the moving linear mass from Eq. (3-1.2) or Eq. (2-2.2), it is much simpler to find it from the gravitational field of the mass. According to Eq. (4-2.4), the cogravitational field **K** of any uniformly moving mass distribution is always

$$\mathbf{K} = (\mathbf{v} \times \mathbf{g})/c^2, \tag{4-2.4}$$

where **g** is the gravitational field of the moving mass distribution. Since $\mathbf{v} = -v\mathbf{i}$, the only non-vanishing component of the cross product in Eq. (4-2.4) is the $z$ component involving $g_y$ only. Substituting $\mathbf{v}$ and Eq. (5-3.22) into Eq. (4-2.4) and denoting $\lambda v$ as the mass current $I$, we obtain

$$\mathbf{K} = \mathbf{k}\,G\frac{I}{c^2R^2}\left[\frac{L_2}{(L_2^2/R^2+1-v^2/c^2)^{1/2}} - \frac{L_1}{(L_1^2/R^2+1-v^2/c^2)^{1/2}}\right]. \quad (5\text{-}3.23)$$

## 5-4. The Gravitational Field of a Point Mass in Arbitrary Motion

As before, we consider a constant mass distribution of total mass $m$ and density $\rho$ confined to a small rectangular prism (Fig. 5.6) whose center is located at the point $x'$, $y'$ in the $xy$ plane of a rectangular system of coordinates, and whose sides $l$, $a$, and $b$ are parallel to the $x$, $y$, and $z$ axis, respectively. The point of observation is at the origin. The distance of the center of the prism from the point of observation (the origin) is $r_0 \gg a,\ b,\ l,$

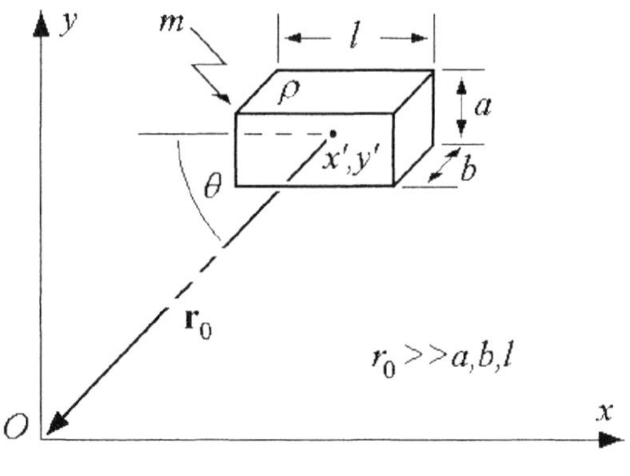


*Fig. 5.6 A mass of uniform density ρ is confined to a small rectangular prism. The mass constitutes a point mass when viewed from a distance large compared to its linear dimensions.*

so that the prism constitutes a point mass.[2] We shall assume that at the retarded time $t'$ the center of the prism moves with velocity $\mathbf{v}$ in the negative $x$ direction and has an acceleration $\dot{\mathbf{v}}$.

For a given present time $t$, the retarded times associated with different points of the prism are different, corresponding to different retarded distances of these points from the point of observation. Therefore the retarded velocities of the different points of the prism are also different. If the prism is sufficiently far from the point of observation, which we assume to be the case, the difference between the retarded times corresponding to different points of the prism is very small, and therefore the retarded acceleration of the prism may be assumed to have the same value $\dot{\mathbf{v}}$ for all points of the prism, even if in reality the acceleration is variable. Therefore the velocities of the different points of the prism can be calculated from velocity formulas for motion with constant acceleration.

As we shall presently see, in addition to the velocity of the center of the prism, we only need the velocities of the right, left, top, and bottom surfaces of the prism. Let the distances of these surfaces from the point of observation be $r_1$, $r_2$, $r_3$, and $r_4$, as shown in Fig. 5.7. The time interval between the retarded time for the center of the prism and for its left or right surface is then approximately $(r_1 - r_2)/2c$ (see Section 4.1), and the time interval between the retarded time for the center of the prism and for its top or bottom surface is approximately $(r_3 - r_4)/2c$. Therefore the (approximate) retarded velocities of the right, left, top, and bottom surfaces of the prism are, respectively, $\mathbf{v}_1 = \mathbf{v} - \dot{\mathbf{v}}(r_1 - r_2)/2c$, $\mathbf{v}_2 = \mathbf{v} + \dot{\mathbf{v}}(r_1 - r_2)/2c$, $\mathbf{v}_3 = \mathbf{v} - \dot{\mathbf{v}}(r_3 - r_4)/2c$, and $\mathbf{v}_4 = \mathbf{v} + \dot{\mathbf{v}}(r_3 - r_4)/2c$.

As was explained in Section 4-1, the apparent size and shape of the prism in its retarded position is not the same as that of the prism when it is at rest. In particular, if the prism moves in the $-x$ direction, the prism appears to be longer, it appears to be slanted, and the effective volume of the prism and of its surface

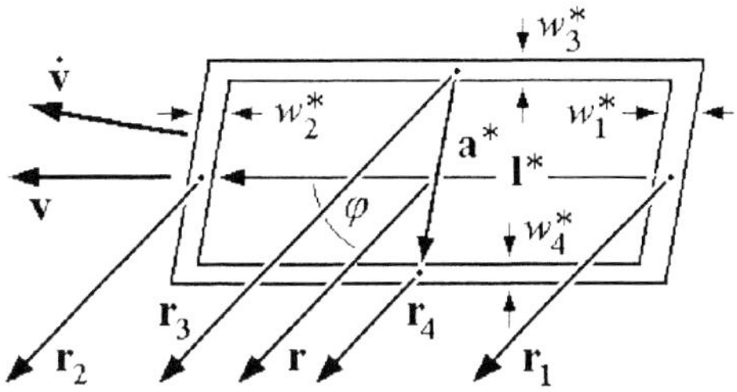


*Fig. 5.7 When the mass shown in Fig. 5.6 is in a state of accelerated motion and is at a retarded position, its apparent length, shape, and thickness of its surface layers are no longer the same as for the stationary mass. The distances from the center of the mass and from the four surface layers to the point of observation are represented by the vectors* $\mathbf{r}$, $\mathbf{r}_1$, $\mathbf{r}_2$, $\mathbf{r}_3$, *and* $\mathbf{r}_4$. *All five* **r***'s meet at the point of observation (origin of coordinates). The acceleration vector is in the xy plane.*

layers changes (Fig. 5.7). As a result, the following geometrical relations hold for the moving prism at its retarded position:

The apparent length of the prism is, by Eq. (4-1.7),

$$l^* = \frac{l}{1 - \mathbf{r} \cdot \mathbf{v}/rc} . \tag{5-4.1}$$

The apparent volume of the prism is, by Eq. (4-1.8),

$$(abl)^* = \frac{abl}{1 - \mathbf{r} \cdot \mathbf{v}/rc} . \tag{5-4.2}$$

By the same equations, the apparent volume of the right surface layer (distance $r_1$ from the origin) is

$$(abw)_1^* = \frac{abw}{1 - \mathbf{r}_1 \cdot \mathbf{v}_1/r_1 c} ; \tag{5-4.3}$$

the apparent volume of the left surface layer (distance $r_2$ from the origin) is

$$(abw)_2^* = \frac{abw}{1 - \mathbf{r}_2 \cdot \mathbf{v}_2 / r_2 c} ; \tag{5-4.4}$$

the apparent volume of the top surface layer (distance $r_3$ from the origin) is

$$(lbw)_3^* = \frac{lbw}{1 - \mathbf{r}_3 \cdot \mathbf{v}_3 / r_3 c} ; \tag{5-4.5}$$

and the apparent volume of the bottom surface layer (distance $r_4$ from the origin) is

$$(lbw)_4^* = \frac{lbw}{1 - \mathbf{r}_4 \cdot \mathbf{v}_4 / r_4 c} . \tag{5-4.6}$$

We shall find the gravitational field of our accelerating point mass by using Eq. (3-1.1)

$$\mathbf{g} = G \int \frac{[\nabla' \rho]}{r} dV + \frac{G}{c^2} \int \frac{1}{r} \left[ \frac{\partial \mathbf{J}}{\partial t} \right] dV' \tag{3-1.1}$$

Consider first the contribution of the gradient of the mass density, $\nabla'\rho$, to the field. Since $\rho$ is constant within the mass, $\nabla'\rho = 0$ within it, so that the only contribution to $\nabla'\rho$ comes from the surface layer of the mass, where $\rho$ changes from 0 (outside the mass) to $\rho$ (inside the mass). Let the actual thickness of the surface layer of the mass be $w$. Taking into account that $\nabla'\rho$ represents the rate of change of $\rho$ in the positive direction of the greatest rate of change, we then have $\nabla'\rho = (\rho/\mathrm{w})\mathbf{n}_{\mathrm{in}}$, where $\mathbf{n}_{\mathrm{in}}$ is a unit vector normal to the surface layer and pointing *into* the mass.[5] Since the center of the mass is in the $xy$ plane ($z' = 0$), the integrals over the two surface layers parallel to the $xy$ plane cancel each other, because $\nabla'\rho$ for one of the layers is opposite to that for the other layer, while $r$ is the same for both layers. Thus, as far as $\nabla'\rho$ is concerned, only the four integrals over the layers parallel to the $xz$ and $yz$ planes remain. Referring to Figs. 5.6 and

5.7, they are the right, left, top, and bottom surface layers, and $\nabla'\rho$ associated with these surface layers is, respectively $-(\rho/w)\mathbf{i}$, $(\rho/w)\mathbf{i}$, $-(\rho/w)\mathbf{j}$, and $(\rho/w)\mathbf{j}$ (these are the same relations that we used for finding the gravitational field of a uniformly moving point mass in Section 5.1).

Assuming that $r_1$, $r_2$, $r_3$, and $r_4$ are much larger than $l^*$, we can replace the integrals over the four layers by the products of the integrands and the retarded volumes of the layers, which gives

$$\mathbf{g} = G\left[\frac{\rho/w}{r_1}(abw)_1^*(-\mathbf{i}) + \frac{\rho/w}{r_2}(abw)_2^*\mathbf{i} + \frac{\rho/w}{r_3}(lbw)_3^*w(-\mathbf{j}) + \frac{\rho/w}{r_4}(lbw)_4^*w\mathbf{j}\right] + \frac{G}{c^2}\int\frac{[\partial\mathbf{J}/\partial t]}{r}dV'. \tag{5-4.7}$$

Let us designate the part of Eq. (5-4.7) which explicitly depends on $\rho$ as $\mathbf{g}_\rho$. Using Eqs. (5-4.3)-(5-4.6) and cancelling $w$, we can write then

$$\mathbf{g}_\rho = G\rho\left[\left(\frac{1}{r_2\{1-\mathbf{r}_2\cdot\mathbf{v}_2/r_2c\}} - \frac{1}{r_1\{1-\mathbf{r}_1\cdot\mathbf{v}_1/r_1c\}}\right)ab\mathbf{i} + \left(\frac{1}{r_4\{1-\mathbf{r}_4\cdot\mathbf{v}_4/r_4c\}} - \frac{1}{r_3\{1-\mathbf{r}_3\cdot\mathbf{v}_3/r_3c\}}\right)bl\mathbf{j}\right]. \tag{5-4.8}$$

The differences of the fractions in this equation are simply the increments of the function $1/(r - \mathbf{r}\cdot\mathbf{v}/c)$ associated with the displacement of the tail of $\mathbf{r}$ over a small distance represented by the vector $\mathbf{l}^*$ [in the $\mathbf{i}$ component of Eq. (5-4.8)] and by the vector $\mathbf{a}^*$ [in the $\mathbf{j}$ component of Eq. (5-4.8)]. Therefore, just as we did in the case of Eq. (5-1.7), we can write Eq. (5-4.8) as

$$\mathbf{g}_\rho = G\rho b\left\{\left[\left(\nabla'\frac{1}{r-\mathbf{r}\cdot\mathbf{v}/c}\right)\cdot\mathbf{l}^*\right]a\mathbf{i} + \left[\left(\nabla'\frac{1}{r-\mathbf{r}\cdot\mathbf{v}/c}\right)\cdot\mathbf{a}^*\right]l\mathbf{j}\right\}. \tag{5-4.9}$$

Using Eqs. (4-1.16), (4-1.14), and (4-1.15), we now have

$$\mathbf{g}_\rho = -G\rho b\left[\left(\frac{\mathbf{r} - r\mathbf{v}/c + (\mathbf{r}\cdot\dot{\mathbf{v}})\mathbf{r}/c^2}{r^3(1-\mathbf{r}\cdot\mathbf{v}/rc)^2}\cdot\mathbf{i}\right)\frac{la}{1-\mathbf{r}\cdot\mathbf{v}/rc}\mathbf{i}\right.$$
$$+\left(\frac{\mathbf{r} - r\mathbf{v}/c + (\mathbf{r}\cdot\dot{\mathbf{v}})\mathbf{r}/c^2}{r^3(1-\mathbf{r}\cdot\mathbf{v}/rc)^2}\cdot\mathbf{i}\right)\frac{y'v/c}{r(1-\mathbf{r}\cdot\mathbf{v}/rc)}al\mathbf{j} \quad (5\text{-}4.10)$$
$$\left.+\left(\frac{\mathbf{r} - r\mathbf{v}/c + (\mathbf{r}\cdot\dot{\mathbf{v}})\mathbf{r}/c^2}{r^3(1-\mathbf{r}\cdot\mathbf{v}/rc)^2}\cdot\mathbf{j}\right)\frac{r-x'v/c}{r(1-\mathbf{r}\cdot\mathbf{v}/rc)}al\mathbf{j}\right].$$

Simplifying and taking into account that $\mathbf{r}\cdot\mathbf{i} = -x'$, $\mathbf{r}\cdot\mathbf{j} = -y'$, $\mathbf{v}\cdot\mathbf{i} = -v$, $\mathbf{v}\cdot\mathbf{j} = 0$, and $\mathbf{r}\cdot\mathbf{v} = x'v$, we obtain

$$\mathbf{g}_\rho = -G\frac{\rho abl}{r^3[1-\mathbf{r}\cdot\mathbf{v}/rc]^3}\left\{[-x' + rv/c - (\mathbf{r}\cdot\dot{\mathbf{v}})x'/c^2]\,\mathbf{i}\right.$$
$$+[-x' + rv/c - (\mathbf{r}\cdot\dot{\mathbf{v}})x'/c^2]\frac{y'v/c}{r}\mathbf{j}$$
$$\left.+[-y' - (\mathbf{r}\cdot\dot{\mathbf{v}})y'/c^2]\frac{r-x'v/c}{r}\mathbf{j}\right\} \quad (5\text{-}4.11)$$
$$= -G\frac{\rho abl}{r^3[1-\mathbf{r}\cdot\mathbf{v}/rc]^3}[-x'\mathbf{i} - r\mathbf{v}/c - (\mathbf{r}\cdot\dot{\mathbf{v}})x'/c^2\mathbf{i}$$
$$+(v^2y'/c^2)\mathbf{j} - y'\mathbf{j} - (\mathbf{r}\cdot\dot{\mathbf{v}})y'/c^2\mathbf{j}].$$

Since we are not interested in the acceleration-independent field $\mathbf{g}_V$ (this field was found in Section 5-1), we shall drop in Eq. (5-4.11) the terms that do not contain the acceleration $\dot{\mathbf{v}}$, and shall designate the rest of the equations as $\mathbf{g}_{A\rho}$, with the subscript "$A$" standing for "acceleration." Noting that $\mathbf{r} = -x'\mathbf{i} - y'\mathbf{j}$, and that $\rho abl = m$, we then obtain

$$\mathbf{g}_{A\rho} = -G\frac{m(\mathbf{r}\cdot\dot{\mathbf{v}})\mathbf{r}}{r^3c^2(1-\mathbf{r}\cdot\mathbf{v}/rc)^3}. \quad (5\text{-}4.12)$$

Consider now the contribution of $\partial\mathbf{J}/\partial t$ to the field. By Eq. (4-1.4), we have

$$\frac{\partial \mathbf{J}}{\partial t} = \frac{\partial(\rho\,\mathbf{v})}{\partial t} = -(\mathbf{v}\cdot\nabla'\rho)\,\mathbf{v} + \rho\frac{\partial \mathbf{v}}{\partial t} = -(\mathbf{v}\cdot\nabla'\rho)\,\mathbf{v} + \rho\,\dot{\mathbf{v}}\,. \qquad (5\text{-}4.13)$$

However, because the retarded velocity is different in different regions (points) of the mass, we must evaluate Eq. (5-4.13) separately for each region under consideration. There are five such regions: the interior of the mass, the right surface, the left surface, the top surface, and the bottom surface.

In the interior of the mass, $\nabla'\rho = 0$. Therefore for the interior we have

$$\frac{\partial \mathbf{J}}{\partial t} = \rho\,\dot{\mathbf{v}}\,. \qquad (5\text{-}4.14)$$

At the right surface, $\nabla'\rho = (\partial\rho/\partial x')\mathbf{i} = -(\rho/w)\mathbf{i}$, and the velocity is $\mathbf{v}_1$. By Eq. (5-4.13), for the right surface we therefore have

$$\frac{\partial \mathbf{J}_1}{\partial t} = -(\mathbf{v}_1\cdot\nabla'\rho)\,\mathbf{v}_1 + \rho\,\dot{\mathbf{v}}_1 = -(v_{1x}\frac{\partial\rho}{\partial x'})\mathbf{v}_1 + \rho\,\dot{\mathbf{v}}_1 = (\rho/w)\,v_{1x}\mathbf{v}_1 + \rho\dot{\mathbf{v}}_1\,. \qquad (5\text{-}4.15)$$

or

$$\frac{\partial \mathbf{J}_1}{\partial t} = (\rho/w)(v_{1x}\mathbf{v}_1 + w\dot{\mathbf{v}}_1)\,, \qquad (5\text{-}4.16)$$

and since we can make $w$ as small as we please,

$$\frac{\partial \mathbf{J}_1}{\partial t} = (\rho/w)\,v_{1x}\mathbf{v}_1\,. \qquad (5\text{-}4.17)$$

At the left surface, $\nabla'\rho = \partial\rho/\partial x'\mathbf{i} = \rho/w\mathbf{i}$, and the velocity is $\mathbf{v}_2$. Therefore, by the same reasoning as in the case of Eq. (5-4.16),

$$\frac{\partial \mathbf{J}_2}{\partial t} = -\,(\rho/w)\,v_{2x}\mathbf{v}_2\,. \qquad (5\text{-}4.18)$$

At the top surface, $\nabla'\rho = \partial\rho/\partial y'\mathbf{j} = -\,\rho/w\mathbf{j}$, and the velocity is $\mathbf{v}_3$. Therefore,

$$\frac{\partial \mathbf{J}_3}{\partial t} = (\rho/w)\,v_{3y}\mathbf{v}_3\,. \qquad (5\text{-}4.19)$$

At the bottom surface, $\nabla'\rho = \partial\rho/\partial y'\mathbf{j} = \rho/w\mathbf{j}$, and the velocity is $\mathbf{v}_4$. Therefore

$$\frac{\partial \mathbf{J}_4}{\partial t} = -\,(\rho/w)\,v_{4y}\mathbf{v}_4\,. \tag{5-4.20}$$

Let us now designate the integral in Eq. (5-4.7) as $\mathbf{g}_\mathbf{J}$. Since, by supposition, all $r$'s for the mass (prism) are much larger than the linear dimensions of the mass, we can replace the integration by the product of the respective integrands and the volumes of the five regions that contribute to $\partial\mathbf{J}/\partial t$. Using Eqs. (5-4.14), (5-4.17)-(5-4.20) and (5-4.2)-(5-4.6), we then have

$$\begin{aligned}\frac{c^2\mathbf{g}_\mathbf{J}}{G} = {} & \frac{\rho\dot{\mathbf{v}}}{r}\left(\frac{abl}{1-\mathbf{r}\cdot\mathbf{v}/rc}\right) \\ & + \frac{\rho}{r_1 w}\left(v_{1x}\mathbf{v}_1\frac{abw}{1-\mathbf{r}_1\cdot\mathbf{v}_1/r_1 c}\right) - \frac{\rho}{r_2 w}\left(v_{2x}\mathbf{v}_2\frac{abw}{1-\mathbf{r}_2\cdot\mathbf{v}_2/r_2 c}\right) \\ & + \frac{\rho}{r_3 w}\left(v_{3y}\mathbf{v}_3\frac{lbw}{1-\mathbf{r}_3\cdot\mathbf{v}_3/r_3 c}\right) - \frac{\rho}{r_4 w}\left(v_{4y}\mathbf{v}_4\frac{lbw}{1-\mathbf{r}_4\cdot\mathbf{v}_4/r_4 c}\right),\end{aligned} \tag{5-4.21}$$

or

$$\begin{aligned}\frac{c^2\mathbf{g}_\mathbf{J}}{G} = {} & \frac{m\dot{\mathbf{v}}}{r(1-\mathbf{r}\cdot\mathbf{v}/rc)} + \rho ab\left(\frac{v_{1x}\mathbf{v}_1}{r_1-\mathbf{r}_1\cdot\mathbf{v}_1/c} - \frac{v_{2x}\mathbf{v}_2}{r_2-\mathbf{r}_2\cdot\mathbf{v}_2/c}\right) \\ & + \rho bl\left(\frac{v_{3y}\mathbf{v}_3}{r_3-\mathbf{r}_3\cdot\mathbf{v}_3/c} - \frac{v_{4y}\mathbf{v}_4}{r_4-\mathbf{r}_4\cdot\mathbf{v}_4/c}\right).\end{aligned} \tag{5-4.22}$$

Since the linear dimensions of the mass are very small compared to the $r$'s, the difference of the fractions in the last two terms of Eq. (5-4.22) can be regarded as the total differential (increment) $df = (\partial f/\partial x')dx' + (\partial f/\partial y')dy'$ of the functions

$$\frac{v_x\mathbf{v}}{r-\mathbf{r}\cdot\mathbf{v}/c} \tag{5-4.23}$$

and

$$\frac{v_y \mathbf{v}}{r - \mathbf{r} \cdot \mathbf{v}/c} \tag{5-4.24}$$

corresponding to the displacements of the tail of $\mathbf{r}$ by $\mathbf{l}^*$ and by $\mathbf{a}^*$, respectively (see Fig. 5.7).

Using Eq. (4-1.16), noting that $\mathbf{r} = -x'\mathbf{i} - y'\mathbf{j}$, noting that $v_y = 0$ (because $\mathbf{v}$ is parallel to the $x$ axis), and remembering that $\mathbf{v}$ and $v$ are functions of the retarded time $t' = t - r/c$, so that $\partial\mathbf{v}/\partial x' = (\partial\mathbf{v}/\partial t')\partial t'/\partial x' = -(\partial\mathbf{v}/\partial t')x'/rc = -\dot{\mathbf{v}}x'/rc$ with similar expressions for $\partial\mathbf{v}/\partial y'$, $\partial v/\partial x'$, and $\partial v/\partial y'$, we have for the needed partial derivatives of the two functions

$$\frac{\partial}{\partial x'}\left(\frac{v_x \mathbf{v}}{r[1-(\mathbf{r}\cdot\mathbf{v})/rc]}\right) = v_x \mathbf{v}\frac{-x' - rv_x/c - (\mathbf{r}\cdot\dot{\mathbf{v}})x'/c^2}{r^3[1-(\mathbf{r}\cdot\mathbf{v})/rc]^2} - \frac{(\dot{v}_x\mathbf{v} + v_x\dot{\mathbf{v}})x'}{r^2c[1-(\mathbf{r}\cdot\mathbf{v})/rc]}, \tag{5-4.25}$$

$$\frac{\partial}{\partial x'}\left(\frac{v_y \mathbf{v}}{r[1-(\mathbf{r}\cdot\mathbf{v})/rc]}\right) = -\frac{\dot{v}_y\mathbf{v}x'}{r^2c[1-(\mathbf{r}\cdot\mathbf{v})/rc]}, \tag{5-4.26}$$

and

$$\frac{\partial}{\partial y'}\left(\frac{v_y \mathbf{v}}{r[1-(\mathbf{r}\cdot\mathbf{v})/rc]}\right) = -\frac{\dot{v}_y\mathbf{v}y'}{r^2c[1-(\mathbf{r}\cdot\mathbf{v})/rc]}. \tag{5-4.27}$$

In evaluating Eq. (5-4.22) with the help of Eqs. (5-4.25)-(5-4.27), we shall omit from Eq. (5-4.25) the terms not containing $\dot{\mathbf{v}}$, since they only contribute to the acceleration-independent field $\mathbf{g}_V$, which we already found in Section 5-1. Combining Eqs. (5-4.22), (5-4.25)-(5-4.27), (4-1.14), and (4-1.15), we then have, denoting the acceleration-dependent field as $\mathbf{g}_{JA}$,

$$\begin{aligned}\frac{c^2\mathbf{g}_{JA}}{G} &= \frac{m\dot{\mathbf{v}}}{r(1-\mathbf{r}\cdot\mathbf{v}/rc)} \\ &+\rho ab\left[\frac{-v_x\mathbf{v}(\mathbf{r}\cdot\dot{\mathbf{v}})x'}{r^3c^2(1-\mathbf{r}\cdot\mathbf{v}/rc)^2}-\frac{(\dot{v}_x\mathbf{v}+v_x\dot{\mathbf{v}})x'}{r^2c(1-\mathbf{r}\cdot\mathbf{v}/rc)}\right]\cdot\frac{l}{(1-\mathbf{r}\cdot\mathbf{v}/rc)} \\ &-\rho bl\left[\frac{\dot{v}_y\mathbf{v}x'}{r^2c(1-\mathbf{r}\cdot\mathbf{v}/rc)}\cdot\frac{ay'v/c}{r(1-\mathbf{r}\cdot\mathbf{v}/rc)}\right. \\ &\left.+\frac{\dot{v}_y\mathbf{v}y'}{r^2c(1-\mathbf{r}\cdot\mathbf{v}/rc)}\cdot\frac{a(r-\mathbf{r}\cdot\mathbf{v}/c)}{r(1-\mathbf{r}\cdot\mathbf{v}/rc)}\right],\end{aligned} \tag{5-4.28}$$

or

$$\begin{aligned}\frac{c^2\mathbf{g}_{JA}}{G} &= \frac{m\dot{\mathbf{v}}}{r(1-\mathbf{r}\cdot\mathbf{v}/rc)} \\ &+\frac{m}{r^2c(1-\mathbf{r}\cdot\mathbf{v}/rc)^2}\left[\frac{-v_x\mathbf{v}(\mathbf{r}\cdot\dot{\mathbf{v}})x'}{rc(1-\mathbf{r}\cdot\mathbf{v}/rc)}-\dot{v}_x\mathbf{v}x'\right. \\ &\left.-v_x\dot{\mathbf{v}}x'-\frac{\dot{v}_y\mathbf{v}x'y'v}{rc}-\dot{v}_y\mathbf{v}y'+\frac{\dot{v}_y\mathbf{v}y'(\mathbf{r}\cdot\mathbf{v})}{rc}\right].\end{aligned} \tag{5-4.29}$$

Since $\mathbf{r}\cdot\mathbf{v} = x'v = -x'v_x$ and since $-\dot{v}_x x' - \dot{v}_y y' = \dot{\mathbf{v}}\cdot\mathbf{r}$ (see Figs. 5.6 and 5.7), Eq. (5-4.29) reduces to

$$\begin{aligned}\frac{c^2\mathbf{g}_{JA}}{G} &= \frac{m\dot{\mathbf{v}}}{r(1-\mathbf{r}\cdot\mathbf{v}/rc)} \\ &+\frac{m}{r^2c(1-\mathbf{r}\cdot\mathbf{v}/rc)^2}\left[\frac{\mathbf{v}(\mathbf{r}\cdot\mathbf{v})(\mathbf{r}\cdot\dot{\mathbf{v}})}{rc(1-\mathbf{r}\cdot\mathbf{v}/rc)}+(\mathbf{r}\cdot\dot{\mathbf{v}})\mathbf{v}+(\mathbf{r}\cdot\mathbf{v})\dot{\mathbf{v}}\right],\end{aligned} \tag{5-4.30}$$

which after elementary simplifications becomes

$$\mathbf{g}_{JA} = G\frac{m\dot{\mathbf{v}}}{c^2r(1-\mathbf{r}\cdot\mathbf{v}/rc)^2} + G\frac{m(\mathbf{r}\cdot\dot{\mathbf{v}})\mathbf{v}}{c^3r^2(1-\mathbf{r}\cdot\mathbf{v}/rc)^3}. \tag{5-4.31}$$

Finally, in accordance with Eq. (5-4.7), adding Eq. (5-4.31) to Eq. (5-4.12), we obtain for $\mathbf{g}_A$

$$\mathbf{g}_A = -\frac{Gm(\mathbf{r}\cdot\dot{\mathbf{v}})\mathbf{r}}{c^2r^3(1-\mathbf{r}\cdot\mathbf{v}/rc)^3} + \frac{Gm(\mathbf{r}\cdot\dot{\mathbf{v}})\mathbf{v}}{c^3r^2(1-\mathbf{r}\cdot\mathbf{v}/rc)^3} + \frac{Gm\dot{\mathbf{v}}}{c^2r(1-\mathbf{r}\cdot\mathbf{v}/rc)^2}, \tag{5-4.32}$$

which can be written in a simpler form as

$$\mathbf{g}_A = -\frac{Gm}{r^3c^2(1-\mathbf{r}\cdot\mathbf{v}/rc)^3}\left\{\mathbf{r}\times\left[\left(\mathbf{r}-\frac{r\mathbf{v}}{c}\right)\times\dot{\mathbf{v}}\right]\right\}. \tag{5-4.33}$$

The total gravitational field is the sum of the acceleration-independent field $\mathbf{g}_V$ given by Eq. (5-1.11) and of $\mathbf{g}_A$ given by Eq. (5-4.33). Adding Eqs. (5-1.11) and (5-4.33), we obtain for the total gravitational field of a point mass in arbitrary motion

$$\mathbf{g} = -G\frac{m}{r^3(1-\mathbf{r}\cdot\mathbf{v}/rc)^3}\left\{\left(\mathbf{r}-\frac{r\mathbf{v}}{c}\right)\left(1-\frac{v^2}{c^2}\right)+\mathbf{r}\times\left[\left(\mathbf{r}-\frac{r\mathbf{v}}{c}\right)\times\frac{\dot{\mathbf{v}}}{c^2}\right]\right\}. \tag{5-4.34}$$

Note that $\mathbf{r}$, $r$, $\mathbf{v}$, $v$, and $\dot{\mathbf{v}}$ in this equation are retarded.

## 5-5. The Cogravitational Field of a Point Mass in Arbitrary Motion

Although by using Eq. (3-1.2) or Eq. (2-2.2) we can find the cogravitational field produced by a point mass in arbitrary motion in the same manner as we found the gravitational field in Section 5-4 (see Example 5-5.1), it is much easier to find it from the known gravitational field by using Eq. (4-2.10).

Applying Eq. (4-2.10) to Eq. (5-4.32), we obtain for the acceleration part of the cogravitational field after elementary simplifications

$$\mathbf{K}_A = G\frac{m}{r^2c^3(1-\mathbf{r}\cdot\mathbf{v}/rc)^2}\left[\frac{(\mathbf{r}\times\mathbf{v})(\mathbf{r}\cdot\dot{\mathbf{v}})}{rc(1-\mathbf{r}\cdot\mathbf{v}/rc)}+\mathbf{r}\times\dot{\mathbf{v}}\right]. \tag{5-5.1}$$

Applying Eq. (4-2.10) to Eq. (5-4.34), we obtain for the total cogravitational field after elementary simplifications

$$\mathbf{K}=G\frac{m}{r^2c^2(1-\mathbf{r}\cdot\mathbf{v}/rc)^2}\left[\frac{1-v^2/c^2+\mathbf{r}\cdot\dot{\mathbf{v}}/c^2}{r(1-\mathbf{r}\cdot\mathbf{v}/rc)}(\mathbf{r}\times\mathbf{v})+\frac{\mathbf{r}\times\dot{\mathbf{v}}}{c}\right]. \tag{5-5.2}$$

▼

**Example 5-5.1** Find the cogravitational field of an accelerating point mass shown in Figs. 5.6 and 5.7 by using Eq. (3-1.2).

Since $\mathbf{J} = \rho\mathbf{v}$, $\nabla' \times \mathbf{J} = \nabla' \times \rho\mathbf{v} = \nabla'\rho \times \mathbf{v} + \rho\nabla' \times \mathbf{v}$. But $\mathbf{v}$ is not a point function (there is no "velocity field"), and therefore $\nabla' \times \mathbf{v} = 0$ and $\nabla' \times \mathbf{J} = \nabla'\rho \times \mathbf{v}$. As we already know from Sections 5-1 and 5-4, $\nabla'\rho$ for our mass is only different from zero at the surface layers of the mass. Therefore the only contribution to the integral in Eq. (3-1.2) comes from the right, left, top, and bottom surface layers, where $\nabla'\rho$ is $-(\rho/w)\mathbf{i}$, $(\rho/w)\mathbf{i}$, $-(\rho/w)\mathbf{j}$, and $(\rho/w)\mathbf{j}$, respectively (by symmetry, the contributions of the front and back surface layers cancel). Since $[\nabla' \times \mathbf{J}]$ in the integral of Eq. (3-1.2) is retarded, the velocity in the expression $[\nabla'\rho \times \mathbf{v}]$ is the retarded velocity of each surface under consideration. By supposition, the distances from the mass to the point of observation is much larger than $l^*$. Therefore the integral in Eq. (3-1.2) can be replaced by the product of the integrand and the volume of integration (the respective volumes of the surface layers). Substituting into $[\nabla' \times \mathbf{J}] = [\nabla'\rho \times \mathbf{v}] = -[\mathbf{v} \times \nabla'\rho]$ the above expressions for $\nabla'\rho$, and using Eqs. (3-1.2) and (5-4.3)-(5-4.6), we then have

$$\begin{aligned}\mathbf{K}=-G\frac{\rho}{wc^2}\bigg[&\frac{abw(\mathbf{v}_1\times\mathbf{i})}{r_1\{1-\mathbf{r}_1\cdot\mathbf{v}_1/r_1c\}}-\frac{abw(\mathbf{v}_2\times\mathbf{i})}{r_2\{1-\mathbf{r}_2\cdot\mathbf{v}_2/r_2c\}}\\&+\frac{blw(\mathbf{v}_3\times\mathbf{j})}{r_3\{1-\mathbf{r}_3\cdot\mathbf{v}_3/r_3c\}}-\frac{blw(\mathbf{v}_4\times\mathbf{j})}{r_4\{1-\mathbf{r}_4\cdot\mathbf{v}_4/r_4c\}}\bigg],\end{aligned} \tag{5-5.3}$$

or

$$\mathbf{K} = -G\frac{\rho}{c^2}\left[\left(\frac{\mathbf{v}_1}{[r_1 - \mathbf{r}_1 \cdot \mathbf{v}_1/c]} - \frac{\mathbf{v}_2}{[r_2 - \mathbf{r}_2 \cdot \mathbf{v}_2/c]}\right) \times \mathbf{i}ab \right.$$
$$\left. + \left(\frac{\mathbf{v}_3}{[r_3 - \mathbf{r}_3 \cdot \mathbf{v}_3/c]} - \frac{\mathbf{v}_4}{[r_4 - \mathbf{r}_4 \cdot \mathbf{v}_4/c]}\right) \times \mathbf{j}bl\right]. \tag{5-5.4}$$

The differences of the fractions in Eq. (5-5.4), just as before in Eq. (5-4.22), are the increments of the functions given by Eqs. (5-4.23) and (5-4.24), except that $v_x$ and $v_y$ in the numerators are now absent. By Eq. (5-4.25), taking into account that $v_y = 0$, the corresponding partial derivatives are

$$\frac{\partial}{\partial x'}\left(\frac{\mathbf{v}}{r - \mathbf{r} \cdot \mathbf{v}/c}\right) = \mathbf{v}\frac{-x' - rv_x/c - (\mathbf{r} \cdot \dot{\mathbf{v}})x'/c^2}{r^3(1 - \mathbf{r} \cdot \mathbf{v}/rc)^2} - \frac{\dot{\mathbf{v}}x'}{r^2c(1 - \mathbf{r} \cdot \mathbf{v}/rc)}, \tag{5-5.5}$$

and

$$\frac{\partial}{\partial y'}\left(\frac{\mathbf{v}}{r - \mathbf{r} \cdot \mathbf{v}/c)}\right) = \mathbf{v}\frac{-y' - (\mathbf{r} \cdot \dot{\mathbf{v}})y'/c^2}{r^3(1 - \mathbf{r} \cdot \mathbf{v}/rc)^2} - \frac{\dot{\mathbf{v}}y'}{r^2c(1 - \mathbf{r} \cdot \mathbf{v}/rc)}. \tag{5-5.6}$$

In evaluating Eq. (5-5.4) with the help of Eqs. (5-5.5) and (5-5.6), we shall omit from Eqs. (5-5.5) and (5-5.6) the terms not containing $\dot{\mathbf{v}}$, since they only contribute to $\mathbf{K}_V$, (the cogravitational field of a uniformly moving mass), which we do not need. Combining Eqs. (5-5.4), (5-5.5), (5-5.6), (4-1.14), and (4-1-15), we then have for the acceleration-dependent field

$$\mathbf{K}_A = -G\frac{\rho}{c^2}\left[\left(\frac{-\mathbf{v}(\mathbf{r} \cdot \dot{\mathbf{v}})x'}{r^3c^2(1 - \mathbf{r} \cdot \mathbf{v}/rc)} - \frac{\dot{\mathbf{v}}x'}{r^2c(1 - \mathbf{r} \cdot \mathbf{v}/rc)}\right) \times \mathbf{i} \cdot \frac{abl}{1 - \mathbf{r} \cdot \mathbf{v}/rc}\right.$$
$$+ \left(\frac{-\mathbf{v}(\mathbf{r} \cdot \dot{\mathbf{v}})x'}{r^3c^2(1 - \mathbf{r} \cdot \mathbf{v}/rc)} - \frac{\dot{\mathbf{v}}x'}{r^2c(1 - \mathbf{r} \cdot \mathbf{v}/rc)}\right) \times \mathbf{j} \cdot \frac{ably'v/c}{r(1 - \mathbf{r} \cdot \mathbf{v}/rc)} \tag{5-5.7}$$
$$\left. + \left(\frac{-\mathbf{v}(\mathbf{r} \cdot \dot{\mathbf{v}})y'}{r^3c^2(1 - \mathbf{r} \cdot \mathbf{v}/rc)} - \frac{\dot{\mathbf{v}}y'}{r^2c(1 - \mathbf{r} \cdot \mathbf{v}/rc)}\right) \times \mathbf{j} \cdot \frac{ably'(1 - \mathbf{r} \cdot \mathbf{v}/rc)}{r(1 - \mathbf{r} \cdot \mathbf{v}/rc)}\right].$$

Expanding Eq. (5-5.7), taking into account that $\mathbf{v} \times \mathbf{i} = 0$, and simplifying, we obtain

$$\mathbf{K}_A = G\frac{m}{r^2c^3(1-\mathbf{r}\cdot\mathbf{v}/rc)^2}\left[(\dot{\mathbf{v}}\times\mathbf{i})x' + \frac{\mathbf{v}(\mathbf{r}\cdot\dot{\mathbf{v}})y'}{rc(1-\mathbf{r}\cdot\mathbf{v}/rc)}\times\mathbf{j} + \dot{\mathbf{v}}\times\mathbf{j}y'\right]. \tag{5-5.8}$$

But $\mathbf{i}x' + \mathbf{j}y' = -\mathbf{r}$, and $\mathbf{v} \times \mathbf{j}y' = -\mathbf{v} \times \mathbf{r}$ (because $\mathbf{v}$ is parallel to the $x$ axis). Therefore Eq. (5-5.8) can be written as

$$\mathbf{K}_A = G\frac{m}{r^2c^3(1-\mathbf{r}\cdot\mathbf{v}/rc)^2}\left[\mathbf{r}\times\dot{\mathbf{v}} + \frac{(\mathbf{r}\times\mathbf{v})(\mathbf{r}\cdot\dot{\mathbf{v}})}{rc(1-\mathbf{r}\cdot\mathbf{v}/rc)}\right]. \tag{5-5.9}$$

The total cogravitational field of an accelerating point mass is the sum of Eq. (5-2.1), representing the cogravitational field of a uniformly moving point mass, and Eq. (5-5.9), representing the effect of the acceleration of the mass on the field. Adding Eqs. (5-2.1) and (5-5.9), we obtain [compare with Eq. (5-5.2)]

$$\mathbf{K} = G\frac{m}{r^2c^2(1-\mathbf{r}\cdot\mathbf{v}/rc)^2}\left[\frac{1-v^2/c^2+(\mathbf{r}\cdot\dot{\mathbf{v}})/c^2}{r(1-\mathbf{r}\cdot\mathbf{v}/rc)}(\mathbf{r}\times\mathbf{v}) + \frac{\mathbf{r}\times\dot{\mathbf{v}}}{c}\right]. \tag{5-5.10}$$

Observe that Eqs. (5-5.9) and (5-5.10) express the cogravitational field in terms of the retarded position of the mass. ▲

## 5-6. Gravitational and Cogravitational Potentials of a Moving Point Mass

Gravitational and cogravitational potentials produced by a moving point mass $m$ can be easily obtained from Eqs. (3-3.5) and (3-3.2).

A "point mass" is a mass distribution viewed from a distance large compared to the linear dimensions of the mass distribution. Therefore, for a point mass, the distance $r$ in the integrals of Eqs. (3-3.5) and (3-3.2) may be considered the same for all volume elements of the mass, and therefore each integral may be replaced by the product of the integrand and the retarded volume of the mass $\Delta V'$.

By Eqs. (3-3.5) and (4-1.8) we then have for the gravitational scalar potential of a moving point mass

$$\varphi = -G\frac{\rho}{r}\Delta V'_{ret} = -G\frac{\rho\Delta V'}{r(1-\mathbf{r}\cdot\mathbf{v}/rc)} \tag{5-6.1}$$

or, replacing $\rho\Delta V'$ by $m$,

$$\varphi = -G\frac{m}{r(1-\mathbf{r}\cdot\mathbf{v}/rc)}. \tag{5-6.2}$$

From Eqs. (3-3.2) and (4-1.8) we similarly have for the cogravitational vector potential of a moving point mass

$$\mathbf{A} = -G\frac{\mathbf{J}}{rc^2}\Delta V'_{ret} = -G\frac{\mathbf{J}\Delta V'}{rc^2(1-\mathbf{r}\cdot\mathbf{v}/rc)}, \tag{5-6.3}$$

and since $\mathbf{J} = \rho\mathbf{v}$,

$$\mathbf{A} = -G\frac{m\mathbf{v}}{rc^2(1-\mathbf{r}\cdot\mathbf{v}/rc)}. \tag{5-6.4}$$

Equations (5-6.2) and (5-6.4) are similar to the *Liénard-Wiechert potentials* of electromagnetic theory.[6,7] They express the potentials of a moving point mass in terms of the retarded position of the mass. If the mass moves with constant velocity, these potentials can be converted to the present position of the mass. Transforming the denominators of Eqs. (5-6.2) and (5-6.4) with the help of Eq. (4-1.26), we obtain for a point mass moving with constant velocity

$$\varphi = -G\frac{m}{r_0[1-(v^2/c^2)\sin^2\theta]^{1/2}} \tag{5-6.5}$$

and

$$\mathbf{A} = -G\frac{m\mathbf{v}}{r_0c^2[1-(v^2/c^2)\sin^2\theta]^{1/2}}, \tag{5-6.6}$$

where $r_0$ is the present position radius vector, and $\theta$ is the angle between $\mathbf{v}$ and $\mathbf{r}_0$.

▼

**Example 5-6.1** Equations (5-6.5) and (5-6.6) represent the "instantaneous" potential of a uniformly moving point mass. Since the mass is moving, the potentials change as time goes by. How should they be written to show explicitly their time dependence?

Assuming that the mass moves in the negative $x$ direction, the $x$ coordinate of the mass diminishes with time according to $x_0' - vt$, where $x_0'$ is the value of the $x$ coordinate at $t = 0$. Expressing the denominators in Eqs. (5-6.5) and (5-6.6) in terms of Cartesian coordinates by means of Eq. (4-1.26) and (4-1.25), and replacing $x_0'$ by $x_0' - vt$, we obtain the time-dependent expressions for the potentials

$$\varphi = -G\frac{m}{[(x_0' - vt)^2 + (1 - v^2/c^2)y'^2]^{1/2}} \tag{5-6.7}$$

and

$$\mathbf{A} = -G\frac{m\mathbf{v}}{c^2[(x_0' - vt)^2 + (1 - v^2/c^2)y'^2]^{1/2}}. \tag{5-6.8}$$

▲

## 5-7. How Accurate are the Equations for the Fields and Potentials Obtained in this Chapter?

In obtaining the expressions for **g** and **K** of moving point masses we used several approximations. Our first approximation was the replacement of the integrals in Eqs. (3-1.1) and (3-1.2) by the products of the integrands and the volumes of integration. This can only be done if the relation $r \gg l^*$ is satisfied. Therefore, by Eq. (4-1.7), our **g** and **K** expressions for moving point masses[8] are subject to the restriction

$$r \gg \frac{l}{1 - \mathbf{r}\cdot\mathbf{v}/rc} = \frac{l}{1 - (v/c)\cos\phi}, \tag{5-7.1}$$

where $l$ is the length of the "point mass," **v** is the velocity of the mass, **r** is the retarded position vector joining the mass with the point of observation, and $\phi$ is the angle between **v** and **r**. Since Eq. (5-7.1) must hold for all values of $\phi$, including $\phi = 0$, the velocity of the mass is subject to the restriction

$$v < c(1 - l/r). \tag{5-7.2}$$

Consider now the approximations that we used for taking into account the acceleration of the mass. The retarded time intervals between the center and the right-left and top-bottom surfaces of the mass are $(r_1 - r_2)/2c \approx (l \cos\phi)/[2c(1 - \mathbf{r}\cdot\mathbf{v}/rc)]$ and $(r_3 - r_4)/2c \approx (a \sin\phi)/[2c(1 - \mathbf{r}\cdot\mathbf{v}/rc)]$, respectively (see Figs. 5.7, 4.2, and 5.3).[9] For Eq. (5-7.1) to hold, the increment in the velocity of the mass during these time intervals must be less than $c - v$. Hence the restrictions on the acceleration of the mass in the direction of the $x$ axis is

$$\dot{v}_x(r_1 - r_2)/2c < c - v, \tag{5-7.3}$$

or

$$\dot{v}_x < \frac{2(c-v)(c - v\cos\phi)}{l\cos\phi}. \tag{5-7.4}$$

A similar restriction applies to the acceleration in any other direction. Since the largest possible value for cos $\varphi$ and sin $\varphi$ is 1, we obtain from Eq. (5-7.4) for the general case of the acceleration $\dot{\mathbf{v}}$

$$\dot{v} < \frac{2(c-v)^2}{L}, \tag{5-7.5}$$

where $L$ is the length of the "point mass" in the direction of the acceleration.

## References and Remarks for Chapter 5

**1.** A "point mass" is by definition any mass distribution viewed from a distance large compared with the linear dimensions of that distribution, similar to the term "light point," which is frequently used in reference to stars. In neither case does the word "point" describe the structure or the constitution of the object; instead, it reflects the attitude of the observer toward this object.

**2.** One may think that by choosing the mass in the shape of a rectangular prism we limit the generality of our derivations. This is not so. Any mass distribution can be regarded as being composed of masses confined to small rectangular prisms: this is exactly what we do when we perform integration over a volume element (rectangular prism!) $dV' = dx'\, dy'\, dz'$.

**3.** The increment $dU$ of any scalar function $U(x, y, z,)$ associated with the displacement $d\mathbf{l} = x\mathbf{i} + y\mathbf{j} + z\mathbf{k}$ is $dU = \nabla U \cdot d\mathbf{l}$ [see, for example, Oleg D. Jefimenko, *Electricity and Magnetism*, 2nd ed., (Electret Scientific, Star City, 1989) pp. 36-38]. Since in the case under consideration the displacements $\mathbf{l}^*$ and $\mathbf{a}^*$ are very small (they represent the length and width of a "point mass"), these displacements can be treated as differentials.

**4.** Oliver Heaviside, "A Gravitational and Electromagnetic Analogy," The Electrician, **31**, 281-282 and 359 (1893). This article is reproduced in modern notation in Oleg D. Jefimenko, *Causality, Electromagnetic Induction and Gravitation,* 2nd ed., (Electret Scientific, Star City, 2000), pp. 189-202.

**5.** Equation (3-1.1) contains a *retarded gradient* of $\rho$ and *retarded time derivative* of $\mathbf{J}$, rather than a gradient of *retarded* $\rho$ and time derivative of *retarded* $\mathbf{J}$. This means that the gradient and the time derivative must be determined for the unretarded (stationary) $\rho$ and $\mathbf{J}$ but must be used at the retarded position of the moving mass.

**6.** A. Liénard, "Champ électrique et magnetique produit par une mass électrique concentré en un point et animé d'un mouvement quelconque," L'Éclairage élect. **16**, 5-14, 53-59, 106-112 (1898).

**7.** E. Wiechert, "Elektrodynamische Elementargesetze," Archives Néerlanaises (2) **5**, 549-573 (1900).

**8.** These expressions also include Eq. (4-2.10).

**9.** To simplify the calculations, we assume here that $\alpha$ in Fig. 5.3 is $\pi/2$.

# 6

# GRAVITATIONAL AND COGRAVITATIONAL FIELDS AND POTENTIALS OF ARBITRARY MASS DISTRIBUTIONS MOVING WITH CONSTANT VELOCITY

Gravitational and cogravitational fields produced by a time-independent stationary mass and mass-current distribution can be calculated with relative ease by a variety of methods. But calculating fields of time-dependent mass and mass-current distributions, and the fields of moving mass distributions in particular, is in general a formidable task. In this chapter we shall obtain formulas that make it possible to determine the fields and potentials of any uniformly moving mass distribution directly and simply in terms of present time integrals that are not much different from the integrals for fields of stationary masses.

## 6-1. Converting Retarded Field Integrals for Uniformly Moving Mass Distributions into Present-Time (Present-Position) Integrals

As we already know from Chapters 3 and 2, gravitational and cogravitational fields of moving mass distributions can be found from the retarded integrals

$$\mathbf{g} = G\int \frac{\left[\nabla'\rho + \frac{1}{c^2}\frac{\partial \mathbf{J}}{\partial t}\right]}{r} dV' \tag{6-1.1}$$

and

$$\mathbf{K} = -\frac{G}{c^2}\int \frac{[\nabla' \times \mathbf{J}]}{r} dV' \tag{6-1.2}$$

or from

$$\mathbf{g} = -G\int\left\{\frac{[\rho]}{r^3} + \frac{1}{r^2 c}\left[\frac{\partial\rho}{\partial t}\right]\right\}\mathbf{r}\,dV' + \frac{G}{c^2}\int \frac{1}{r}\left[\frac{\partial \mathbf{J}}{\partial t}\right] dV' \tag{6-1.3}$$

and

$$\mathbf{K} = -\frac{G}{c^2}\int\left\{\frac{[\mathbf{J}]}{r^3} + \frac{1}{r^2 c}\left[\frac{\partial \mathbf{J}}{\partial t}\right]\right\}\times \mathbf{r}\,dV'. \tag{6-1.4}$$

We shall presently show that for time-independent mass distributions moving with constant velocity, these integrals can be converted to the "present" position of the mass distribution, so that the integration is performed not over the retarded, or effective, volume (see Section 4-1), but over the real volume that the mass distribution occupies at the moment $t$ for which the fields are being determined.

The conversion is based on certain properties and relations involving retarded integrals and retarded quantities which are reviewed below.

Although in the retarded integrals the retardation symbol [ ] usually appears only in the numerators of the integrands, all quantities in the integrals are retarded. In particular, the volume element $dV'$ stands for the retarded volume element $dV_{ret}' = [dV'] = d[x']d[y']d[z']$, $r$ stands for the retarded distance $[r]$, and $\mathbf{r}$ stands for the retarded position vector $[\mathbf{r}]$. Note that $[\nabla\rho]$ means "ordinary $\nabla\rho$ used at retarded position," $[\partial\rho/\partial t]$ means "derivative of ordinary $\rho$ with respect to ordinary time used at retarded position," and $[\partial\mathbf{J}/\partial t]$ means "derivative of ordinary $\mathbf{J}$ with respect to ordinary time used at retarded position."

In the derivations that follow, we shall assume that the point of observation is at $x = y = z = 0$, and we shall only consider a time-independent mass distribution moving with constant velocity in the $-x$ direction. For such a mass distribution, because the mass density is not a function of time, $[\rho] = \rho$, and, because $\mathbf{v}$ is constant, $[\mathbf{v}] = \mathbf{v}$. Also, as explained in Section 4-1 [see Eqs. (4-1.8), (4-1.3), (4-1.4), (4-1.25), and (4-1.26)], the following relations hold for such a mass distribution

$$[dV'] = \frac{dV'}{1 - [\mathbf{r}\cdot\mathbf{v}]/rc}, \tag{6-1.5}$$

$$\frac{\partial\rho}{\partial t} = -\,\mathbf{v}\cdot\nabla'\rho = v\frac{\partial\rho}{\partial x'}, \tag{6-1.6}$$

$$\frac{\partial\mathbf{J}}{\partial t} = -\,(\mathbf{v}\cdot\nabla'\rho)\mathbf{v} = -v^2\frac{\partial\rho}{\partial x'}\mathbf{i}, \tag{6-1.7}$$

$$\begin{aligned}[r] - [\mathbf{r}\cdot\mathbf{v}]/c &= \{x_0'^2 + y'^2 + z'^2 - (y'^2 + z'^2)v^2/c^2\}^{1/2} \\ &= \{x_0'^2 + (y'^2 + z'^2)(1 - v^2/c^2)\}^{1/2} = \{x_0'^2 + (y'^2 + z'^2)/\gamma^2\}^{1/2},\end{aligned} \tag{6-1.8}$$

[we are using the standard abbreviation $\gamma = 1/(1 - \mathrm{v}^2/\mathrm{c}^2)^{1/2}$], and

$$[r] - [\mathbf{r}\cdot\mathbf{v}]/c = r_0\{1 - (v^2/c^2)\sin^2\theta\}^{1/2}, \tag{6-1.9}$$

where $\sin^2\theta = (y'^2 + z'^2)/(x_0'^2 + y'^2 + z'^2)$ and $\theta$ is the angle between the velocity vector $\mathbf{v}$ and the vector $\mathbf{r}_0$ joining $dV'$ with the point of observation. For clarity, all retarded quantities and expressions in the above equations are placed between square brackets; the quantities without brackets, and the quantities between braces in Eq. (6-1.8) and (6-1.9) in particular, are present-time quantities. Observe that Eq. (6-1.8) is obtained from Eq. (4-1.25) by replacing $y'^2$ by $y'^2 + z'^2$; the replacement is needed because we no longer deal with a point mass and therefore cannot assume that the mass is confined to the $xy$ plane.

We can now proceed with the conversion of Eqs. (6-1.1)-(6-1.4). Once again, we shall only consider a time-independent mass distribution moving with constant velocity $\mathbf{v} = -v\mathbf{i}$.

***Converting Eq. (6-1.1).*** Using Eqs. (6-1.5) and (6-1.7) and remembering that $\rho$ and $\mathbf{v}$ are not affected by retardation and that $\nabla'\rho$ in Eq. (6-1.1) is the ordinary gradient, we can write Eq. (6-1.1) as

$$\begin{aligned}\mathbf{g} &= G\int \frac{\nabla'\rho - (\mathbf{v}\cdot\nabla'\rho)\mathbf{v}/c^2}{[r - \mathbf{r}\cdot\mathbf{v}/c]}\, dV' \\ &= G\int \frac{\nabla'\rho - \mathbf{i}(v^2/c^2)(\partial\rho/\partial x')}{[r - \mathbf{r}\cdot\mathbf{v}/c]}\, dV',\end{aligned} \tag{6-1.10}$$

where only the denominator is retarded. Converting the retarded denominator in Eq. (6-1.10) with the help of Eq. (6-1.8), we obtain the desired equation (we are omitting the subscript "0" at $x'$ for simplicity)

$$\mathbf{g} = G\int \frac{\nabla'\rho - \mathbf{i}(v^2/c^2)(\partial\rho/\partial x')}{\{x'^2 + (y'^2 + z'^2)/\gamma^2\}^{1/2}}\, dV', \tag{6-1.11}$$

where the integral is a "present position" integral, and where all quantities are present-time quantities.

Equation (6-1.11) can be written in an alternative form. Using Eq. (6-1.9) for converting the denominator of the integrand in Eq. (6-1.10), we obtain (omitting the subscript "0" at $r$ for simplicity)

$$\mathbf{g} = G\int \frac{\nabla'\rho - \mathbf{i}(v^2/c^2)(\partial\rho/\partial x')}{r\{1 - (v^2/c^2)\sin^2\theta\}^{1/2}}\, dV'. \tag{6-1.12}$$

An even simpler expression for $\mathbf{g}$ of a moving mass distribution can be obtained from Eq. (6-1.1) if the density of the mass under consideration is constant within the volume occupied by the mass. As was shown in Section 3-2, in this case the mass gradient exists only at the surface of the mass, and the volume integral reduces to a surface integral. Equation (6-1.12) becomes then

$$\mathbf{g} = -G\rho \oint \frac{d\mathbf{S}' \pm \mathbf{i}(v^2/c^2)\,dy'\,dz'}{r\{1-(v^2/c^2)\sin^2\theta\}^{1/2}}, \tag{6-1.13}$$

where the surface element vector $d\mathbf{S}'$ is directed from the mass distribution into the surrounding space, and the sign in front of $\mathbf{i}$ is the same as that of $\partial\rho/\partial x'$ .

▼

**Example 6-1.1**. A thin ring of width $w$, thickness $b$, and radius $a \gg b$ has a uniformly distributed mass $m$ and moves with velocity $\mathbf{v} = -v\mathbf{i}$ along the $x$ axis, which is also the symmetry axis of the ring (Fig. 6.1). Find the gravitational field produced by the ring at the origin of coordinates when the center of the ring is at a distance $x'$ from the origin.

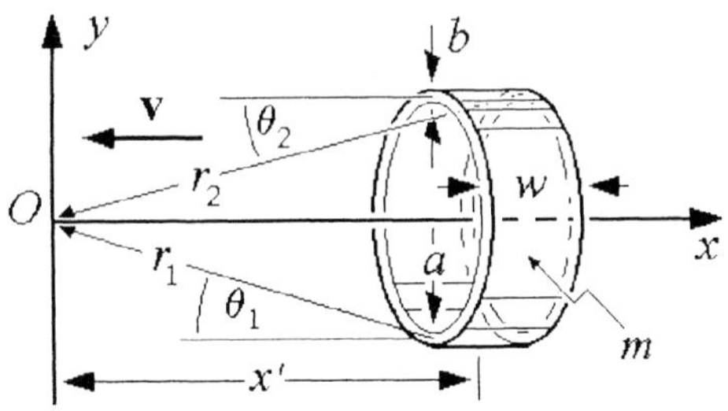


*Fig. 6.1 A thin ring of mass m moves with velocity* $\mathbf{v} = -v\mathbf{i}$ *along the x axis. Find the gravitational field at the origin.*

We can solve this problem by using Eq. (6-1.13). By symmetry, only the front (leading) and the back (trailing) surface of the ring contribute to the gravitational field at the origin. Let the distances from the front and the back surface of the ring to the origin be $r_1$ and $r_2$. We then have $r_1 = [(x' - w/2)^2 + a^2]^{1/2}$, $r_2 = [(x' + w/2)^2 + a^2]^{1/2}$, $\sin\theta_1 = a/[(x' - w/2)^2 + a^2]^{1/2}$, $\sin\theta_2 = a/[(x' + w/2)^2 + a^2]^{1/2}$. Equation (6-1.13) becomes therefore

$$\mathbf{g} = -G\rho\left(\int \frac{-\{1-v^2/c^2\}dy'dz'\mathbf{i}}{r_1\{1-(v^2/c^2)\sin^2\theta_1\}^{1/2}} + \int \frac{\{1-v^2/c^2\}dy'dz'\mathbf{i}}{r_2\{1-(v^2/c^2)\sin^2\theta_2\}^{1/2}}\right), \tag{6-1.14}$$

where the integration is over the two flat surfaces of the ring. Substituting the above values for $r_1$, $r_2$, $\sin\theta_1$, and $\sin\theta_2$ and taking into account that the area of each flat surface of the ring is $2\pi ab$, we then have

$$\mathbf{g} = -\mathbf{i}G\rho(1-v^2/c^2)2\pi ab\left(\frac{-1}{\{(x'-w/2)^2+a^2-v^2a^2/c^2\}^{1/2}} + \frac{1}{\{(x'+w/2)^2+a^2-v^2a^2/c^2\}^{1/2}}\right), \tag{6-1.15}$$

or

$$\mathbf{g} = \mathbf{i}G\frac{m(1-v^2/c^2)}{w}\left(\frac{1}{\{(x'-w/2)^2+(1-v^2/c^2)a^2\}^{1/2}} - \frac{1}{\{(x'+w/2)^2+(1-v^2/c^2)a^2\}^{1/2}}\right). \tag{6-1.16}$$

**Example 6-1.2**. A very long, thin, straight ribbon of width $a$ and thickness $b$ has a mass of uniform density $\rho$ and moves along its length with velocity $\mathbf{v} = -v\mathbf{i}$ (Fig. 6.2). The plane of the ribbon is in the $xz$ plane of rectangular coordinates and the center line of the ribbon is on the $x$ axis. Find the gravitational and cogravitational fields produced by the ribbon at the point $P(0, 0, R)$.

We can solve this problem by using Eqs. (6-1.13) and (4-2.4). According to Eq. (6-1.13), the only contribution to the gravitational

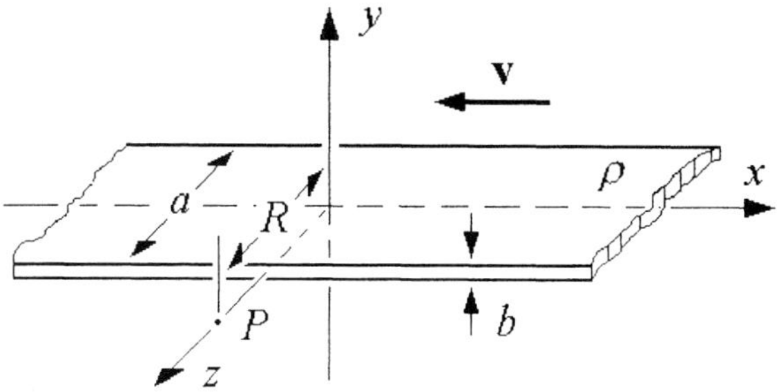


*Fig. 6.2 A very long thin ribbon of mass density ρ moves with uniform velocity* $\mathbf{v} = -v\mathbf{i}$ *along the x axis of rectangular coordinates. Find the gravitational and cogravitational fields produced by the ribbon at the point P.*

field of the ribbon at $P$ comes from the edges of the ribbon located at $z' = a/2$ and $z' = -a/2$ (by symmetry, the flat surfaces make no contribution). Let us assume that the ends of the ribbon are at $x' = -L_1$ and $x' = L_2$. By Eqs. (6-1.13), (6-1-9), and (6-1.8), we then have

$$\mathbf{g} = -G\rho\left(\int_{-L_1}^{L_2} \frac{\mathbf{k}b\,dx'}{\{x'^2+(R-a/2)^2/\gamma^2\}^{1/2}} - \int_{-L_1}^{L_2} \frac{\mathbf{k}b\,dx'}{\{x'^2+(R+a/2)^2/\gamma^2\}^{1/2}}\right)$$
$$= -\mathbf{k}G\rho b\Big\{\ln\left(x' + \{x'^2+(R-a/2)^2/\gamma^2\}^{1/2}\right) \tag{6-1.17}$$
$$- \ln\left(x' + \{x'^2+(R+a/2)^2/\gamma^2\}^{1/2}\right)\Big\}\Big|_{-L_1}^{L_2},$$

or

$$\mathbf{g} = \mathbf{k}G\rho b\left[\ln\frac{L_2+\{L_2^2+(R-a/2)^2/\gamma^2\}^{1/2}}{-L_1+\{L_1^2+(R-a/2)^2/\gamma^2\}^{1/2}}\right.$$
$$\left. - \ln\frac{L_2+\{L_2^2+(R+a/2)^2/\gamma^2\}^{1/2}}{-L_1+\{L_1^2+(R+a/2)^2/\gamma^2\}^{1/2}}\right]. \tag{6-1.18}$$

Since $R - a \ll L_1, L_2$ and $R + a \ll L_1, L_2$, we can expand the expressions in the braces and keep only the leading terms, obtaining

$$\mathbf{g} = -\mathbf{k}G\rho b\left\{\ln\frac{L_2+L_2+(R-a/2)^2/2L_2\gamma^2}{-L_1+L_1+(R-a/2)^2/2L_2\gamma^2}\right.$$

$$\left.-\ln\frac{L_2+L_2+(R+a/2)^2/2L_2\gamma^2}{-L_1+L_1+(R+a/2)^2/2L_2\gamma^2}\right\} \qquad (6\text{-}1.19)$$

$$= -\mathbf{k}G\rho b\left\{\ln\frac{2L_2+(R-a/2)^2/2L_2\gamma^2}{(R-a/2)^2/2L_1\gamma^2} - \ln\frac{2L_2+(R+a/2)^2/2L_2\gamma^2}{(R+a/2)^2/2L_1\gamma^2}\right\}$$

and, finally,

$$\mathbf{g} = -\mathbf{k}G2\rho b\ln\frac{(R+a/2)}{(R-a/2)}. \qquad (6\text{-}1.20)$$

To find the cogravitational field, we will use Eq. (4-2.4). By Eqs. (4-2.4) and (6-1.20), we have

$$\mathbf{K} = \frac{1}{c^2}\mathbf{v}\times\mathbf{g} = -(-\mathbf{i}\times\mathbf{k})\frac{G2\rho v b}{c^2}\ln\frac{(R+a/2)}{(R-a/2)} \qquad (6\text{-}1.21)$$

or

$$\mathbf{K} = -\mathbf{j}\frac{G2Jb}{c^2}\ln\frac{(R+a/2)}{(R-a/2)} = -\mathbf{j}\frac{2GI}{ac^2}\ln\frac{(R+a/2)}{(R-a/2)}, \qquad (6\text{-}1.22)$$

where $J$ is the mass-current density and $I = \rho vab$ is the mass-current formed by the ribbon.

Observe that Eq. (6-1.22) becomes the same as Eq. (5-3.23) if $R \gg a$ and, in Eq. (5-3.23), $L_1 = \infty = -L_2$. Taking into account the difference of the methods used for obtaining Eq. (6-1.22) and Eq. (5-3.23), this result is quite remarkable.

▲

***Converting Eq. (6-1.3).*** As before, we assume that the mass is time independent and moves with constant velocity $\mathbf{v} = -v\mathbf{i}$. Using Eqs. (6-1.6) and (6-1.7), we can write Eq. (6-1.3) as

$$\mathbf{g} = -G\int \frac{[\rho]}{r^3}\mathbf{r}\,dV' - \frac{G}{c}\int \frac{-[\mathbf{v}\cdot\nabla'\rho]\mathbf{r} + \mathbf{v}[\mathbf{v}\cdot\nabla'\rho]r/c}{r^2}dV' . \tag{6-1.23}$$

Note that $\nabla'\rho$ in this equation represents the ordinary gradient, that is, the gradient with respect to the ordinary source-point coordinates. For the calculations that follow, we need to convert $\nabla'\rho$ into the gradient with respect to the *retarded* coordinates. According to Eq. (4-1.7),

$$d[x'] = \frac{dx'}{1-[\mathbf{r}\cdot\mathbf{v}]/[r]c}, \tag{6-1.24}$$

and therefore

$$\frac{\partial}{\partial x'} = \frac{1}{1-[\mathbf{r}\cdot\mathbf{v}]/[r]c}\frac{\partial}{\partial[x']}. \tag{6-1.25}$$

Since $\mathbf{v}$ is along the $x$ axis, the $y'$ and $z'$ are not affected by retardation, so that $\partial/\partial y' = \partial/\partial[y']$ and $\partial/\partial z' = \partial/\partial[z']$. Hence

$$[\mathbf{v}\cdot\nabla'\rho] = \frac{[\mathbf{v}]\cdot[\nabla'][\rho]}{1-[\mathbf{r}\cdot\mathbf{v}]/[r]c}. \tag{6-1.26}$$

Substituting this expression into Eq. (6-1.23), we obtain

$$\mathbf{g} = -G\int_{ret} \frac{\rho}{r^3}\mathbf{r}\,dV' - \frac{G}{c}\int_{ret} \frac{(\mathbf{v}\,r/c - \mathbf{r})\mathbf{v}\cdot\nabla'\rho}{r^2(1-\mathbf{v}\cdot\mathbf{r}/rc)}dV', \tag{6-1.27}$$

where all the quantities under the integral signs are retarded, and where we have replaced the retardation brackets in the integrands by the subscript "*ret*" at the integral signs.

Let us designate the last term in Eq. (6-1.27) as $\mathbf{g}_2$. We have

$$\mathbf{g}_2 = -\frac{G}{c}\int_{ret} \frac{(\mathbf{v}\,r/c - \mathbf{r})\mathbf{v}\cdot\nabla'\rho}{r^2(1-\mathbf{v}\cdot\mathbf{r}/rc)}dV' . \tag{6-1.28}$$

To convert this integral to the present position of the mass, we shall first eliminate $\nabla'\rho$ from it. To do so, we shall write Eq. (6-1.28) in terms of its Cartesian components. For the $x$ component we have, remembering that $\mathbf{v} = -v\mathbf{i}$ and that $\mathbf{r} = -(x'\mathbf{i} + y'\mathbf{j} + z'\mathbf{k})$,

$$g_{2x} = \frac{G}{c}\int_{ret} \frac{(vr/c - x')\mathbf{v}\cdot\nabla'\rho}{r^2(1-\mathbf{v}\cdot\mathbf{r}/rc)}dV'. \tag{6-1.29}$$

Let us now factor out $\mathbf{v}\cdot$ and, using vector identity (V-5), let us write the integral as a difference of two integrals

$$\begin{aligned} g_{2x} &= G\frac{\mathbf{v}\cdot}{c}\int_{ret} \frac{(vr/c - x')\nabla'\rho}{r^2(1-\mathbf{v}\cdot\mathbf{r}/rc)}dV' \\ &= G\frac{\mathbf{v}\cdot}{c}\left\{\int_{ret}\nabla'\frac{(vr/c - x')\rho}{r^2(1-\mathbf{v}\cdot\mathbf{r}/rc)}dV' - \int_{ret}\rho\nabla'\frac{(vr/c - x')}{r^2(1-\mathbf{v}\cdot\mathbf{r}/rc)}dV'\right\}. \end{aligned} \tag{6-1.30}$$

The first integral in the last expression can be converted into a surface integral by means of Gauss's theorem of vector analysis [vector identity (V-19)], and since there is no mass outside the mass distribution under consideration, the integral vanishes. Differentiating the integrand in the second integral, collecting terms, reintroducing $\mathbf{v}\cdot$ under the integral sign, and simplifying, we obtain

$$g_{2x} = -G\int_{ret}\rho\frac{\{v^2/c^2 - 2\mathbf{v}\cdot\mathbf{r}/rc + (\mathbf{v}\cdot\mathbf{r}/rc)^2\}x' - (v^2/c^2 - 1)vr/c}{r^3(1-\mathbf{v}\cdot\mathbf{r}/rc)^2}dV'. \tag{6-1.31}$$

Proceeding in the same manner with the $y$ and $z$ components of Eq. (6-1.28), we obtain

$$g_{2y} = -G\int_{ret}\rho\frac{\{v^2/c^2 - 2\mathbf{v}\cdot\mathbf{r}/rc + (\mathbf{v}\cdot\mathbf{r}/rc)^2\}y'}{r^3(1-\mathbf{v}\cdot\mathbf{r}/rc)^2}dV' \tag{6-1.32}$$

and

$$g_{2z} = -G \int_{ret} \rho \frac{\{v^2/c^2 - 2\mathbf{v} \cdot \mathbf{r}/rc + (\mathbf{v} \cdot \mathbf{r}/rc)^2\} z'}{r^3(1 - \mathbf{v} \cdot \mathbf{r}/rc)^2} dV'. \quad (6\text{-}1.33)$$

Multiplying Eqs. (6-1.31)-(6-1.33), respectively, by **i**, **j**, and **k** and then adding them together, we again obtain a single vector equation for $\mathbf{g}_2$:

$$\mathbf{g}_2 = -G \int_{ret} \rho \frac{\{2\mathbf{v} \cdot \mathbf{r}/rc - (\mathbf{v} \cdot \mathbf{r}/rc)^2 - v^2/c^2\}\mathbf{r} + (v^2/c^2 - 1)\mathbf{v}r/c}{r^3(1 - \mathbf{v} \cdot \mathbf{r}/rc)^2} dV'. \quad (6\text{-}1.34)$$

Let us now rewrite Eq. (6-1.27) using Eq. (6-1.34) for the second integral of Eq. (6-1.27). We then have

$$\mathbf{g} = -G \int_{ret} \frac{\rho}{r^3} \mathbf{r}\, dV' \quad (6\text{-}1.35)$$
$$-G \int_{ret} \rho \frac{\{2\mathbf{v} \cdot \mathbf{r}/rc - (\mathbf{v} \cdot \mathbf{r}/rc)^2 - v^2/c^2\}\mathbf{r} + (v^2/c^2 - 1)\mathbf{v}r/c}{r^3(1 - \mathbf{v} \cdot \mathbf{r}/rc)^2} dV'.$$

Adding the two integrals, we obtain

$$\mathbf{g} = -G \int_{ret} \rho \frac{(1 - v^2/c^2)(\mathbf{r} - \mathbf{v}r/c)}{r^3(1 - \mathbf{v} \cdot \mathbf{r}/rc)^2} dV'. \quad (6\text{-}1.36)$$

We shall now convert the retarded integral in Eq. (6-1.36) to the present position of the mass. Replacing the retarded $dV'$ in Eq. (6-1.36) by the ordinary $dV'$ with the help of Eq. (6-1.5) and writing $1/\gamma^2$ for $1 - v^2/c^2$, we have

$$\mathbf{g} = -\frac{G}{\gamma^2} \int \frac{\rho([\mathbf{r}] - \mathbf{v}[r]/c)}{[r]^3(1 - \mathbf{v} \cdot [\mathbf{r}]/[r]c)^3} dV', \quad (6\text{-}1.37)$$

where, since $\rho$, **v**, $v$, and $c$ do not depend on time, only **r** and $r$ are retarded. But according to Eq. (4-1.19), the present-position vector $\mathbf{r}_0$ and the retarded position vector $\mathbf{r}$ are connected by the relation

$$\mathbf{r}_0 = [\mathbf{r}] - \mathbf{v}[r]/c, \tag{6-1.38}$$

so that the numerator in the integrand of Eq. (6-1.37) is simply $\rho$ multiplied by the present-position vector $\mathbf{r}_0$. Furthermore, according to Eq. (6-1.9), the denominator is simply

$$r_0^3\{1-(v^2/c^2)\sin^2\theta\}^{3/2}, \tag{6-1.39}$$

where $r_0$ is the distance from the present-position volume element $dV'$ to the point of observation, and $\theta$ is the angle between $\mathbf{v}$ and $\mathbf{r}_0$. Hence Eq. (6-1.37) can be written as

$$\mathbf{g} = -\frac{G}{\gamma^2}\int\frac{\rho\mathbf{r}_0}{r_0^3\{1-(v^2/c^2)\sin^2\theta\}^{3/2}}dV', \tag{6-1.40}$$

where the integration is over the volume of the mass at its present position [compare with Eq. (5-1.13)].[1]

▼

**Example 6-1.3.** An irregularly shaped mass distribution of total mass $m$ moves with constant velocity $\mathbf{v} = v\mathbf{i}$. The longest linear dimension of the distribution is $a$. Find the gravitational field of the distribution at a distance $r \gg a$ from the distribution.

We can solve the problem by using Eq. (6-1.40). Since $r \gg a$, we can assume $r$ and $\theta$ to be the same for all points of the mass. Therefore we can factor out $\mathbf{r}$ and the denominator of the integrand in Eq. (6-1.40), obtaining [compare with Eq. (5-1.13)]

$$\begin{aligned}\mathbf{g} &= -\frac{G\mathbf{r}_0}{\gamma^2 r_0^3\{1-(v^2/c^2)\sin^2\theta\}^{3/2}}\int\rho\, dV' \\ &= -\frac{Gm\mathbf{r}_0}{\gamma^2 r_0^3\{1-(v^2/c^2)\sin^2\theta\}^{3/2}}.\end{aligned} \tag{6-1.41}$$

▲

***Converting Eqs. (6-1.2) and (6-1.4).*** The retarded integrals for the cogravitational fields in Eq. (6-1.2) and (6-1.4) can be converted to the present position of the mass in the same manner as the integrals in Eqs. (6-1.1) and (6-1.3) for the gravitational field. However, there is no need to resort to this conversion process, because by Eq. (4-2.4) the gravitational and cogravitational fields of any uniformly moving mass distribution are connected by the relation

$$\mathbf{K} = (\mathbf{v} \times \mathbf{g})/c^2. \quad (6\text{-}1.42)$$

From Eqs. (6-1.12) and (6-1.42) we then have, noting that $\mathbf{v} \times \mathbf{i} = 0$,

$$\mathbf{K} = \frac{G}{c^2} \int \frac{\mathbf{v} \times \nabla' \rho}{r_0\{1 - (v^2/c^2)\sin^2\theta\}^{1/2}} dV'. \quad (6\text{-}1.43)$$

From Eqs. (6-1.13) and (6-1.42) we have

$$\mathbf{K} = -G\frac{\rho}{c^2} \oint \frac{\mathbf{v} \times d\mathbf{S}'}{r_0\{1 - (v^2/c^2)\sin^2\theta\}^{1/2}}. \quad (6\text{-}1.44)$$

And from Eqs. (6-1.40) and (6-1.42) we have

$$\mathbf{K} = -\frac{G}{c^2\gamma^2} \int \frac{\rho \mathbf{v} \times \mathbf{r}_0}{r_0^3\{1 - (v^2/c^2)\sin^2\theta\}^{3/2}} dV'. \quad (6\text{-}1.45)$$

## 6-2. Converting Retarded Potential Integrals for Uniformly Moving Mass Distributions into Present-Time (Present Position) Integrals

We know from Chapter 3, Eqs. (3-3.5) and (3-3.2), that the gravitational potential $\varphi$ and the cogravitational vector potential $\mathbf{A}$ of time-variable mass and mass-current distributions in a vacuum can be found from the retarded integrals

$$\varphi = -G\int \frac{[\rho]}{r} dV' \tag{6-2.1}$$

and

$$\mathbf{A} = -\frac{G}{c^2}\int \frac{[\mathbf{J}]}{r} dV'. \tag{6-2.2}$$

As we shall presently see, for time-independent mass distributions moving with constant velocity, these integrals can be converted to the present position of the mass, so that the integration is performed not over the retarded volume, but over the volume that the mass distribution occupies at the moment $t$ for which the potentials are being determined.

***Converting Eq. (6-2.1).*** Using Eq. (6-1.5) and remembering that $\rho$ and $\mathbf{v}$ are not affected by retardation, we can write Eq. (6-2.1) as

$$\varphi = -G\int \frac{\rho}{[r - \mathbf{r}\cdot\mathbf{v}/c]} dV', \tag{6-2.3}$$

where only the denominator is retarded. Converting the retarded denominator in Eq. (6-2.3) with the help of Eq. (6-1.8), we obtain the desired equation (omitting the subscript "0" for simplicity)

$$\varphi = -G\int \frac{\rho}{\{x'^2 + (y'^2 + z'^2)/\gamma^2\}^{1/2}} dV', \tag{6-2.4}$$

where the integral is a "present position" integral, and where all quantities are present-time quantities.

Equation (6-2.4) can be written in an alternative form. Using Eqs. (6-1.8) and (6-1.9) for converting the denominator of the integrand in Eq. (6-2.4), we obtain

$$\varphi = -G\int \frac{\rho}{r\{1 - (v^2/c^2)\sin^2\theta\}^{1/2}} dV'. \tag{6-2.5}$$

Equations (6-2.4) and (6-2.5) can be further modified so that the potential is expressed not in terms of the mass density $\rho$ as such, but in terms of $\nabla\rho$ (that is, in terms of the "mass inhomogeneities"). This can be done as follows.

Taking into account that the position vector $\mathbf{r}$ is directed toward the point of observation, so that $\mathbf{r} = -x'\mathbf{i} - y'\mathbf{j} - z'\mathbf{k}$ and $\nabla' \cdot \mathbf{r} = -3$, we write [see vector identity (V-8)]

$$\nabla' \cdot \frac{\mathbf{r}\rho}{\{x'^2+(y'^2+z'^2)/\gamma^2\}^{1/2}} = \frac{\mathbf{r}}{\{x'^2+(y'^2+z'^2)/\gamma^2\}^{1/2}} \cdot \nabla'\rho$$
$$+\rho\nabla' \cdot \frac{\mathbf{r}}{\{x'^2+(y'^2+z'^2)/\gamma^2\}^{1/2}}$$
$$= \frac{\mathbf{r}}{\{x'^2+(y'^2+z'^2)/\gamma^2\}^{1/2}} \cdot \nabla'\rho - \frac{3\rho}{\{x'^2+(y'^2+z'^2)/\gamma^2\}^{1/2}}$$
$$- \frac{\mathbf{r} \cdot \{x'\mathbf{i}+(y'\mathbf{j}+z'\mathbf{k})/\gamma^2\}\rho}{\{x'^2+(y'^2+z'^2)/\gamma^2\}^{3/2}} \tag{6-2.6}$$
$$= \frac{\mathbf{r} \cdot \nabla'\rho}{\{x'^2+(y'^2+z'^2)/\gamma^2\}^{1/2}} - \frac{2\rho}{\{x'^2+(y'^2+z'^2)/\gamma^2\}^{1/2}} .$$

Using Eq. (6-2.6) and Eq. (6-2.4), we can now express the potential as

$$\varphi = \frac{G}{2}\int \nabla' \cdot \frac{\mathbf{r}\rho}{\{x'^2+(y'^2+z'^2)/\gamma^2\}^{1/2}}\, dV'$$
$$- \frac{G}{2}\int \frac{\mathbf{r} \cdot \nabla\rho}{\{x'^2+(y'^2+z'^2)/\gamma^2\}^{1/2}}\, dV' . \tag{6-2.7}$$

The first integral in this equation can be transformed into a surface integral over all space by means of Gauss's theorem of vector analysis [vector identity (V-19)], and, since there are no masses at infinity, the integral vanishes. Hence the potential can be written as

$$\varphi = -\frac{G}{2}\int \frac{\mathbf{r}\cdot\boldsymbol{\nabla}'\rho}{\{x'^2+(y'^2+z'^2)/\gamma^2\}^{1/2}}\,dV', \tag{6-2.8}$$

or, by using Eqs. (6-1.8) and (6-1.9), as

$$\varphi = -\frac{G}{2}\int \frac{\mathbf{r}\cdot\boldsymbol{\nabla}'\rho}{r\{1-(v^2/c^2)\sin^2\theta\}^{1/2}}\,dV'. \tag{6-2.9}$$

Equations (6-2.8) and (6-2.9) can be written in a much simpler form, if $\rho$ is constant within the mass distribution. In this case $\boldsymbol{\nabla}'\rho$ is different from zero only in the surface layer of the mass distribution, where the mass changes from $\rho$ within the distribution to zero outside the distribution. We then have $\boldsymbol{\nabla}'\rho = (\rho/\tau)\mathbf{n}_u$, where $\tau$ is the thickness of the surface layer of the distribution, and $\mathbf{n}_u$ is a unit vector normal to the surface of the distribution and directed into the distribution. The volume element $dV'$ in Eqs. (6-2.8) and (6-2.9) becomes then $\tau dS'$, where $dS'$ is a surface area element of the distribution, and therefore Eqs. (6-2.8) and (6-2.9) reduce to

$$\varphi = \frac{G\rho}{2}\oint \frac{\mathbf{r}\cdot d\mathbf{S}'_{out}}{\{x'^2+(y'^2+z'^2)/\gamma^2\}^{1/2}}, \tag{6-2.10}$$

and

$$\varphi = \frac{G\rho}{2}\oint \frac{\mathbf{r}\cdot d\mathbf{S}'_{out}}{r\{1-(v^2/c^2)\sin^2\theta\}^{1/2}}, \tag{6-2.11}$$

where $d\mathbf{S}_{out}$ is a surface element vector directed from the mass distribution into the surrounding space.

***Converting Eq. (6-2.2).*** The mass-current density produced by a uniformly moving mass distribution is $\mathbf{J} = \rho\mathbf{v}$ with $\mathbf{v} =$ const. The vector potential $\mathbf{A}$ for such a mass distribution is, by Eqs. (6-2.2) and (6-2.1),

$$\mathbf{A} = \frac{\mathbf{v}\varphi}{c^2}, \tag{6-2.12}$$

Hence, using Eqs. (6-2.5), (6-2.9), and (6-2.11), we have

$$\mathbf{A} = -\frac{G\mathbf{v}}{c^2}\int \frac{\rho}{r\{1-(v^2/c^2)\sin^2\theta\}^{1/2}}\,dV', \tag{6-2.13}$$

$$\mathbf{A} = -\frac{G\mathbf{v}}{2c^2}\int \frac{\mathbf{r}\cdot\nabla'\rho}{r\{1-(v^2/c^2)\sin^2\theta\}^{1/2}}\,dV', \tag{6-2.14}$$

and

$$\mathbf{A} = \frac{G\mathbf{v}\rho}{2c^2}\oint \frac{\mathbf{r}\cdot d\mathbf{S}'_{out}}{r\{1-(v^2/c^2)\sin^2\theta\}^{1/2}}, \tag{6-2.15}$$

and similar expressions corresponding to Eqs. (6-2.4), (6-2.8), and (6-2.10):

$$\mathbf{A} = -\frac{G\mathbf{v}}{c^2}\int \frac{\rho}{\{x'^2+(y'^2+z'^2)/\gamma^2\}^{1/2}}\,dV', \tag{6-2.16}$$

$$\mathbf{A} = -\frac{G\mathbf{v}}{2c^2}\int \frac{\mathbf{r}\cdot\nabla'\rho}{\{x'^2+(y'^2+z'^2)/\gamma^2\}^{1/2}}\,dV', \tag{6-2.17}$$

$$\mathbf{A} = \frac{G\mathbf{v}\rho}{2c^2}\oint \frac{\mathbf{r}\cdot d\mathbf{S}'_{out}}{\{x'^2+(y'^2+z'^2)/\gamma^2\}^{1/2}}. \tag{6-2.18}$$

▼

**Example 6-2.1**. An irregularly shaped mass distribution of total mass $m$ moves with constant velocity $\mathbf{v} = -v\mathbf{i}$. The longest linear dimension of the mass distribution is $a$. Find the gravitational and cogravitational potentials produced by the mass at a distance $r \gg a$ from the mass.

We can solve the problem by using Eqs. (6-2.5) and (6-2.13). Since $r \gg a$, we can assume $r$ and $\theta$ to be the same for all points of the mass. Therefore we can factor out the denominator of the integrands in Eq. (6-2.5) and (6-2.13), obtaining [compare with Eqs. (5-6.5) and (5-6.6)]

$$\varphi = -G\frac{m}{r\{1-(v^2/c^2)\sin^2\theta\}^{1/2}}, \qquad (6\text{-}2.19)$$

$$\mathbf{A} = -G\frac{\mathbf{v}m}{c^2 r\{1-(v^2/c^2)\sin^2\theta\}^{1/2}}. \qquad (6\text{-}2.20)$$

▲

## 6-3. Some Peculiarities of the Expressions for the Fields and Potentials Derived in this Chapter

Three peculiarities of the equations for the gravitational and cogravitational fields and potentials derived in this chapter should be noted.

First, in the equations developed in the preceding chapters we used both retarded and present-time (present position) coordinates, and therefore we needed to use different notation for the two types of coordinates. In particular, we designated the present position vector as $\mathbf{r}_0$ and the $x$ component of this vector as $x_0'$, while we designated the retarded position vector as $\mathbf{r}$ and its $x$ component as $x'$. However, since all the resulting expressions for the fields and potentials developed in this chapter are for the present position of the mass distributions, there is no longer a need to use the subscript "0" at $\mathbf{r}$ or $x'$. Therefore, in the field and potential equations obtained in this chapter, $\mathbf{r}$ and $x'$ stand for the present-time (present position) coordinates.

Second, in deriving our equations for the potentials of moving mass distributions, we assumed that the field point (the point for which the potentials are determined) was at the origin. However, in practical application of the potentials it is usually necessary to differentiate the potentials with respect to the field point. In particular, for finding gravitational and cogravitational fields from potentials it is necessary to operate upon the gravitational and cogravitational potentials with the operator $\nabla$ (which operates upon the field point coordinates). Therefore, in general, the field point must be allowed to vary.

We can easily convert our equations for the potentials (and fields) into equations with a variable field point. Let us designate the coordinates of this point as $x$, $y$, and $z$. If we then replace the $x'$, $y'$, and $z'$ coordinates appearing explicitly or implicitly in our equations for potentials or fields by $(x - x')$, $(y - y')$, and $(z - z')$, respectively, the new equations will apply to fields and potentials determined for the field point $x$, $y$, $z$. However, if the mass density $\rho$ within the mass distribution under consideration is constant, we can differentiate the potentials with respect to the field point without actually replacing the $x'$, $y'$, $z'$ coordinates at all, because in this case, by vector identity (V-27), the only difference between the differentiation of the integrands with respect to $x'$, $y'$, $z'$ and with respect to $x$, $y$, $z$ is in the sign of the resulting expression. Thus, in the case of constant $\rho$, we can compute gravitational and cogravitational fields from the potentials derived in this chapter without changing the coordinates, provided that after placing $\nabla$ under the integral sign we replace it by $-\nabla'$ (see Example 6-3.1).

Third, all the fields and potentials derived in this chapter are "snapshots" representing only the instantaneous values of the observed fields and potentials. In reality the fields and potentials of a moving mass distribution vary as the mass distribution moves relative to the point of observation. For practical applications it may be necessary to determine time derivatives of the fields and potentials. Therefore, in general, the fields and potentials must be expressed as a function of time. This can be easily done by noting that when a mass distribution moves with constant speed parallel to the $x$ axis, the present position of $dV'$ (or $d\mathbf{S}'$) is $x' \mp vt$ (the minus applies to motion against the $x$ axis, the plus applies to the motion in the direction of the $x$ axis). Thus all we need to do for introducing the time dependence into the fields and potentials derived in this chapter is to replace $x'$ appearing explicitly or implicitly in our field and potential equations by $x' \mp vt$ (see Example 6-3.1, see also Examples 5-1.2 and 5-6.1).

▼

**Example 6-3.1** A very long hollow cylinder of wall thickness $b$ and radius $a \gg b$ has a uniformly distributed mass of density $\rho$ and moves with velocity $\mathbf{v} = -v\mathbf{i}$ along the $x$ axis, which is also the symmetry axis of the cylinder (Fig. 6.3). Find the gravitational field produced by the cylinder at the origin of coordinates when the leading end of the cylinder is at a distance $x'$ from the origin.

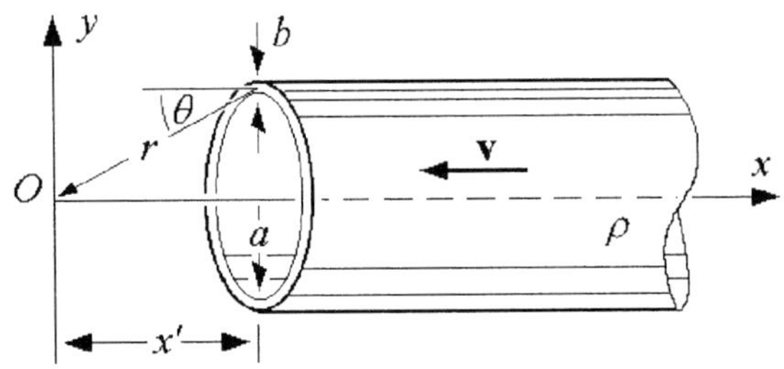


*Fig. 6.3 A very long cylinder of mass density ρ moves with uniform velocity* $\mathbf{v} = -v\mathbf{i}$ *along the x axis. Find the gravitational field produced by the cylinder at the origin.*

We shall solve this problem by using Eqs. (6-2.4) and (6-2.16). Applying the relation $\mathbf{g} = -\nabla\varphi - \partial\mathbf{A}/\partial t$ [this is Eq. (3-3.4) derived in Section 3.3] to Eqs. (6-2.4) and (6-2.16), we obtain

$$\mathbf{g} = -\nabla\left(-G\int\frac{\rho}{\{x'^2+(y'^2+z'^2)/\gamma^2\}^{1/2}}dV'\right) - \frac{\partial}{\partial t}\left(-\frac{G\mathbf{v}}{c^2}\int\frac{\rho}{\{x'^2+(y'^2+z'^2)/\gamma^2\}^{1/2}}dV'\right). \tag{6-3.1}$$

In Eq. (6-3.1), $\nabla$ operates upon the field point coordinates $x$, $y$, $z$, which do not appear in Eq. (6-3.1). However, as explained above, for constant $\rho$ we can leave the first integral in Eq. (6-3.1) as it now is, provided that for the actual differentiation we replace $\nabla$ by $-\nabla'$. Placing $\nabla$ under the integral sign and replacing it by $-\nabla'$, we have for the part of the gravitational field due to $\varphi$ (using $\mathbf{g} = \mathbf{g}_\varphi + \mathbf{g}_A$)

$$\mathbf{g}_\varphi = -G\int\nabla'\frac{\rho}{\{x'^2+(y'^2+z'^2)/\gamma^2\}^{1/2}}dV'. \tag{6-3.2}$$

To differentiate the second integral in Eq. (6-3.1), we must first express the integrand as a function of $t$. Replacing $x'$ in the integrand by $x' - vt$, placing $\partial/\partial t$ under the integral sign, and differentiating the integrand, we then have for the part of the gravitational field due to $\mathbf{A}$

$$\mathbf{g}_A = G\frac{\mathbf{v}}{c^2}\int\frac{\rho(x'-vt)v}{\{(x'-vt)^2+(y'^2+z'^2)/\gamma^2\}^{3/2}}dV', \tag{6-3.3}$$

or, setting $t = 0$,

$$\mathbf{g}_A = G\frac{\mathbf{v}}{c^2}\int\frac{vx'\rho}{\{x'^2+(y'^2+z'^2)/\gamma^2\}^{3/2}}dV', \tag{6-3.4}$$

which, as one can easily verify by direct differentiation, is the same as

$$\mathbf{g}_A = \frac{G\mathbf{v}}{c^2}\mathbf{v}\cdot\int\nabla'\frac{\rho}{\{x'^2+(y'^2+z'^2)/\gamma^2\}^{1/2}}dV'. \tag{6-3.5}$$

The total field is therefore

$$\mathbf{g} = -G\int \nabla' \frac{\rho}{\{x'^2 + (y'^2 + z'^2)/\gamma^2\}^{1/2}} dV' + G\frac{\mathbf{v}}{c^2}\mathbf{v}\cdot\int \nabla' \frac{\rho}{\{x'^2 + (y'^2 + z'^2)/\gamma^2\}^{1/2}} dV'. \tag{6-3.6}$$

Using now Gauss's theorem of vector analysis [vector identity (V-19)], we can convert the two integrals into integrals over the surface of the cylinder, obtaining[2]

$$\mathbf{g} = -G\rho\left\{\oint \frac{d\mathbf{S}_{out}}{\{x'^2 + (y'^2 + z'^2)/\gamma^2\}^{1/2}} - \frac{\mathbf{v}}{c^2}\mathbf{v}\cdot\oint \frac{d\mathbf{S}_{out}}{\{x'^2 + (y'^2 + z'^2)/\gamma^2\}^{1/2}}\right\}, \tag{6-3.7}$$

where $d\mathbf{S}_{out}$ is a surface element vector directed outward from the volume of the cylinder.

By the symmetry of the system, the gravitational field at the point of observation has only the $x$ component. The only surfaces of the cylinder contributing to that component are the surfaces of the leading and trailing ends of the cylinder. However, since the cylinder is very long, the contribution of the trailing end is negligible. Furthermore, since the cylinder's wall is thin, the integration over the leading end can be replaced by the multiplication of the integrand by the surface area $S = 2\pi ab$ of the leading end's wall. Taking into account that $\mathbf{v} = -v\mathbf{i}$, that for the leading end $y'^2 + z'^2 = a^2$, $d\mathbf{S}_{out} = -dS\mathbf{i}$, and $\mathbf{v}\cdot d\mathbf{S}_{out} = vdS$, we finally obtain for the "snapshot" of the gravitational field produced by the cylinder at the point of observation

$$\mathbf{g} = G\frac{2\pi\rho ab(1 - v^2/c^2)}{\{x'^2 + a^2(1 - v^2/c^2)\}^{1/2}}\mathbf{i}. \tag{6-3.8}$$

**Example 6-3.2** A line mass of length $2L$ and linear mass density $\lambda$ moves along its length with constant velocity $\mathbf{v} = -v\mathbf{i}$ in the $xy$ plane of a rectangular system of coordinates at a distance $y = R$ above the $x$ axis. The point of observation is at the origin. Find the gravitational potential, the gravitational field, and the cogravitational field at the origin at the moment when the two ends of the mass are at equal distances $L$ from the $y$ axis and then obtain the limiting value of the fields for a very long mass.

To find the gravitational potential, we use Eq. (6-2.4) with $\rho dV'$ replaced by $\lambda dx'$. Integrating over the length of the line mass we then have

$$\varphi = -G\int_{-L}^{L}\frac{\lambda}{(x'^2+y'^2/\gamma^2)^{1/2}}dx' \qquad (6\text{-}3.9)$$
$$= -G\lambda\ln\{x'+(x'^2+y'^2/\gamma^2)^{1/2}\}\Big|_{-L}^{L},$$

or

$$\varphi = -G\lambda\ln\frac{\{L+(L^2+y'^2/\gamma^2)^{1/2}\}}{\{-L+(L^2+y'^2/\gamma^2)^{1/2}\}}. \qquad (6\text{-}3.10)$$

To find the gravitational field, we differentiate Eq. (6-3.10) with respect to $y'$, using the *positive* derivative (by symmetry, the vector potential makes no contribution to the gravitational field at the origin). The result is

$$\mathbf{g} = G\frac{2\lambda}{y'(1+y'^2/\gamma^2L^2)^{1/2}}\mathbf{j} = G\frac{2\lambda}{R(1+R^2/\gamma^2L^2)^{1/2}}\mathbf{j}. \qquad (6\text{-}3.11)$$

The cogravitational field of the line mass is, by Eqs. (6-3.11) and (4-2.4),

$$\mathbf{K} = -G\frac{2\lambda v}{c^2R(1+R^2/\gamma^2L^2)^{1/2}}\mathbf{k}. \qquad (6\text{-}3.12)$$

For a very long mass, $L \gg R$, so that Eqs. (6-3.11) and (6-3.12) reduce to

$$\mathbf{g} = G\frac{2\lambda}{R}\mathbf{j} \tag{6-3.13}$$

and

$$\mathbf{K} = -G\frac{2\lambda v}{c^2 R}\mathbf{k}. \tag{6-3.14}$$

It is interesting to note that the gravitational field given by Eq. (6-3.13) is the same as that of a *stationary* infinitely long line mass, and that the cogravitational field given by Eq. (6-3.14) is the same as the cogravitational field produced by a mass-current $I = \lambda v$ (compare with Example 6-1.2 and Eq. 5-3.23).

▲

## References and Remarks for Chapter 6

**1.** As was mentioned in Section 5-1, Eq. (5-1.13) for a moving point mass was first derived (in a different form) by Oliver Heaviside. Heaviside noted that his equation provided an explanation for the absence of gravitational aberration even if gravitation propagated at a finite speed. Equation (6-1.40) shows that also in the case of a mass distribution of any shape moving with a finite speed $v$ there is no gravitational aberration, because the gravitational field appears to originate at the present position of the moving mass.

**2.** When using this method, the volume of integration must be inside the mass distribution, because only there $\rho$ is constant. The surface of integration remains *within* the mass distribution, just touching the surface layer of the mass, but not stepping out of the mass distribution into the space where there is no mass. See Section 3-2.

# 7

# DIFFERENTIAL EQUATIONS FOR GRAVITATIONAL AND COGRAVITATIONAL FIELDS; ELECTROMAGNETIC ANALOGY

In this chapter we shall derive differential equations for gravitational and cogravitational fields. As we shall see, these equations are similar to Maxwell's electromagnetic equations, which makes the generalized theory of gravitation very similar to Maxwellian electrodynamics. An important consequence of this similarity is that many methods and techniques originally developed for solving electromagnetic problems can be used for solving problems involving gravitational and cogravitational interactions.

## 7-1. Differential Equations for Gravitational and Cogravitational Fields; Analogy with Maxwell's Electromagnetic Equations

Practical applications of the principal field equations, Eqs. (2-2.1) and (2-2.2), as well as of their special forms derived in Chapter 3 are rather difficult because they involve retarded integrals, in which the integrands must be evaluated for a past time $t'$, rather than for the present time $t$ (the time for which the fields **g** and **K** are determined). Therefore, for practical applications, Eqs. (2-2.1), (2-2.2) and their equivalents should be preferably converted into equations where all the quantities are evaluated for the present time $t$. In general, converting equations involving retarded integrals into equations with present-time integrals is not possible (see Chapters 5 and 6 for several exceptions). However, as we shall presently see, some equations containing retarded integrals can be converted into differential and integral equations involving present-time quantities only.

From the theoretical point of view, particularly important is the fact that Eqs. (2-2.1) and (2-2.2) can be converted into the following present-time differential equations:

$$\nabla\cdot\mathbf{g} = -4\pi G\rho, \tag{7-1.1}$$

$$\nabla\cdot\mathbf{K} = 0, \tag{7-1.2}$$

$$\nabla\times\mathbf{g} = -\frac{\partial\mathbf{K}}{\partial t}, \tag{7-1.3}$$

and

$$\nabla\times\mathbf{K} = -\frac{4\pi G}{c^2}\mathbf{J} + \frac{1}{c^2}\frac{\partial\mathbf{g}}{\partial t}. \tag{7-1.4}$$

By applying vector identity (V-19) to Eqs. (7-1.1) and (7-1.2) and by applying vector identity (V-17) to Eqs. (7-1.3) and (7-1.4), Eqs. (7-1.1)-(7-1.4) can be further converted into the following present-time integral equations:

$$\oint\mathbf{g}\cdot d\mathbf{S} = -4\pi G\int\rho\, dV. \tag{7-1.5}$$

$$\oint\mathbf{K}\cdot d\mathbf{S} = 0. \tag{7-1.6}$$

$$\oint\mathbf{g}\cdot d\mathbf{l} = -\frac{\partial}{\partial t}\int\mathbf{K}\cdot d\mathbf{S}. \tag{7-1.7}$$

and

$$\oint \mathbf{K} \cdot d\mathbf{l} = -\frac{1}{c^2}\int\left(4\pi G\mathbf{J} - \frac{\partial \mathbf{g}}{\partial t}\right)\cdot d\mathbf{S}. \qquad (7\text{-}1.8)$$

Readers familiar with electromagnetic theory will immediately recognize that Eqs. (7-1.1)-(7-1.8) are similar to Maxwell's electromagnetic equations in their differential and integral forms,[1] except that the symbols and constants in Eqs. (7-1.1)-(7-1.8) are different from the symbols and constants appearing in Maxwell's equations. It is important to note that, as early as in 1893, Oliver Heaviside suggested the possibility of an analogy between gravitation and electromagnetics and the possibility of expressing gravitational fields and the gravitational "analog of magnetic" fields by equations similar to Maxwell's equations.[2] Heaviside's suggestion was entirely intuitive and was not supported by substantive physical or mathematical arguments. However, as is now clear, this analogy is actually a rigorous consequence of the fundamental premises of the generalized theory of gravitation and of Eqs. (2-2.1) and (2-2.2) in particular. It is also clear therefore that the similarity of Eqs. (7-1.1)-(7-1.8) with Maxwell's electromagnetic equations is more than a mere analogy: Eqs. (7-1.1)-(7-1.8) are completely autonomous equations for gravitational and cogravitational fields reflecting intrinsic properties of these fields.

The analogy between gravitational-cogravitational and electrodynamic equations is not perfect, of course. In particular, whereas the electric field may be directed to or from the electric charge by which it is created (depending on whether the charge is negative or positive), the gravitational field is always directed to the mass by which it is created. Also, whereas the magnetic field is always right-handed relative to the electric current by which it is created, the cogravitational field is always left-handed relative to the mass current by which it is created, and whereas like electric currents attract each other and opposite electric currents repel each other, like mass currents repel each other and opposite mass currents attract each other. Furthermore, whereas electric charges may attract or repel each other, masses always attract each other (see, however, Chapter 19).

There may also be a difference in the interpretation of the physical significance of Eq. (7-1.3) and of its electromagnetic counterpart. Maxwell's electromagnetic equation similar to Eq. (7-1.3) was in the past interpreted as representing the phenomenon of electromagnetic induction and was thought to show that a changing magnetic field creates an electric field. One may think therefore that Eq. (7-1.3) likewise represents a gravitational-cogravitational induction phenomenon. However, it has now been proved that Maxwell's electromagnetic equation similar to Eq. (7-1.3) does not represent electromagnetic induction and that electric fields are not created by changing magnetic fields.[3,4] Clearly then, Eq. (7-1.3) does not represent an induction effect either. Moreover, since the gravitational and cogravitational fields in this equation are simultaneous in time, the equation does not reveal any causal relation between these fields. Similar considerations apply to Eq. (7-1.4), where the cogravitational and gravitational fields are also simultaneous in time and therefore are not causally connected with each other.

**Derivation of Eq. (7-1.1).** We start with Eq. (3-1.1) [which, as is shown in Chapter 3, is a consequence of Eq. (2-2.1)]

$$\mathbf{g} = G\int \frac{[\nabla'\rho]}{r}dV' + \frac{G}{c^2}\int \frac{1}{r}\left[\frac{\partial \mathbf{J}}{\partial t}\right]dV'. \qquad (3\text{-}1.1)$$

Multiplying Eq. (3-1.1) by $\nabla\cdot$ , we have

$$\nabla\cdot\mathbf{g} = G\nabla\cdot\int \frac{[\nabla'\rho]}{r}dV' + \frac{G}{c^2}\nabla\cdot\int \frac{1}{r}\left[\frac{\partial \mathbf{J}}{\partial t}\right]dV'. \qquad (7\text{-}1.9)$$

The operator $\nabla\cdot$ can be placed under the integral sign because it operates on the field-point coordinates $x$, $y$, $z$, while the integration is over the source-point coordinates $x'$, $y'$, $z'$. This gives

$$\nabla\cdot\mathbf{g} = G\int\nabla\cdot\frac{[\nabla'\rho]}{r}dV' + \frac{G}{c^2}\int\nabla\cdot\frac{1}{r}\left[\frac{\partial\mathbf{J}}{\partial t}\right]dV'. \quad (7\text{-}1.10)$$

Applying vector identity (V-34) to the first integral of Eq. (7-1.10), we have

$$\int\nabla\cdot\frac{[\nabla'\rho]}{r}dV' = -\int\nabla'\cdot\frac{[\nabla'\rho]}{r}dV' + \int\frac{[\nabla'\cdot\nabla'\rho]}{r}dV'. \quad (7\text{-}1.11)$$

The first integral on the right of Eq. (7-1.11) can be transformed into a surface integral by means of vector identity (V-19). But this surface integral vanishes, because $\rho$ is confined to a finite region of space, while the surface of integration is at infinity (as was explained in Section 2-2, unless stated otherwise, all integrals in this book are over all space). Thus we obtain for the first integral in Eq. (7-1.10)

$$\int\nabla\cdot\frac{[\nabla'\rho]}{r}dV' = \int\frac{[\nabla'\cdot\nabla'\rho]}{r}dV'. \quad (7\text{-}1.12)$$

Using the same considerations, we obtain for the second integral in Eq. (7-1.10)

$$\int\nabla\cdot\frac{1}{r}\left[\frac{\partial\mathbf{J}}{\partial t}\right]dV' = \int\frac{1}{r}\left[\nabla'\cdot\frac{\partial\mathbf{J}}{\partial t}\right]dV' = \int\frac{1}{r}\left[\frac{\partial\nabla'\cdot\mathbf{J}}{\partial t}\right]dV'. \quad (7\text{-}1.13)$$

Thus, Eq. (7-1.10) can be written as

$$\nabla\cdot\mathbf{g} = G\int\frac{[\nabla'\cdot\nabla'\rho]}{r}dV' + \frac{G}{c^2}\int\frac{1}{r}\left[\frac{\partial\nabla'\cdot\mathbf{J}}{\partial t}\right]dV'. \quad (7\text{-}1.14)$$

Using now the continuity law, Eq. (2-2.4), with **J** substituted for $\rho\mathbf{v}$,

$$\nabla' \cdot \mathbf{J} = - \frac{\partial \rho}{\partial t}, \tag{2-2.4}$$

we can rewrite Eq. (7-1.14) as

$$\nabla \cdot \mathbf{g} = G \int \frac{\left[ \nabla' \cdot (\nabla' \rho) - \frac{1}{c^2} \frac{\partial^2 \rho}{\partial t^2} \right]}{r} dV'. \tag{7-1.15}$$

But according to vector identity (V-30), the right side of Eq. (7-1.15) is simply $- 4\pi G\rho$, and therefore

$$\nabla \cdot \mathbf{g} = - 4\pi G\rho. \tag{7-1.1}$$

**Derivation of Eq. (7-1.2).** We start with Eq. (3-1.2) [which, as is shown in Chapter 3, is a consequence of Eq. (2-2.2)]

$$\mathbf{K} = - \frac{G}{c^2} \int \frac{[\nabla' \times \mathbf{J}]}{r} dV', \tag{3-1.2}$$

Multiplying Eq. (3-1.2) by $\nabla \cdot$ , we have

$$\nabla \cdot \mathbf{K} = - \frac{G}{c^2} \nabla \cdot \int \frac{[\nabla' \times \mathbf{J}]}{r} dV'. \tag{7-1.16}$$

As in the case of Eq. (7-1.9), the operator $\nabla \cdot$ can be placed under the integral sign because it operates on the field-point coordinates $x$, $y$, $z$, while the integration is over the source-point coordinates $x'$, $y'$, $z'$. This gives

$$\nabla \cdot \mathbf{K} = - \frac{G}{c^2} \int \nabla \cdot \frac{[\nabla' \times \mathbf{J}]}{r} dV'. \tag{7-1.17}$$

Transforming now the integral in Eq. (7-1.17) just as we transformed the first integral in Eq. (7-1.11), we obtain

$$\nabla \cdot \mathbf{K} = -\frac{G}{c^2}\int \frac{[\nabla' \cdot (\nabla' \times \mathbf{J})]}{r} dV', \qquad (7\text{-}1.18)$$

and, since $\nabla' \cdot \nabla' \times = 0$,

$$\nabla \cdot \mathbf{K} = 0. \qquad (7\text{-}1.2)$$

**Derivation of Eq. (7-1.3).** We again start with Eq. (3-1.1)

$$\mathbf{g} = G\int \frac{[\nabla' \rho]}{r} dV' + \frac{G}{c^2}\int \frac{1}{r}\left[\frac{\partial \mathbf{J}}{\partial t}\right] dV'. \qquad (3\text{-}1.1)$$

Multiplying Eq. (3-1.1) by $\nabla \times$ , we have

$$\nabla \times \mathbf{g} = G\nabla \times \int \frac{[\nabla' \rho]}{r} dV' + \frac{G}{c^2}\nabla \times \int \frac{1}{r}\left[\frac{\partial \mathbf{J}}{\partial t}\right] dV'. \qquad (7\text{-}1.19)$$

The operator $\nabla \times$ can be placed under the integral sign because it operates on the field-point coordinates $x$, $y$, $z$, while the integration is over the source-point coordinates $x'$, $y'$, $z'$. We then have

$$\nabla \times \mathbf{g} = G\int \nabla \times \frac{[\nabla' \rho]}{r} dV' + \frac{G}{c^2}\int \nabla \times \frac{1}{r}\left[\frac{\partial \mathbf{J}}{\partial t}\right] dV'. \qquad (7\text{-}1.20)$$

Applying vector identity (V-34) to the first integral of Eq. (7-1.20), we have

$$\int \nabla \times \frac{[\nabla' \rho]}{r} dV' = -\int \nabla' \times \frac{[\nabla' \rho]}{r} dV' + \int \frac{[\nabla' \times \nabla' \rho]}{r} dV'. \qquad (7\text{-}1.21)$$

The first integral on the right of Eq. (7-1.21) can be transformed into a surface integral by means of vector identity (V-21). But this surface integral vanishes, because $\rho$ is confined to a finite region of space, while the surface of integration is at infinity. Thus we obtain for the first integral in Eq. (7-1.20)

$$\int \nabla \times \frac{[\nabla' \rho]}{r} dV' = \int \frac{[\nabla' \times \nabla' \rho]}{r} dV' . \tag{7-1.22}$$

However, $\nabla' \times \nabla' = 0$, and therefore the last integral vanishes and so does the first integral in Eq. (7-1.20). Equation (7-1.20) reduces therefore to

$$\nabla \times \mathbf{g} = \frac{G}{c^2} \int \nabla \times \frac{1}{r} \left[ \frac{\partial \mathbf{J}}{\partial t} \right] dV' . \tag{7-1.23}$$

Let us now differentiate Eq. (3-1.2) with respect to time. We have

$$\frac{\partial \mathbf{K}}{\partial t} = - \frac{G}{c^2} \frac{\partial}{\partial t} \int \frac{[\nabla' \times \mathbf{J}]}{r} dV' . \tag{7-1.24}$$

Applying vector identity (V-34) to the integral in Eq. (7-1.24) and eliminating $\nabla' \times \{[\nabla' \times \mathbf{J}]/r\}$ by means of vector identity (V-21) [see the explanation below Eq. (7-1.21); note that $\mathbf{J} = \rho \mathbf{v}$ and therefore $\mathbf{J}$ is confined to a finite region of space], we obtain

$$\frac{\partial \mathbf{K}}{\partial t} = - \frac{G}{c^2} \frac{\partial}{\partial t} \int \nabla \times \frac{[\mathbf{J}]}{r} dV' . \tag{7-1.25}$$

Differentiating under the integral sign, we obtain

$$\frac{\partial \mathbf{K}}{\partial t} = -\frac{G}{c^2}\int \boldsymbol{\nabla}\times\frac{1}{r}\left[\frac{\partial \mathbf{J}}{\partial t}\right]dV', \tag{7-1.26}$$

which together with Eq. (7-1.23) yields

$$\nabla\times\mathbf{g} = -\frac{\partial \mathbf{K}}{\partial t}, \tag{7-1.3}$$

**Derivation of Eq. (7-1.4).** We start once again with Eq. (3-1.2)

$$\mathbf{K} = -\frac{G}{c^2}\int \frac{[\nabla'\times\mathbf{J}]}{r}dV', \tag{3-1.2}$$

Multiplying Eq. (3-1.2) by $\boldsymbol{\nabla}\times$ , we have

$$\boldsymbol{\nabla}\times\mathbf{K} = -\frac{G}{c^2}\boldsymbol{\nabla}\times\int \frac{[\nabla'\times\mathbf{J}]}{r}dV'. \tag{7-1.27}$$

In Eq. (7-1.27) the operator $\boldsymbol{\nabla}\times$ can be placed under the integral sign because it operates on the field-point coordinates $x$, $y$, $z$, while the integration is over the source-point coordinates $x'$, $y'$, $z'$. We then have

$$\boldsymbol{\nabla}\times\mathbf{K} = -\frac{G}{c^2}\int \boldsymbol{\nabla}\times\frac{[\nabla'\times\mathbf{J}]}{r}dV'. \tag{7-1.28}$$

Transforming now the integral in Eq. (7-1.28) just as we transformed the first integral in Eq. (7-1.20), we obtain

$$\boldsymbol{\nabla}\times\mathbf{K} = -\frac{G}{c^2}\int \frac{[\boldsymbol{\nabla}'\times(\boldsymbol{\nabla}'\times\mathbf{J})]}{r}dV'. \tag{7-1.29}$$

Let us now find the time derivative of **g** by differentiating Eq. (3-1.1). We have

$$\frac{\partial \mathbf{g}}{\partial t} = G \int \frac{\partial}{\partial t} \frac{[\nabla' \rho]}{r} dV' + \frac{G}{c^2} \int \frac{\partial}{\partial t} \frac{1}{r} \left[ \frac{\partial \mathbf{J}}{\partial t} \right] dV'$$
$$= G \int \frac{1}{r} \left[ \nabla' \frac{\partial \rho}{\partial t} \right] dV' + \frac{G}{c^2} \int \frac{1}{r} \left[ \frac{\partial^2 \mathbf{J}}{\partial t^2} \right] dV' \tag{7-1.30}$$

and, making use of the continuity law, Eq. (2-2.4), we obtain

$$\frac{\partial \mathbf{g}}{\partial t} = - G \int \frac{[\nabla' (\nabla' \cdot \mathbf{J})]}{r} dV' + \frac{G}{c^2} \int \frac{1}{r} \left[ \frac{\partial^2 \mathbf{J}}{\partial t^2} \right] dV' . \tag{7-1.31}$$

Next, let us divide Eq. (7-1.31) by $c^2$ and subtract it from Eq. (7-1.29). Combining the three integrals into a single integral, we obtain

$$\nabla \times \mathbf{K} - \frac{1}{c^2} \frac{\partial \mathbf{g}}{\partial t} = \frac{G}{c^2} \int \frac{\left[ \nabla' (\nabla' \cdot \mathbf{J}) - \nabla' \times (\nabla' \times \mathbf{J}) - \frac{1}{c^2} \frac{\partial^2 \mathbf{J}}{\partial t^2} \right]}{r} dV' . \tag{7-1.32}$$

But, according to vector identity (V-28), the integral on the right in Eq. (7-1.32) is simply $-4\pi G\mathbf{J}/c^2$. Replacing this integral by $-4\pi G\mathbf{J}/c^2$ and transferring $(1/c^2)(\partial \mathbf{g}/\partial t)$ to the right, we obtain

$$\nabla \times \mathbf{K} = - \frac{4\pi G}{c^2} \mathbf{J} + \frac{1}{c^2} \frac{\partial \mathbf{g}}{\partial t} . \tag{7-1.4}$$

## 7-2. Corresponding Gravitational-Cogravitational and Electromagnetic Equations

The similarity of differential equations for gravitational-cogravitational fields, Eqs. (7-1.1)-(7-1.8), with Maxwell's electromagnetic equations indicates that many methods and techniques originally developed for solving problems involving electromagnetic fields can be used for solving problems involving gravitational and cogravitational fields. This similarity also indicates that it is possible to convert many equations originally derived for electromagnetic systems into the corresponding equations for gravitational-cogravitational systems. The corresponding equations are identical except for the symbols and constants occurring in them. Therefore, in order to convert an appropriate electromagnetic equation into a gravitational-cogravitational equation, one only needs to replace symbols and constants appearing in the electromagnetic equation by the corresponding gravitational-cogravitational symbols and constants. The relations between the corresponding symbols and constants are shown in Table 7-1.[5]

**Table 7-1**

**Corresponding Electromagnetic and Gravitational-Cogravitational Symbols and Constants**

| Electric | Gravitational |
|---|---|
| $q$ (charge) | $m$ (mass) |
| $\rho$ (volume charge density) | $\rho$ (volume mass density) |
| $\sigma$ (surface charge density) | $\sigma$ (surface mass density) |
| $\lambda$ (line charge density) | $\lambda$ (line mass density) |
| $\varphi$ (scalar potential) | $\varphi$ (scalar potential) |
| $\mathbf{A}$ (vector potential) | $\mathbf{A}$ (vector potential) |
| $\mathbf{J}$ (convection current density) | $\mathbf{J}$ (mass-current density) |
| $I$ (electric current) | $I$ (mass current) |
| $\mathbf{m}$ (magnetic dipole moment) | $\mathbf{d}$ (cogravitational moment) |
| $\mathbf{E}$ (electric field) | $\mathbf{g}$ (gravitational field) |
| $\mathbf{B}$ (magnetic field) | $\mathbf{K}$(cogravitational field) |
| $\varepsilon_0$ (permittivity of space) | $-1/4\pi G$ |
| $\mu_0$ (permeability of space) | $-4\pi G/c^2$ |
| $-1/4\pi\varepsilon_0$ or $-\mu_0 c^2/4\pi$ | $G$ (gravitational constant) |

Symbols that are not specific to electromagnetism, such as those for force, energy, momentum, etc., need not be replaced.

It is important to keep in mind, however, that only electromagnetic equations for fields in a vacuum have their gravitational counterparts, and only the electromagnetic symbols listed in Table 7-1 can be directly replaced by the corresponding gravitational symbols. In all other cases the following conversion procedure should be used:

(1) If an electromagnetic equation is for fields in the presence of material media, reduce the equation to fields in a vacuum.

(2) If electromagnetic equations contain field vectors **D** or **H**, replace them by **E** or **B**, using the relations $\mathbf{D} = \varepsilon_0\mathbf{E}$ and $\mathbf{B} = \mu_0\mathbf{H}$.

(3) Use Table 7-1 to replace electromagnetic constants by the corresponding gravitational-cogravitational constants.

## 7-3. Gravitational-Cogravitational Equations Obtained by Analogy with Electromagnetic Equations

Listed below are gravitational-cogravitational equations that have been obtained by converting electromagnetic equations in accordance with the procedure explained in Section 7-2. The electromagnetic equations used for conversion were taken from the author's book *Electricity and Magnetism.*[1] Some readers may want to examine these electromagnetic equations and their derivations. For this purpose each gravitational-cogravitational equation appearing below is provided with the number of the page where the corresponding electromagnetic equation appears in *Electricity and Magnetism* (hereafter abbreviated as EM). The equations are arranged in three categories: equations for calculating fields and potentials, equations for calculating energy and forces, and wave equations. Note that traditionally by far the majority of electromagnetic equations are derived for static fields or for fields involving slowly moving charges (conduction currents). In such fields the retardation does not exist or is ignored. Therefore some of the gravitational-cogravitational equations listed below do not apply to systems involving time-dependent or rapidly moving mass distributions; consult the derivations presented in this book if in doubt. Several equations listed below have already been directly derived in the preceding chapters; nevertheless, it is instructive to introduce them also as equations analogous to electromagnetic equations. (Primed fields and potentials indicate fields and potentials created by external sources.)

**(1) Equations for calculating gravitational fields and potentials:**

*Basic gravitational laws in present-time differential notation, EM502*

$$\nabla \cdot \mathbf{g} = -4\pi G\rho , \tag{7-3.1}$$

$$\nabla \cdot \mathbf{K} = 0 , \tag{7-3.2}$$

$$\nabla \times \mathbf{g} = -\frac{\partial \mathbf{K}}{\partial t} , \tag{7-3.3}$$

$$\nabla \times \mathbf{K} = -\frac{4\pi G}{c^2}\mathbf{J} + \frac{1}{c^2}\frac{\partial \mathbf{g}}{\partial t} . \tag{7-3.4}$$

*Basic gravitational laws in present-time integral notation, EM502*

$$\oint \mathbf{g} \cdot d\mathbf{S} = -4\pi G \int \rho \, dV . \tag{7-3.5}$$

$$\oint \mathbf{K} \cdot d\mathbf{S} = 0 . \tag{7-3.6}$$

$$\oint \mathbf{g}\cdot d\mathbf{l} = -\frac{\partial}{\partial t}\int \mathbf{K}\cdot d\mathbf{S}. \tag{7-3.7}$$

$$\oint \mathbf{K}\cdot d\mathbf{l} = -\frac{1}{c^2}\int\left(4\pi G\mathbf{J} - \frac{\partial \mathbf{g}}{\partial t}\right)\cdot d\mathbf{S}. \tag{7-3.8}$$

*Gravitational field of a point mass, EM96*

$$\mathbf{g} = -G\frac{m}{r^2}\mathbf{r}_u. \tag{7-3.9}$$

*Gravitational field of a mass distribution, EM93*

$$\mathbf{g} = -G\int \frac{\rho}{r^2}\mathbf{r}_u dV'. \tag{7-3.10}$$

*Gravitational field in terms of mass inhomogeneities (constant interior mass), EM103*

$$\mathbf{g} = -G\rho\oint \frac{d\mathbf{S}'}{r}. \tag{7-3.11}$$

*Gravitational scalar potential (with respect to ∞), EM120*

$$\varphi = -G\int \frac{\rho}{r}dV'. \tag{7-3.12}$$

*Gravitational potential of a point mass, EM121*

$$\varphi = -G\frac{m}{r}. \tag{7-3.13}$$

*Gravitational field in terms of scalar potential, EM111*

$$\mathbf{g} = -\nabla\varphi. \tag{7-3.14}$$

*Gravitational potential in terms of the field, EM112*

$$\varphi_a = \int_a^c \mathbf{g} \cdot d\mathbf{l} + \varphi_c . \tag{7-3.15}$$

*Poisson's equation for scalar potential, EM142*

$$\nabla^2 \varphi = 4\pi G \rho . \tag{7-3.16}$$

*Gravitational field in terms of vector potential*[6]

$$\mathbf{g} = \boldsymbol{\nabla} \times \mathbf{A}_g . \tag{7-3.17}$$

*Cogravitational field of a moving point mass, EM390*

$$\mathbf{K} = -G \frac{m(\mathbf{v} \times \mathbf{r}_u)}{c^2 r^2} . \tag{7-3.18}$$

*Cogravitational field of a current distribution, EM344*

$$\mathbf{K} = -\frac{G}{c^2} \int \frac{\mathbf{J} \times \mathbf{r}_u}{r^2} dV' . \tag{7-3.19}$$

*Cogravitational field in terms of current inhomogeneities (constant mass-current density)*, EM352

$$\mathbf{K} = -\frac{G}{c^2} \oint \frac{\mathbf{J} \times d\mathbf{S}'}{r} . \tag{7-3.20}$$

*Cogravitational vector potential, EM364*

$$\mathbf{A} = -\frac{G}{c^2} \int \frac{\mathbf{J}}{r} dV' . \tag{7-3.21}$$

*Cogravitational field in terms of vector potential, EM363*

$$\mathbf{K} = \nabla \times \mathbf{A} . \tag{7-3.22}$$

*Poisson's equation for cogravitational vector potential, EM364*

$$\nabla^2 \mathbf{A} = \frac{4\pi G}{c^2}\mathbf{J}. \tag{7-3.23}$$

*Cogravitational field in terms of scalar potential, EM373*

$$\mathbf{K} = \frac{4\pi G}{c^2}\nabla \varphi_c. \tag{7-3.24}$$

*Cogravitational dipole moment of filamentary mass current I (***S**′ *is right-handed relative to I), EM381*

$$\mathbf{d} = -\frac{4\pi G}{c^2} I\mathbf{S}'. \tag{7-3.25}$$

*Cogravitational dipole field, EM381*

$$\mathbf{K} = \frac{d}{2\pi r^3}\cos\theta\,\mathbf{r}_u + \frac{d}{4\pi r^3}\sin\theta\,\boldsymbol{\theta}_u. \tag{7-3.26}$$

**(2) Equations for calculating gravitational forces and energy:**

*Gravitational force on a mass distribution, EM208*

$$\mathbf{F} = \int \rho \mathbf{g}'\, dV. \tag{7-3.27}$$

*Gravitational force in terms of scalar potential*[7] *(single mass of constant density)*, EM211

$$\mathbf{F} = -\rho \oint \varphi'\, d\mathbf{S}. \tag{7-3.28}$$

*Gravitational force in terms of vector potential*[6,7] *(single mass of constant density)*

$$\mathbf{F} = -\rho \oint \mathbf{A}'_g \times d\mathbf{S}. \tag{7-3.29}$$

*Maxwell's stress integral for the gravitational field, EM215*

$$\mathbf{F} = \frac{1}{8\pi G}\oint \mathbf{g}^2 d\mathbf{S} - \frac{1}{4\pi G}\oint \mathbf{g}(\mathbf{g}\cdot d\mathbf{S}). \qquad (7\text{-}3.30)$$

*Cogravitational force on a mass current, EM440*

$$\mathbf{F} = \int \mathbf{J}\times\mathbf{K}' dV. \qquad (7\text{-}3.31)$$

*Cogravitational force on a mass-current dipole, EM446*

$$\mathbf{F} = -\frac{c^2}{4\pi G}(\mathbf{d}\cdot\nabla)\mathbf{K}'. \qquad (7\text{-}3.32)$$

*Cogravitational torque on a mass-current dipole, EM446*

$$\mathbf{T} = -\frac{c^2}{4\pi G}\mathbf{d}\times\mathbf{K}'. \qquad (7\text{-}3.33)$$

*Cogravitational force in terms of vector potential*[7] (*constant mass-current density*), EM453

$$\mathbf{F} = \oint \mathbf{A}'\cdot\mathbf{J}\, d\mathbf{S}. \qquad (7\text{-}3.34)$$

*Cogravitational force in terms of scalar potential*[7] (*constant mass-current density*), EM453

$$\mathbf{F} = \frac{4\pi G}{c^2}\oint \varphi_c'\mathbf{J}\times d\mathbf{S}. \qquad (7\text{-}3.35)$$

*Maxwell's stress integral for the cogravitational field, EM447*

$$\mathbf{F} = \frac{c^2}{8\pi G}\oint \mathbf{K}^2 d\mathbf{S} - \frac{c^2}{4\pi G}\oint \mathbf{K}(\mathbf{K}\cdot d\mathbf{S}). \qquad (7\text{-}3.36)$$

*Gravitational field energy, EM186*

$$U = -\frac{1}{8\pi G}\int \mathbf{g}^2 dV. \tag{7-3.37}$$

*Gravitational energy in terms of potential, EM190*

$$U = \frac{1}{2}\int \varphi\rho\, dV. \tag{7-3.38}$$

*Energy of a system of point masses, EM192*

$$U = -\frac{G}{2}\sum_i \sum_k{}' \frac{m_i m_k}{r_{ik}} + U_s. \tag{7-3.39}$$

*Energy of a mass distribution in an external field, EM195*

$$U' = \int \rho\varphi'\, dV. \tag{7-3.40}$$

*Energy of a point mass in an external field, EM195*

$$U' = m\varphi'. \tag{7-3.41}$$

*Cogravitational field energy, EM427*

$$U = -\frac{c^2}{8\pi G}\int \mathbf{K}^2 dV. \tag{7-3.42}$$

Cogravitational energy in terms of vector potential, EM430

$$U = \frac{1}{2}\int \mathbf{A}\cdot\mathbf{J}\, dV. \tag{7-3.43}$$

*Cogravitational energy of a mass current in an external field, EM432*

$$U' = \int \mathbf{J}\cdot\mathbf{A}'\, dV. \tag{7-3.44}$$

*Gravitational Poynting's vector, EM509*

$$\mathbf{P} = \frac{c^2}{4\pi G}\mathbf{K}\times\mathbf{g}. \tag{7-3.45}$$

*Gravitational field momentum, EM513*

$$\mathbf{G} = \frac{1}{4\pi G}\int \mathbf{K}\times\mathbf{g}\,dV. \tag{7-3.46}$$

*Gravitational field angular momentum*

$$\mathbf{L} = \frac{1}{4\pi G}\int \mathbf{r}\times(\mathbf{K}\times\mathbf{g})\,dV. \tag{7-3.47}$$

**(3) Equations for gravitational waves (see also Chapter 18):**

*Direction of field vectors in a plane wave propagating in the z-direction, EM531*

$$\mathbf{K} = \frac{1}{c}\mathbf{k}\times\mathbf{g}. \tag{7-3.48}$$

*Energy density in a gravitational wave, EM533*

$$U_v = -\frac{1}{4\pi G}\mathbf{g}^2 = -\frac{c^2}{4\pi G}\mathbf{K}^2. \tag{7-3.49}$$

The analogy between electromagnetic and gravitational-cogravitational equations is, of course, not limited to the equations listed above.[8] Not only basic electromagnetic equations, but also most equations representing a solution of an electromagnetic problem for fields or forces not involving conducting, dielectric, or magnetic bodies have their gravitational counterparts. However, if the propagation velocity of gravitation is not equal to the velocity of light (see Section 9-1), then $c$ appearing in the gravitational-cogravitational equations should be, in general, the velocity of the propagation of gravitation rather than the velocity of light.

Observe, however, that gravitational equations depicting "nonlinear" gravitational effects (see Chapter 19) do not have their electromagnetic counterparts.

## References and Remarks for Chapter 7

**1.** See, for example, Oleg D. Jefimenko, *Electricity and Magnetism*, 2nd ed., (Electret Scientific, Star City, 1989), p. 500.
**2.** The existence of a second gravitational field, analogous to the magnetic field, was first suggested by Oliver Heaviside in his two-part article "A Gravitational and Electromagnetic Analogy," The Electrician **31**, 281-282 and 359 (1893). This article is reproduced in modern notations in Ref. 3, pp. 189-202.
**3.** See Oleg D. Jefimenko, *Causality, Electromagnetic Induction, and Gravitation*, 2nd ed., (Electret Scientific, Star City, 2000), pp. 16-33.
**4.** Nevertheless, gravitational and cogravitational effects similar to electromagnetic induction do exist. See Chapter 12 for details.
**5.** Table 7-1 was originally published in 1991 in the 1st edition of Ref. 3, p. 104.
**6.** The possibility of representing a gravitational field by vector potential is not well known. As in the case of the electrostatic field, such a representation is only possible for the gravitational field in a mass-free space, where $\nabla \cdot \mathbf{g} = 0$. See Chapter 10 and Ref. 7 for details.
**7.** See Oleg D. Jefimenko, "Direct calculation of electric and magnetic forces from potentials," Am. J. Phys. **58**, 625-631 (1990).
**8.** See, for example, Oleg D. Jefimenko, "Presenting electromagnetic theory in accordance with the principle of causality," Eur. J. Phys. **25**, 287-296 (2004).

# 8

# ENERGY, ACTION-REACTION, AND MOMENTUM IN GRAVITATIONAL AND COGRAVITATIONAL FIELDS

In this chapter we shall discuss energy and momentum relations in gravitational and cogravitational fields. We shall present the proof that energy is always conserved in closed gravitational-cogravitational systems and is always conserved in gravitational and cogravitational interactions. We shall also present the proof that momentum in closed gravitational-cogravitational systems is always conserved. And we shall prove that, although the law of action and reaction does not always hold in gravitational-cogravitational interactions, the law of momentum conservation is always fulfilled in such interactions.

## 8-1. Conservation of Energy in Gravitational and Cogravitational Systems

Let us consider a closed gravitational-cogravitational system. In such a system there is no inflow or outflow of field energy to or from the system. By Eq. (2-2.9) we then have

$$\oint \mathbf{P}\cdot d\mathbf{S} = \frac{c^2}{4\pi G}\oint \mathbf{K}\times\mathbf{g}\cdot d\mathbf{S} = 0, \tag{8-1.1}$$

where the surface of integration encloses the system under consideration. Transforming the last surface integral in Eq. (8-1.1) into a volume integral by using vector identity (V-19), we obtain

$$\frac{c^2}{4\pi G}\oint \mathbf{K}\times\mathbf{g}\cdot d\mathbf{S} = \frac{c^2}{4\pi G}\int \nabla\cdot(\mathbf{K}\times\mathbf{g})dV = 0. \tag{8-1.2}$$

Expanding the integrand of the last integral in Eq. (8-1.2) by means of vector identity (V-9), we obtain

$$\nabla\cdot(\mathbf{K}\times\mathbf{g}) = \mathbf{g}\cdot(\nabla\times\mathbf{K}) - \mathbf{K}\cdot(\nabla\times\mathbf{g}). \tag{8-1.3}$$

Using now Eqs. (7-1.4) and (7-1.3), we can write the two terms on the right of Eq. (8-1.3) as

$$\mathbf{g}\cdot(\nabla\times\mathbf{K}) = -\mathbf{g}\cdot\frac{4\pi G}{c^2}\mathbf{J} + \mathbf{g}\cdot\frac{1}{c^2}\frac{\partial\mathbf{g}}{\partial t} \tag{8-1.4}$$

and

$$\mathbf{K}\cdot(\nabla\times\mathbf{g}) = -\mathbf{K}\cdot\frac{\partial\mathbf{K}}{\partial t}. \tag{8-1.5}$$

Substituting Eqs.(8-1.4) and (8-1.5) into Eq. (8-1.3), we obtain

$$\nabla\cdot(\mathbf{K}\times\mathbf{g}) = -\mathbf{g}\cdot\frac{4\pi G}{c^2}\mathbf{J} + \mathbf{g}\cdot\frac{1}{c^2}\frac{\partial\mathbf{g}}{\partial t} + \mathbf{K}\cdot\frac{\partial\mathbf{K}}{\partial t}, \tag{8-1.6}$$

which can be written as

$$\nabla\cdot(\mathbf{K}\times\mathbf{g}) = -\frac{4\pi G}{c^2}\mathbf{g}\cdot\mathbf{J} + \frac{1}{c^2}\left[\frac{1}{2}\frac{\partial(\mathbf{g}\cdot\mathbf{g})}{\partial t} + \frac{c^2}{2}\frac{\partial(\mathbf{K}\cdot\mathbf{K})}{\partial t}\right], \tag{8-1.7}$$

or

$$\nabla\cdot(\mathbf{K}\times\mathbf{g}) = -\frac{4\pi G}{c^2}\mathbf{g}\cdot\mathbf{J} + \frac{1}{c^2}\left[\frac{1}{2}\frac{\partial\mathbf{g}^2}{\partial t} + \frac{c^2}{2}\frac{\partial\mathbf{K}^2}{\partial t}\right]. \tag{8-1.8}$$

By Eq. (2-2.8), Eq. (8-1.8) can be written as

$$\nabla \cdot (\mathbf{K} \times \mathbf{g}) = -\frac{4\pi G}{c^2}\mathbf{g} \cdot \mathbf{J} - \frac{4\pi G}{c^2}\frac{\partial U_v}{\partial t}, \tag{8-1.9}$$

where $U_v$ is the energy density in the gravitational and cogravitational field.

Substituting Eq. (8-1.9) into Eq. (8-1.2) and noting that $\mathbf{J} = \boldsymbol{\rho}\mathbf{v}$, we obtain

$$\int \mathbf{g} \cdot \rho\mathbf{v}\, dV + \int \frac{\partial U_v}{\partial t} dV = 0, \tag{8-1.10}$$

or, for constant $\mathbf{g}$,

$$m\mathbf{g} \cdot \mathbf{v} + \frac{\partial U}{\partial t} = 0, \tag{8-1.11}$$

where $m$ is the mass contained in the system under consideration and $U$ is the gravitational-cogravitational energy of the system.

The first term in Eqs. (8-1.10) and (8-1.11) represents the rate at which the energy of the moving mass $m$ (its kinetic energy) increases under the actions of the gravitational field $\mathbf{g}$. The second term represents the *decrease* [note that, by Eqs. (2-2.7) and (2-2.8), $U$ is negative] of the gravitational-cogravitational field energy. Thus Eqs.(8-1.10) and (8-1.11) show that the total energy in a closed gravitational-cogravitational system is conserved: kinetic energy of a mass (or masses) in the system increases at the expense of the field energy and vice-versa [the latter is true because Eqs. (8-1.10) and (8-1.11) remain valid when multiplied by $-1$].[1]

## 8-2. Conservation of Momentum in Gravitational and Cogravitational Systems

Let us again consider a closed gravitational-cogravitational system. In such a system there is no inflow or outflow of field momentum to or from the system. By Eq. (2-2.11) we then have

$$\frac{1}{4\pi G}\left[\frac{1}{2}\oint(\mathbf{g}^2+c^2\mathbf{K}^2)d\mathbf{S}-\oint\mathbf{g}(\mathbf{g}\cdot d\mathbf{S})-c^2\oint\mathbf{K}(\mathbf{K}\cdot d\mathbf{S})\right]=0 \quad (8\text{-}2.1)$$

and therefore

$$\frac{d\mathbf{G}_M}{dt}=-\frac{1}{4\pi G}\int\frac{\partial}{\partial t}(\mathbf{K}\times\mathbf{g})dV. \quad (8\text{-}2.2)$$

By Eq. (2-2.11) we then have

$$\frac{d\mathbf{G}_M}{dt}=-\frac{\partial\mathbf{G}_f}{\partial t}. \quad (8\text{-}2.3)$$

Thus in a closed gravitational-cogravitational system the mechanical momentum of the system increases at the expense of the field momentum, and vice versa [the latter is true because Eq. (8-2.3) remains valid when multiplied by $-1$].[1]

## 8-3. Action and Reaction in Gravitational-Cogravitational Systems

Let us consider a closed gravitational-cogravitational system consisting of two mass distributions $\rho_1$ and $\rho_2$ producing, respectively, gravitational fields $\mathbf{g}_1$ and $\mathbf{g}_2$ and cogravitational fields $\mathbf{K}_1$ and $\mathbf{K}_2$. If we apply vector identity (V-22) to the fields $\mathbf{g}_1$ and $\mathbf{g}_2$, we obtain

$$\oint(\mathbf{g}_1\cdot\mathbf{g}_2)d\mathbf{S}-\oint\mathbf{g}_2(\mathbf{g}_1\cdot d\mathbf{S})-\oint\mathbf{g}_1(\mathbf{g}_2\cdot d\mathbf{S})=$$
$$\int[\mathbf{g}_1\times(\nabla\times\mathbf{g}_2)+\mathbf{g}_2\times(\nabla\times\mathbf{g}_1)-\mathbf{g}_1(\nabla\cdot\mathbf{g}_2)-\mathbf{g}_2(\nabla\cdot\mathbf{g}_1)]dV, \quad (8\text{-}3.1)$$

where the surface of integration encloses the system. Since the system is closed, the surface integrals vanish (we can assume that the surface of integration is at infinity, where there are no gravitational and cogravitational fields). We are thus left with

$$\int [\mathbf{g}_1 \times (\nabla \times \mathbf{g}_2) + \mathbf{g}_2 \times (\nabla \times \mathbf{g}_1) - \mathbf{g}_1(\nabla \cdot \mathbf{g}_2) - \mathbf{g}_2(\nabla \cdot \mathbf{g}_1)]dV = 0. \quad (8\text{-}3.2)$$

Let us now assume that the two mass distributions are stationary and are time independent. In this case, by Eq. (7-1.3), $\nabla \times \mathbf{g} = 0$, so that the first two terms in the integrand of Eq. (8-3.2) vanish. Using Eq. (7-1.1) for replacing $\nabla \cdot \mathbf{g}_1$ and $\nabla \cdot \mathbf{g}_2$ by $(-4\pi G\rho_1)$ and $(-4\pi G\rho_2)$, respectively, canceling $(-4\pi G)$, and expressing the resulting integral as two integrals, we then obtain

$$\int \rho_1 \mathbf{g}_2 dv = -\int \rho_2 \mathbf{g}_1 dV. \quad (8\text{-}3.3)$$

But, by Eq. (2-2.6), $\int \rho_1 \mathbf{g}_2 dV$ is the force exerted by $\rho_2$ upon $\rho_1$ and $\int \rho_2 \mathbf{g}_1 dV$ is the force exerted by $\rho_1$ upon $\rho_2$. Hence the forces acting on the two mass distributions are equal in magnitude and opposite in direction, as required by the law of action and reaction. Thus the law of action and reaction holds for gravitational interactions between constant stationary mass distributions.

Let us now assume that $\rho_2$ is moving and/or is time dependent. In this case $\nabla \times \mathbf{g}_1 = 0$, but, by Eq. (7-1.3), $\nabla \times \mathbf{g}_2 = -\partial \mathbf{K}_2/\partial t$. Substituting in Eq. (8-3.2) $\nabla \cdot \mathbf{g}_1$, $\nabla \cdot \mathbf{g}_2$, $\nabla \times \mathbf{g}_1$, and $\nabla \times \mathbf{g}_2$, we find that now only the second term in the integrand of Eq. (8-3.2) vanishes. Simplifying, we then obtain

$$\int \rho_1 \mathbf{g}_2 dV = -\int \rho_2 \mathbf{g}_1 dV + \frac{1}{4\pi G}\int \mathbf{g}_1 \times \frac{\partial \mathbf{K}_2}{\partial t} dV. \quad (8\text{-}3.4)$$

Hence, if one of the two interacting mass distributions is time variable or is moving, the action-reaction law does not hold: the two forces differ by the value of the integral containing $\mathbf{K}_2$.

Let us now assume that both mass distributions are moving and/or are time dependent. In this case $\nabla \times \mathbf{g}_1 = -\partial \mathbf{K}_1/\partial t$ and $\nabla \times \mathbf{E}_2 = -\partial \mathbf{K}_2/\partial t$. Substituting in Eq. (8-3.2) $\nabla \cdot \mathbf{g}_1$, $\nabla \cdot \mathbf{g}_2$, $\nabla \times \mathbf{K}_1$, and $\nabla \times \mathbf{K}_2$, we obtain

$$\int \rho_1 \mathbf{g}_2 dV - \frac{1}{4\pi G}\int \mathbf{g}_2 \times \frac{\partial \mathbf{K}_1}{\partial t} dV$$
$$= - \int \rho_2 \mathbf{g}_1 dV + \frac{1}{4\pi G}\int \mathbf{g}_1 \times \frac{\partial \mathbf{K}_2}{\partial t} dV. \tag{8-3.5}$$

Thus, when both mass distributions are time variable or are in motion, the law of action and reaction, in general, does not hold: the two forces differ by the value of the two integrals containing $\mathbf{K}_1$ and $\mathbf{K}_2$. However, if the two integrals happen to be equal in magnitude but have opposite signs, they cancel each other, so that in this case the law of action and reaction does hold even when the two masses vary or move.

It should be pointed out that although the law of action and reaction does not hold for certain types of gravitational interactions, the law of conservation of momentum is valid for all gravitational and cogravitational interactions, without exceptions. This will be shown later in this chapter.

We shall now examine what happens to the action and reaction law in cogravitational interactions.

Consider two constant, stationary mass-current distributions $\mathbf{J}_1$ and $\mathbf{J}_2$ in a closed gravitational-cogravitational system. The cogravitational fields produced by $\mathbf{J}_1$ and $\mathbf{J}_2$ are $\mathbf{K}_1$ and $\mathbf{K}_2$, respectively. The force exerted by $\mathbf{J}_2$ on $\mathbf{J}_1$ is, by Eq. (2-2.6), $\int \mathbf{J}_1 \times \mathbf{K}_2\, dV$ and the force exerted by $\mathbf{J}_1$ on $\mathbf{J}_2$ is $\int \mathbf{J}_2 \times \mathbf{K}_1\, dV$.

Applying vector identity (V-22) to $\mathbf{K}_1$ and $\mathbf{K}_2$, we have

$$\oint (\mathbf{K}_1 \cdot \mathbf{K}_2) d\mathbf{S} - \oint \mathbf{K}_2(\mathbf{K}_1 \cdot d\mathbf{S}) - \oint \mathbf{K}_1(\mathbf{K}_2 \cdot d\mathbf{S}) =$$
$$\int [\mathbf{K}_1 \times (\nabla \times \mathbf{K}_2) + \mathbf{K}_2 \times (\nabla \times \mathbf{K}_1) - \mathbf{K}_1(\nabla \cdot \mathbf{K}_2) - \mathbf{K}_2(\nabla \cdot \mathbf{K}_1)] dV. \tag{8-3.6}$$

As before, the surface integrals vanish. By Eq. (7-1.2), $\nabla \cdot \mathbf{K} = 0$, so that the last two terms in the volume integral vanish also. Taking into account that, by Eq. (7-1.4), for stationary (time-independent) systems $\nabla\times\mathbf{K} = -(4\pi G/c^2)\mathbf{J}$, and simplifying the two remaining terms in Eq. (8-3.6), we then obtain

$$\int \mathbf{J}_1 \times \mathbf{K}_2 dV = -\int \mathbf{J}_2 \times \mathbf{K}_1 dV. \tag{8-3.7}$$

Thus, for cogravitational interactions between two constant stationary mass-currents, the two forces are equal in magnitude and opposite in direction, and the law of action and reaction holds.

Let us now assume that $\mathbf{J}_2$ is variable or is in a state of motion. In this case, by Eq. (7-1.4), $\nabla\times\mathbf{K}_2 = -(4\pi G/c^2)\mathbf{J}_2 + (1/c^2)\partial\mathbf{g}_2/\partial t$. Noting that the surface integrals of Eq. (8-3.6) still vanish, and simplifying the volume integral as before, we obtain

$$\int \mathbf{J}_1 \times \mathbf{K}_2 dV = -\int \mathbf{J}_2 \times \mathbf{K}_1 dV - \frac{1}{4\pi G}\int \mathbf{K}_1 \times \frac{\partial \mathbf{g}_2}{\partial t} dV. \tag{8-3.8}$$

Hence, when one of the currents is changing or is in motion, the two forces are not equal and differ by the amount of the integral containing $\mathbf{g}_2$. The law of action and reaction does not hold.

Let us now assume that $\mathbf{J}_1$ is also variable. In this case $\nabla\times\mathbf{K}_1 = -(4\pi G/c^2)\mathbf{J}_1 + (1/c^2)\partial\mathbf{g}_1/\partial t$ and $\nabla\times\mathbf{K}_2 = -(4\pi G/c^2)\mathbf{J}_2 + (1/c^2)\partial\mathbf{g}_2/\partial t$. From Eq. (8-3.6) we now obtain (noting that the surface integrals vanish as before)

$$\begin{aligned}\int \mathbf{J}_1 \times \mathbf{K}_2 dV &+ \frac{1}{4\pi G}\int \mathbf{K}_2 \times \frac{\partial \mathbf{g}_1}{\partial t} dV \\ &= -\int \mathbf{J}_2 \times \mathbf{K}_1 dV - \frac{1}{4\pi G}\int \mathbf{K}_1 \times \frac{\partial \mathbf{g}_2}{\partial t} dV.\end{aligned} \tag{8-3.9}$$

Thus, when both mass-currents are changing or are moving, the law of action and reaction, in general, does not hold: the forces differ by the values of the integrals containing $\mathbf{g}_1$ and $\mathbf{g}_2$. However, if the two integrals are equal in magnitude but have opposite signs, they cancel, and then the law of action and reaction does hold.

## 8-4. The Law of Action and Reaction and the Law of Conservation of Momentum

As we have seen, Newton's third law, the law of action and reaction, has only a limited validity in the domain of gravitational and cogravitational interactions. In general, it holds only for interactions between constant stationary masses and for interactions between constant stationary mass currents. However, it is not necessary to state Newton's third law as the law of action and reaction. One can state this law more accurately as the law of conservation of momentum. If we examine the time dependent terms appearing in Eqs. (8-3.4), (8-3.5), (8-3.8), and (8-3.9), we recognize that these terms represent rates of change of gravitational-cogravitational field momentum, Eq. (2-2.11),

$$\mathbf{G}_f = \frac{1}{4\pi G}\int \mathbf{K}\times\mathbf{g}\,dV, \tag{2-2.11}$$

Therefore these equations show that, although the forces are different, the total momentum (mechanical plus gravitational-cogravitational) of the system is always conserved. An exchange of momentum between a mass or a mass current and the surrounding field is, of course, necessary since gravitational and cogravitational fields propagate with finite speed, so that no direct interactions between field-producing and field-experiencing masses or mass currents are possible (see Section 16-2).

It is important to note that Eqs. (8-3.4), (8-3.5), (8-3.8), and (8-3.9) involve only the interaction, or mutual, momentum rather than the total gravitational-cogravitational momentum of the systems under consideration. Specifically, in the case of gravitational systems, the rate of momentum change is expressed as the cross product of the gravitational field and the time derivative of the cogravitational field. And in the case of cogravitational systems, this rate of momentum change is expressed as the cross product of the cogravitational field and the time derivative of the gravitational field. A remarkable feature of these equations is that they only involve partial fields: $\mathbf{g}_1$ and $\mathbf{K}_2$ or $\mathbf{g}_2$ and $\mathbf{K}_1$. This means that even in a region of space where the total field $\mathbf{g} = \mathbf{g}_1+\mathbf{g}_2$ or $\mathbf{K} = \mathbf{K}_1+\mathbf{K}_2$ is zero, there still can be an exchange of the gravitational-cogravitational and mechanical momentum.

The apparent simplicity of Eq. (8-3.3) for the gravitational interactions is misleading. Assumed by Newton to hold for all gravitational interactions, and for the interactions in the Solar system in particular, it actually holds only for interactions between stationary time-independent masses. According to the generalized theory of gravitation, it certainly does not hold for planetary interactions. As we shall see later (Sections 14.3 and 15.1), the force with which the Sun acts on a planet is not equal to the force with which the planet acts on the Sun. In fact, according to the generalized theory of gravitation, the interaction is actually not between the Sun and the planet, but between the planet and the field created by the Sun at the location of the planet and between the Sun and the field created by the planet at the locations of the Sun. However, although the law of action and reaction does not hold for planetary interaction, the law of momentum conservation does hold without exception.[2]

## References and Remarks for Chapter 8

**1.** See also Chapter 16.

**2.** For a related discussion of the action-reaction law and of the momentum conservation law in electromagnetic systems see O. D. Jefimenko, *Causality, Electromagnetic Induction, and Gravitation*, 2nd ed., (Electret Scientific, Star City, 2000), pp. 67-79.

# 9

# GENERALIZED THEORY OF GRAVITATION AND THE SPECIAL RELATIVITY THEORY

As we saw in Chapter 7, many electromagnetic equations have their gravitational and cogravitational counterparts. In this chapter we shall explore the analogy between electromagnetism and gravitation even further, and, on the basis of this analogy, shall develop a relativistic theory of gravitation analogous to relativistic electrodynamics. We shall present illustrative examples demonstrating the use and power of relativistic transformations applied to gravitational and cogravitational equations. Then we shall briefly discuss the so-called "covariant formulation" of gravitational and cogravitational equations.

## 9-1. Relativistic Transformation Equations for Gravitational and Cogravitational Fields

Until recently it was believed that the analogy between electromagnetic and gravitational equations could not apply to fast moving systems, because the electric charge is not affected by velocity, but the mass of a moving body was thought to vary with velocity. It is now generally accepted that mass, just like the electric charge, does not depend on velocity.[1] This also means that transformation equations of the special relativity theory developed for electromagnetic systems[2] have their gravitational and cogravitational counterparts. In agreement with considerations presented in Section 7-2, the only essential difference between the relativistic gravitational-cogravitational equations and the corresponding electromagnetic equations is in the symbols and constants appearing in these equations.

Thus there is no need to *derive* relativistic gravitational-cogravitational transformation equations, because we can easily obtain them by replacing symbols and constants appearing in relativistic electromagnetic equations by the corresponding gravitational-cogravitational symbols and constants with the help of Table 7-1. The basic relativistic gravitational-cogravitational transformation equations obtained in this way[3] are listed below. In these equations, the unprimed quantities are those measured in the stationary reference frame Σ ("laboratory"), and the primed quantities are those measured in the moving reference frame Σ′.

**Transformation equations correlating quantities measured in Σ with quantities measured in Σ′:**

(a) Equations for space and time coordinates

$$x = \gamma(x' + vt'), \tag{9-1.1}$$

$$y = y', \tag{9-1.2}$$

$$z = z', \tag{9-1.3}$$

$$t = \gamma(t' + vx'/c^2). \tag{9-1.4}$$

(b) Equations for the gravitational field

$$g_x = g_x', \tag{9-1.5}$$

$$g_y = \gamma(g_y' + vK_z'), \tag{9-1.6}$$

$$g_z = \gamma(g_z' - vK_y'). \tag{9-1.7}$$

(c) Equations for the cogravitational field

$$K_x = K'_x , \tag{9-1.8}$$

$$K_y = \gamma(K'_y - vg'_z/c^2), \tag{9-1.9}$$

$$K_z = \gamma(K'_z + vg'_y/c^2). \tag{9-1.10}$$

(d) Equations for the mass and mass-current densities

$$\rho = \gamma[\rho' + (v/c^2)J'_x], \tag{9-1.11}$$

$$J_x = \gamma(J'_x + v\rho'), \tag{9-1.12}$$

$$J_y = J'_y , \tag{9-1.13}$$

$$J_z = J'_z . \tag{9-1.14}$$

(e) Equations for gravitational and cogravitational potentials

$$\varphi = \gamma(\varphi' + vA'_x), \tag{9-1.15}$$

$$A_x = \gamma[A'_x + (v/c^2)\varphi'], \tag{9-1.16}$$

$$A_y = A'_y , \tag{9-1.17}$$

$$A_z = A'_z . \tag{9-1.18}$$

**Transformation equations correlating quantities measured in $\Sigma'$ with quantities measured in $\Sigma$:**

(a) Equations for space and time coordinates

$$x' = \gamma(x - vt), \tag{9-1.19}$$

$$y' = y, \tag{9-1.20}$$

$$z' = z, \tag{9-1.21}$$

$$t' = \gamma(t - vx/c^2). \tag{9-1.22}$$

(b) Equations for the gravitational field

$$g_x' = g_x, \tag{9-1.23}$$

$$g_y' = \gamma(g_y - vK_z), \tag{9-1.24}$$

$$g_z' = \gamma(g_z + vK_y). \tag{9-1.25}$$

(c) Equations for the cogravitational field

$$K_x' = K_x, \tag{9-1.26}$$

$$K_y' = \gamma(K_y + vg_z/c^2), \tag{9-1.27}$$

$$K_z' = \gamma(K_z - vg_y/c^2). \tag{9-1.28}$$

(d) Equations for the mass and mass current densities

$$\rho' = \gamma[\rho - (v/c^2)J_x], \tag{9-1.29}$$

$$J_x' = \gamma(J_x - v\rho), \tag{9-1.30}$$

$$J_y' = J_y, \tag{9-1.31}$$

$$J_z' = J_z. \tag{9-1.32}$$

(e) Equations for gravitational and cogravitational potentials

$$\varphi' = \gamma(\varphi - vA_x), \tag{9-1.33}$$

$$A_x' = \gamma[A_x - (v/c^2)\varphi], \tag{9-1.34}$$

$$A_y' = A_y, \tag{9-1.35}$$

$$A_z' = A_z. \tag{9-1.36}$$

$$g_x = -G\frac{m}{[(\gamma x)^2 + y^2]^{3/2}}\gamma x = -G\frac{m}{\gamma^2(x^2 + y^2/\gamma^2)^{3/2}}x. \quad (9\text{-}1.40)$$

To obtain the $y$ component of the gravitational field of the moving mass, we shall use Eqs. (9-1.6), (9-1.19), and again Eq. (9-1.20). Since $\mathbf{K}' = 0$ for the stationary mass, Eq. (9-1.6) tells us that, to find $g_y$ of the moving mass, we must replace $g_y'$ on the left of Eq. (9-1.39) by $g_y/\gamma$, while Eqs. (9-1.19) and (9-1.20) tell us that we must replace $x'$ in Eq. (9-1.39) by $\gamma x$ and $y'$ by $y$. Making the substitutions, we then obtain for $g_y$ of the moving point mass

$$g_y/\gamma = -G\frac{m}{[(\gamma x)^2 + y^2]^{3/2}}y, \quad (9\text{-}1.41)$$

or

$$g_y = -G\frac{m}{\gamma^2(x^2 + y^2/\gamma^2)^{3/2}}y. \quad (9\text{-}1.42)$$

Replacing now $\gamma$ in Eqs. (9-1.40) and (9-1.42) by $1/(1 - v^2/c^2)^{1/2}$, factoring out $x^2 + y^2$ from the denominator, taking into account that $\mathbf{r} = x\mathbf{i} + y\mathbf{j}$, where $\mathbf{i}$ and $\mathbf{j}$ are unit vectors in the direction of the $x$ and $y$ axes, and noting that $y^2/(x^2 + y^2) = \sin^2\theta$, where $\theta$ is the angle between $\mathbf{v}$ and $\mathbf{r}$, we finally obtain

$$\mathbf{g} = -G\frac{m(1 - v^2/c^2)}{r^3[1 - (v^2/c^2)\sin^2\theta]^{3/2}}\mathbf{r}. \quad (9\text{-}1.43)$$

Equation (9-1.43) is the same as Eq. (5-1.13) (the "Heaviside equation") that we obtained in Chapter 5 directly from the fundamental gravitational field equation. Note that $r$ and $\mathbf{r}$ in Eq. (9-1.43) represent the present position of the mass and are therefore the same as $r_0$ and $\mathbf{r}_0$ in Eq. (5-1.13). Note also that in applying relativistic transformations we did not transform the mass $m$. Just like the electric charge $q$, the mass of a body is invariant under relativistic transformations.

**Example 9-1.2** The gravitational scalar potential of a stationary mass distribution $\rho$ is represented by the Newtonian Eq. (1-1.11) [which is the same as Eq. (3-3.7)]

$$\varphi = -G\int \frac{\rho}{r}dV. \tag{1-1.11}$$

(we omit the prime on $dV$ so as not to confuse $dV$ referring to the reference frame $\Sigma$ with $dV'$ referring to the reference frame $\Sigma'$). Using relativistic transformation equations, convert Eq. (1-1.11) into the equation for the cogravitational vector potential produced by a mass distribution moving with constant velocity $\mathbf{v} = v\,\mathbf{i}$.

Let a mass distribution $\rho'$ be at rest in the moving reference frame $\Sigma'$ and let the point of observation be at the origin. By Eq. (1-1.11), the gravitational potential $\varphi'$ produced by $\rho'$ in this reference frame is

$$\varphi' = -G\int \frac{\rho'}{r'}dV'. \tag{9-1.44}$$

Observed from the laboratory (reference frame $\Sigma$), the mass distribution $\rho'$ moves with velocity $v$ along a line parallel to the $x$ axis. Like all moving masses, it creates a cogravitational field. To find the associated cogravitational vector potential, we transform Eq. (9-1.44) by using appropriate transformation equations listed in Section 9-1. However, first we express $r'$ and $dV'$ appearing in Eq. (9-1.44) in terms of $x'$, $y'$, and $z'$:

$$\varphi' = -G\int\int\int \frac{\rho'}{(x'^2 + y'^2 + z'^2)^{1/2}}dx'dy'dz'. \tag{9-1.45}$$

Since we are free to choose the time of observation in the laboratory, we choose $t = 0$ for simplicity.[6] By Eqs. (9-1.19), (9-1.20) and (9-1.21) we then have

$$x' = \gamma x,\ y' = y,\ z' = z. \tag{9-1.46}$$

By Eq. (9-1.11) (noting that there is no mass current in $\Sigma'$ where the mass is stationary) we have

$$\rho' = \rho/\gamma. \tag{9-1.47}$$

By Eq. (9-1.16) (noting that there is no cogravitational vector potential in $\Sigma'$ where the mass is stationary), we have

$$\varphi' = A_x c^2/\gamma v. \tag{9-1.48}$$

Substituting Eqs. (9-1.46)-(9-1.48) into Eq. (9-1.45), we obtain

$$\begin{aligned} A_x &= -G\frac{\gamma v}{c^2}\int\int\int \frac{\rho/\gamma}{(\gamma x^2 + y^2 + z^2)^{1/2}} d(\gamma x)dydz \\ &= -G\frac{v}{c^2}\int\int\int \frac{\rho}{[x^2 + (y^2 + z^2)/\gamma]^{1/2}} dxdydz. \end{aligned} \tag{9-1.49}$$

and since

$$1/\gamma^2 = 1 - v^2/c^2, \tag{9-1.50}$$

we obtain upon simplifying the denominator in the last integral of Eq. (9-1.49) and replacing *dxdydz* by *dV*

$$A_x = -G\frac{v}{c^2}\int \frac{\rho}{r[1 - (y^2 + z^2)v^2/r^2c^2]^{1/2}} dV, \tag{9-1.51}$$

or

$$A_x = -G\frac{v}{c^2}\int \frac{\rho}{r[1 - (v^2/c^2)\sin^2\theta]^{1/2}} dV, \tag{9-1.52}$$

where $\theta$ is the angle between the velocity vector $\mathbf{v}$ of the moving mass distribution and the radius vector $\mathbf{r}$ connecting $dV$ with the point of observation. For the $y$ and $z$ components of the vector potential we obtain from Eqs. (9-1.17) and (9-1.18)

$$A_y = A_z = 0. \tag{9-1.53}$$

Observe that Eqs. (9-1.52) and (9-1.53) are the $x$, $y$, and $z$ components of Eq. (6-2.13) that we previously derived from the fundamental gravitational and cogravitational laws. It is quite remarkable that, by applying relativistic transformations to the Newtonian equation for the gravitational potential, we have obtained equations for the cogravitational vector potential, although at first sight there appears to be no connection whatsoever between the Newtonian gravitational potential and the cogravitational vector potential of the generalized theory of gravitation.

**Example 9-1.3** The Newtonian gravitational field of a stationary mass distribution $\rho$ is represented by Eq. (1-1.7)

$$\mathbf{g} = -G\int \frac{\rho}{r^3}\mathbf{r}dV, \tag{1-1.7}$$

Applying relativistic transformation equations to Eq. (1-1.7), find the cogravitational field produced by a mass distribution moving with constant velocity $\mathbf{v} = v\,\mathbf{i}$.

As in the preceding example, let a mass distribution $\rho'$ be at rest in the moving reference frame $\Sigma'$. Rewriting Eq. (1-1.7) in terms of its Cartesian components and prime coordinates, we have for the gravitational field produced by $\rho'$ in $\Sigma'$

$$g'_x = -G\int\int\int \frac{\rho' x'}{(x'^2 + y'^2 + z'^2)^{3/2}} dx'dy'dz', \tag{9-1.54}$$

$$g'_y = -G\int\int\int \frac{\rho' y'}{(x'^2 + y'^2 + z'^2)^{3/2}} dx'dy'dz', \tag{9-1.55}$$

$$g'_z = -G\int\int\int \frac{\rho' z'}{(x'^2 + y'^2 + z'^2)^{3/2}} dx'dy'dz'. \tag{9-1.56}$$

For the time of observation in the laboratory we choose as before $t = 0$, so that Eq. (9-1.46) applies again. Also, since there is no mass current in $\Sigma'$, Eq. (9-1.47) applies. By Eqs. (9-1.8)-(9-1.10) (noting that there is no cogravitational field in $\Sigma'$ where the mass is stationary) we have

$$K_x = 0, \tag{9-1.57}$$

$$K_y = -\gamma v g_z'/c^2, \tag{9-1.58}$$

$$K_z = \gamma v g_y'/c^2. \tag{9-1.59}$$

Substituting Eqs. (9-1.46), (9-1.47), (9-1.58), and (9-1.59) into Eqs. (9-1.56) and (9-1.55), we obtain

$$K_y = G\frac{\gamma v}{c^2}\int\int\int \frac{(\rho/\gamma)z}{(\gamma^2x^2 + y^2 + z^2)^{3/2}} d(\gamma x)dydz, \tag{9-1.60}$$

$$K_z' = -G\frac{\gamma v}{c^2}\int\int\int \frac{(\rho/\gamma)y}{(\gamma^2x^2 + y^2 + z^2)^{3/2}} d(\gamma x)dydz. \tag{9-1.61}$$

Rewriting Eq. (9-1.57) and simplifying Eqs. (9-1.60) and (9-1.61) just as we simplified Eq. (9-1.49) in Example 9-1.2, we obtain for the cogravitational field produced by a moving mass distribution

$$K_x = 0, \tag{9-1.62}$$

$$K_y = G\frac{v}{c^2}\int \frac{\rho(1 - v^2/c^2)z}{r^2[1 - (v^2/c^2)\sin^2\theta]^{3/2}} dV, \tag{9-1.63}$$

$$K_z = -G\frac{v}{c^2}\int \frac{\rho(1 - v^2/c^2)y}{r^2[1 - (v^2/c^2)\sin^2\theta]^{3/2}} dV. \tag{9-1.64}$$

Observe that Eqs. (9-1.62), (9-1.63) and (9-1.64) are the $x$, $y$, and $z$ components of Eq. (6-1.45) that we previously derived from the fundamental gravitational and cogravitational laws. One cannot help but be impressed by the fact that, by applying relativistic transformations to the Newtonian equation for the gravitational field, we have obtained equations for the cogravitational field, although at first sight there appears to be no connection whatsoever between the Newtonian gravitational field and the cogravitational field of the generalized theory of gravitation.

▲

## 9-2. Covariant Formulation of the Generalized Theory of Gravitation

Covariant formulation of physical formulas and equations is considered by some authors to be the most appropriate formulation for expressing the laws of physics in a frame-independent form. It is also believed by some authors to be more concise and occasionally more informative than the conventional formulation. Since any equation invariant under relativistic transformations should be expressible in a covariant form, and since the principle of relativity is considered to be a fundamental law of nature, the laws of physics that cannot be expressed in a covariant form are considered by some authors to be incomplete or incorrect.[7]

Newton's gravitational law is an example of a physical law that cannot be expressed in a covariant form. The problem of finding an invariant form of the law of gravitation was first considered by Poincaré, but without success.[8] It is interesting to note that Poincaré attempted to solve the problem on the basis of just one gravitational field (the gravitational analog of the electrostatic field). But even if the theory of gravitation is built upon two fields, a covariant theory of gravitation is not possible unless the gravitational mass, just like the electric charge, does not depend on the velocity with which the mass moves.

As already mentioned in Section 9-1, until recently it was believed that the mass of a moving body was a function of the velocity of the body and thus was not invariant under relativistic transformations. This was the most important reason for questioning the possibility of a relativistic theory of gravitation analogous to relativistic electromagnetism. If mass, unlike the electric charge, is not invariant, then the analogy between electromagnetism and gravitation is not sufficiently complete to allow a construction of a relativistic gravitational theory similar to relativistic electrodynamics based on the gravitational field vector, with or without the addition of a second (the cogravitational) field vector.

However, it is now generally accepted that mass does not depend on the velocity with which a body moves.[1] Therefore a covariant formulation of the theory of gravitation based on gravitational-cogravitational fields is not only possible but can be constructed straightaway from the covariant theory of electromagnetism by a mere substitution of symbols and constants in accordance with Table 7-1.

In particular, from electromagnetic equations[9] we directly obtain for the covariant "position 4-vector"

$$\boldsymbol{r} = (x_1, x_2, x_3, x_4) = (x, y, z, ict), \tag{9-2.1}$$

where $i$ is $\sqrt{-1}$. From the 4-vector electric current[9] we obtain by substitutions the covariant expression for the 4-vector mass current

$$\boldsymbol{J} = (J_1, J_2, J_3, J_4) = (J_x, J_y, J_z, ic\rho), \tag{9-2.2}$$

where $J_x$, $J_y$, and $J_z$ are the $x$, $y$, and $z$ components of mass-current density. From the electromagnetic field tensor[9] we obtain the gravitational-cogravitational field tensor by replacing, with the help of Table 7-1, the $x$, $y$, and $z$ components of **E** by the corresponding components of **g** and the $x$, $y$, and $z$ components of **B** by the corresponding components of **K**

$$\boldsymbol{F}_{\mu\nu} = \begin{bmatrix} 0 & K_z & -K_y & -ig_x/c \\ -K_z & 0 & K_x & -ig_y/c \\ K_y & -K_x & 0 & -ig_z/c \\ ig_x/c & ig_y/c & ig_z/c & 0 \end{bmatrix}. \tag{9-2.3}$$

where the subscript $\mu$ indicates the row (1, 2, 3, 4 top to bottom) and the subscript $\nu$ indicates the column (1, 2, 3, 4 left to right).

Finally, in the same manner, we obtain covariant expressions of the present-time differential equations for gravitational-cogravitational fields:

$$\sum_{\nu=1}^{4} \frac{\partial F_{\mu\nu}}{\partial x_\nu} = -\frac{4\pi G}{c^2} J_\mu \tag{9-2.4}$$

and

$$\frac{\partial F_{\mu\nu}}{\partial x_\lambda} + \frac{\partial F_{\nu\lambda}}{\partial x_\mu} + \frac{\partial F_{\lambda\mu}}{\partial x_\nu} = 0 . \tag{9-2.5}$$

It should be kept in mind, however, that $c$ in the gravitational-cogravitational equations stands for the speed of propagation of gravitational-cogravitational fields, which is generally assumed to be the same as the speed of light, but has never been actually measured.[10]

▼

**Example 9-2.1** Show that Eq. (9-2.4) is equivalent to Eqs. (7-1.1) and (7-1.4), and that Eq. (9-2.5) is equivalent to Eqs. (7-1.2) and (7-1.3).

Replacing in Eq. (9-2.4) $F_{\mu\nu}$ by $F_{4\nu}$, substituting $x$, $y$, $z$, and $ict$ for $x_1$, $x_2$, $x_3$, and $x_4$, respectively, using, according to Eq. (9-2.3), $F_{41} = ig_x/c$, $F_{42} = ig_y/c$, $F_{43} = ig_z/c$, and $F_{44} = 0$, and using, according to Eq. (9-2.2), $J_4 = ic\rho$, we have

$$\frac{\partial(ig_x/c)}{\partial x} + \frac{\partial(ig_y/c)}{\partial y} + \frac{\partial(ig_z/c)}{\partial z} + \frac{\partial 0}{\partial(ict)} = -\frac{4\pi G}{c^2} ic\rho , \tag{9-2.6}$$

which, after cancelling $i$ and $c$, becomes the same as Eq. (7-1.1).

Setting in Eq. (9-2.4) $\mu = 1$, and using, according to Eq. (9-2.3), $F_{11} = 0$, $F_{12} = K_z$, $F_{13} = -K_y$, $F_{14} = -ig_x/c$, we similarly obtain

$$\frac{\partial 0}{\partial x} + \frac{\partial K_z}{\partial y} - \frac{\partial K_y}{\partial z} - \frac{\partial (ig_x/c)}{\partial (ict)} = -\frac{4\pi G}{c^2}J_x, \tag{9-2.7}$$

or

$$\frac{\partial K_z}{\partial y} - \frac{\partial K_y}{\partial z} = -\frac{4\pi G}{c^2}J_x + \frac{\partial g_x}{c^2 \partial t}, \tag{9-2.8}$$

which is the $x$ component of Eq. (7-1.4). Likewise, setting $\mu = 2$ and then $\mu = 3$ in Eq. (9-2.4) and using Eq. (9-2.3), we obtain the $y$ and $z$ components of Eq. (7-1.4).

Setting in Eq. (9-2.5) $\mu = 1$, $\nu = 2$, $\lambda = 3$, and using Eq. (9-2.3), we obtain

$$\frac{\partial K_z}{\partial z} + \frac{\partial K_x}{\partial x} + \frac{\partial K_y}{\partial y} = 0, \tag{9-2.9}$$

which is the same as Eq. (7-1.2).

Setting in Eq. (9-2.5) $\mu = 2$, $\nu = 3$, $\lambda = 4$, and using Eq. (9-2.3), we similarly obtain

$$\frac{\partial g_z}{\partial y} - \frac{\partial g_y}{\partial z} = -\frac{\partial K_x}{\partial t}, \tag{9-2.10}$$

which is the $x$ component of Eq. (7-1.3). The remaining two components are obtained in the same manner by setting $\mu = 1$, $\nu = 3$, $\lambda = 4$ and $\mu = 1$, $\nu = 2$, $\lambda = 4$.

▲

## References and Remarks for Chapter 9

**1**. For a discussion of the history and use of the concept of relativistic mass, see Carl G. Adler, "Does mass really depend on velocity, dad?" Am. J. Phys. **55**, 739-743 (1987); L. B. Okun, "The concept of mass," Phys. Today **42**, June, 31-36 (1989) and letters in response to this article in Phys. Today **43**, May, 13-15, 115-117 (1990); T. R. Sandin, "In defense of relativistic mass," Am J. Phys. **59**, 1032-1036 (1991).

**2.** See Oleg D. Jefimenko, *Electromagnetic Retardation and Theory of Relativity*, 2nd ed., (Electret Scientific, Star City, 2004), pp. 148-206.

**3.** It is important to note that these equations can be derived directly, without using the analogy between electromagnetic and gravitational-cogravitational systems. See Oleg D. Jefimenko, "Derivation of Relativistic Transformations for Gravitational Fields from Retarded Field Integrals," Galilean Electrodynamics **6**, 23-30 (1995).

**4.** For a detailed discussion of this effect, including the possibility of antigravitational mass distributions arising from it, see Chapter 19.

**5.** Contrary to the prevailing belief, equations of relativistic electrodynamics and the entire theory of special relativity is also only approximately correct, since it is valid only for inertial systems ("inertial frames of reference"). In reality such systems do not exist, because everywhere in the Universe there is a gravitational force field, making all systems and locations in the Universe non-inertial.

**6.** One may think that since the values of $g'$, $\varphi'$, $\rho'$, and $r'$ in $\Sigma'$ correspond to some time of observation $t'$ = *constant*, they cannot be transformed to $t$ = *constant* in $\Sigma$ by Lorentz transformations, and therefore using $t = 0$ in $\Sigma$ is in violation of the theory of relativity. However, by the principle of relativity, if a physical law (Coulomb's law, for example) does not depend on time in one inertial reference frame, then it (or its equivalent) cannot depend on time in any other inertial reference frame. Moreover, if a quantity does not explicitly appear in the formula under consideration, then this quantity is not subject to Lorentz transformation in the course of transforming this formula to a different reference frame. Therefore, since Eq. (9-1.44) in $\Sigma'$ is time-independent, and since $t$ does not appear in Eq. (9-1.44) (the time of observation in $\Sigma'$ is unspecified), we are free to choose any convenient time of observation in $\Sigma$. Similarly, since the actual volume of integration in Eq. (9-1.44) in $\Sigma'$ is arbitrary (as long as it encloses the mass distribution under consideration), we need not transform it to some particular volume in Σ. More generally, if the domain of integration (volume, surface, contour) in one inertial reference frame is unspecified, it is not subject to Lorentz transformation in the course of transforming the corresponding integral to a different reference frame. See also Oleg D. Jefimenko, "A neglected topic in relativistic electrodynamics: transformation of electromagnetic integrals," http://arxiv.org/ftp/physics/papers/0509/0509159.pdf

**7.** This view is unquestionably wrong, since according to it, even Maxwell's equations in their vector form should be classified as "incomplete" or "incorrect" (see Ref. 2, Section 7.4). Note also that covariant formulation changes the form of equations but does not create new physical laws and thus is of very limited utility.

**8.** H. Poincaré, "Sur la Dynamique de L'Électron," Rend. Circ. mat. Palermo **21**, 129-176 (1906) and .H. Poincaré, "La Dynamique de L'Électron," in *Revue général des Sciences pures et appliquées* **19**, 386-402 (1908).

**9.** See Ref. 2, pp. 284-292.

**10.** In 2002 Fomalont and Kopeikin attempted an indirect measurement of the velocity of gravitation and reported that the velocity of gravitation was found to be equal to the velocity of light. See E. B. Fomalont and S. M. Kopeikin, "The measurement of the light deflection from Jupiter: Experimental results," Astrophysical Journal **598** (1), 704-711 (2003).

# 10

# CALCULATION OF GRAVITATIONAL AND COGRAVITATIONAL FORCES FROM POTENTIALS

One of the main problems in the generalized theory of gravitation is the determination of forces involved in gravitational and cogravitational interactions. In this chapter we shall present a new method for calculating these forces based on using gravitational and cogravitational potentials rather than fields. Since potentials are frequently easier to compute than the corresponding fields, this method provides an effective alternative for force calculations. From the theoretical point of view, this method reveals a physical significance of gravitational and cogravitational potentials not heretofore apparent.

## 10-1. Calculation of Gravitational Forces in Time-Independent Systems from Scalar Potentials

The gravitational force on a mass distribution $\rho$ located in a gravitational field $\mathbf{g}$ is, according to Eq. (2-2.6),

$$\mathbf{F} = \int \rho \mathbf{g} dV. \tag{10-1.1}$$

Let us write this equation as[1]

$$\mathbf{F} = \int_{S.\ layer} \rho \mathbf{g} dV + \int_{Interior} \rho \mathbf{g} dV, \tag{10-1.2}$$

where the first integral is extended over the surface layer of the mass distribution, and the second integral is extended over the interior of the mass distribution. The volume of the surface layer may be assumed as small as we please, so that the first integral may be disregarded. We then have

$$\mathbf{F} = \int_{Interior} \rho \mathbf{g} dV, \tag{10-1.3}$$

In accordance with Eq. (1-1.9) [which for time-independent systems is the same as Eq. (3-3.4)], let us now replace **g** in Eq. (10-1.3) by $-\nabla\varphi$, and let us then transform the integrand by means of vector identity (V-5). We have

$$\begin{aligned} \mathbf{F} &= \int_{Interior} \rho \mathbf{g} dV = - \int_{Interior} \rho \nabla \varphi dV \\ &= \int_{Interior} \varphi \nabla \rho dV - \int_{Interior} \nabla (\rho \varphi) dV . \end{aligned} \tag{10-1.4}$$

If we now transform the last integral by means of vector identity (V-20), we obtain

$$\mathbf{F} = \int_{Interior} \varphi \nabla \rho dV - \oint_{Surface} \rho \varphi d\mathbf{S}, \tag{10-1.5}$$

where the second integral is extended over the surface of the mass distribution.[2, 3]

A remarkable feature of Eq. (10-1.5) is that it associates the force on a mass distribution directly with the potential rather than with the field. The equation is immediately suspect, because the potential is determined only to within an additive constant, while the force must be a single-valued quantity. However, a closer examination of Eq. (10-1.5) shows that any additive constant appearing in $\varphi$ integrates out and has no effect on the force.[4]

If the mass distribution is constant, the first integral in Eq.(10-1.5) vanishes, and we have

$$\mathbf{F} = -\rho \oint \varphi d\mathbf{S}. \tag{10-1.6}$$

If the mass is confined to a thin layer, the surface integral in Eqs. (10-1.5) and (10-1.6) can be split into the integrals over the broad surface of the layer and over the rim of the layer. The latter integral contributes to the total force an amount

$$\mathbf{F}_{rim} = -\oint \rho \varphi t d\mathbf{l}_{out} = -\oint \sigma \varphi d\mathbf{l}_{out}. \tag{10-1.7}$$

where $t$ is the thickness of the layer, $\sigma$ is the surface mass density of the layer, and $d\mathbf{l}_{out}$ is a vector representing a length element of the rim directed out of the mass distribution (normal to the rim).

It should be noted that the external potential $\varphi$ appearing in the above equations can be replaced by the total potential because a self-potential cannot produce a net force on a mass distribution.

▼

**Example 10-1.1** A point mass $m$ is located on the axis ($x$ axis) of a thin-walled cylinder of uniform surface mass (mass per unit surface area) $\sigma$, length $2L$, and radius $a$. The distance between $m$ and the center of the cylinder (assumed to be to the right of $m$) is $x$. Find the force exerted on the cylinder by the point mass.

Since the mass distribution is uniform, only the surfaces of the cylinder contribute to the force experienced by the cylinder, and, by the symmetry of the system, the only contribution comes from the two end surfaces (rims) of the cylinder. By Eq. (1-1.12), the potential produced by $m$ at the rim of the cylinder closest to the mass is[5] $\varphi = -Gm[(x-L)^2 + a^2]^{1/2}$, and that at the other end is

$\varphi = -Gm[(x+L)^2 + a^2]^{1/2}$. By Eq. (10-1.7), taking into account that the integrand is a constant, the force is then (compare with Example 6-1.1 for $v = 0$)

$$\mathbf{F} = -\mathbf{i}Gm\sigma 2\pi a\left[\frac{1}{(x-L)^2+a^2]^{1/2}} - \frac{1}{[(x+L)^2+a^2]^{1/2}}\right]. \qquad (10\text{-}1.8)$$

**Example 10-1.2** A spherical mass $m$ of uniform density $\rho$ and radius $a$ consists of two separate hemispheres. Find the force between the hemispheres.

Since $\rho$ is constant, we can use Eq. (10-1.6). We shall use it with the total potential $\varphi$, because the external potential is difficult to compute in this particular case. The total potential at a distance $r \le a$ from the center of the sphere is (see Example 13-1.3)

$$\varphi = -G\frac{m}{2a^3}(3a^2 - r^2), \qquad (10\text{-}1.9)$$

where $r$ is the distance from the center of the sphere. Let us assume that the hemispheres are separated by a horizontal plane, and let us calculate the force on the upper hemisphere. The surface integral in Eq. (10-1.6) can be split into a part over the hemispherical surface and a part over the flat base of the upper hemisphere. Since the magnitude of $\int d\mathbf{S}$ over a hemispherical surface is just the area of the projection of the hemisphere on its base, the contribution of the hemispherical surface to the force is

$$\mathbf{F}_1 = -\mathbf{i}G\rho m\pi a^2/a = -\mathbf{i}G\rho m\pi a, \qquad (10\text{-}1.10)$$

where $\mathbf{i}$ is a unit vector normal to the base and directed downward. The contribution of the base of the hemisphere is

$$\mathbf{F}_2 = \mathbf{i}G\rho\int_0^a \frac{m}{2a^3}(3a^2 - r^2)2\pi r dr = \mathbf{i}G\rho\pi\left(\frac{3ma}{2} - \frac{ma}{4}\right)$$
$$= \mathbf{i}G\frac{5\pi\rho ma}{4}. \qquad (10\text{-}1.11)$$

The total force $\mathbf{F}_1 + \mathbf{F}_2$ is therefore

$$\mathbf{F} = \mathbf{i}G\frac{\rho m \pi a}{4} = \mathbf{i}G\frac{3m^2}{16a^2}. \qquad (10\text{-}1.12)$$

▲

## 10-2. Calculation of Gravitational Forces in Time-Independent Systems from Vector Potentials

As we know from Section 3-3, gravitational fields in mass-free regions can be represented not only as gradients of scalar potentials but also as curls of vector potentials.

Let us replace $\mathbf{g}$ in Eq. (10-1.1) by $\nabla \times \mathbf{A}_g$, where $\mathbf{A}_g$ is the vector potential due to the sources producing $\mathbf{g}$ [the presence of $\rho$ in Eq. (10-1.1) does not preclude the existence of the external vector potential $\mathbf{A}_g$ at the location of $\rho$, since all sources of $\mathbf{A}_g$ are outside of $\rho$). We have

$$\mathbf{F} = \int \rho \nabla \times \mathbf{A}_g dV. \qquad (10\text{-}2.1)$$

Splitting the integral into an integral over the surface layer of the mass and an integral over the interior of the mass, and ignoring the first integral as before in Eq. (10-1.2), we have

$$\mathbf{F} = \int_{Interior} \rho \nabla \times \mathbf{A}_g dV. \qquad (10\text{-}2.2)$$

Using now vector identity (V-11), we can write

$$\mathbf{F} = \int_{Interior} \rho \nabla \times \mathbf{A}_g dV = \int_{Interior} \nabla \times (\rho \mathbf{A}_g) dV - \int_{Interior} \nabla \rho \times \mathbf{A}_g dV, \qquad (10\text{-}2.3)$$

and, using vector identity (V-21), we obtain

$$\mathbf{F} = \int_{Interior} \mathbf{A}_g \times \nabla \rho \, dV - \oint_{Surface} \rho \mathbf{A}_g \times d\mathbf{S}. \tag{10-2.4}$$

For constant $\rho$, Eq. (10-2.4) simplifies to

$$\mathbf{F} = -\rho \oint \mathbf{A}_g \times d\mathbf{S}. \tag{10-2.5}$$

For a mass layer the contribution of the rim of the layer to the total force is

$$\mathbf{F}_{rim} = -\int \rho \mathbf{A}_g \times t \, d\mathbf{l}_{out} = -\int \sigma \mathbf{A}_g \times d\mathbf{l}_{out}, \tag{10-2.6}$$

where $t$, $\sigma$, and $d\mathbf{l}_{out}$ are the same as in Eq. (10-1.7).

Note that, in contrast to the similar equations for scalar potentials, only the external vector potentials can be used in Eqs. (10-2.1)-(10-2.6), because a gravitational vector potential is defined only for regions of space external to the masses that produce the vector potential.

▼

**Example 10-2.1** A point mass $m$ is at the origin of coordinates. A thin disk of mass $M$ and radius $a$ has its center on the $x$ axis at a distance $x$ from the origin. The density $\sigma$ (mass per unit area) of the disk is uniform. The surface of the disk is perpendicular to the $x$ axis. Find the force with which the point mass attracts the disk.

The vector potential of the point mass is, in spherical coordinates centered at the point mass,[6]

$$\mathbf{A}_g = -G\frac{m}{r}\left[\frac{1 - \cos\theta}{\sin\theta}\right]\boldsymbol{\phi}_u \tag{10-2.7}$$

where $\theta$ is the angle between $r$ and the $x$ axis, and $\boldsymbol{\phi}_u$ is a circular unit vector right-handed relative to the $x$ axis.

Only the rim of the disk makes a non-vanishing contribution to the force (the contributions of the two flat surfaces, being in opposite directions, cancel each other). Substituting $\cos\theta = x/(x^2 + a^2)^{1/2}$ and $r\sin\theta = a$, we obtain, by Eq. (10-2.6) (compare with Example 3-2.1),

$$\begin{aligned}\mathbf{F} &= -\mathbf{i}G\oint\frac{\sigma m[1 - x/(x^2 + a^2)^{1/2}]}{a}dl \\ &= -\mathbf{i}G\frac{\sigma m[1 - x/(x^2 + a^2)^{1/2}]}{a}2\pi a \qquad (10\text{-}2.8) \\ &= -\mathbf{i}G\frac{2mM}{a^2}\left[1 - \frac{x}{(x^2 + a^2)^{1/2}}\right].\end{aligned}$$

**Example 10-2.2** An infinitely long line mass of density $\lambda$ is placed along the $z$ axis of rectangular coordinates. An infinite plane sheet of surface mass density

$$\sigma = \sigma_0[a^2/(a^2 + y^2)] \qquad (10\text{-}2.9)$$

is placed parallel to the $yz$ plane at the distance $x = a$ from the line mass; the center of the sheet being on the $x$ axis. Find the force per unit length exerted by the line mass on the sheet.

Since the density of the sheet is not constant, we must use Eq. (10-2.4). Assuming that the thickness of the sheet is $t$, we have for $\nabla\rho$

$$\nabla\rho = \frac{1}{t}\nabla\sigma = -\frac{1}{t}\sigma_0\frac{2ya^2}{(a^2 + y^2)^2}\mathbf{j}. \qquad (10\text{-}2.10)$$

The vector potential produced by the line mass is, in cylindrical coordinates,[6]

$$\mathbf{A}_g = -G2\lambda\theta\mathbf{k}, \qquad (10\text{-}2.11)$$

where $\theta$ is the angle around the $z$ axis in the $xy$ plane. By the symmetry of the system, the surface integral in Eq. (10-2.4) makes no contribution to the force (the contributions of the front and back surfaces cancel, and there is no contribution from the edges of the sheet at infinity).

Expressing the gradient of the mass density in terms of the angle $\theta$, we have

$$\nabla\rho = -(2\sigma_0/ta)\sin\theta\cos^3\theta\,\mathbf{j}. \tag{10-2.12}$$

Using the first integral in Eq. (10-2.4), we obtain for the force per unit length exerted by the line mass on the sheet

$$\begin{aligned}\mathbf{F}_l &= -\mathbf{i}G\int_{-\infty}^{\infty} 2\lambda\theta\frac{2\sigma_0}{ta}\sin\theta\cos^3\theta\, t\,dy \\ &= -\mathbf{i}G2\lambda\sigma_0\int_{-\pi/2}^{\pi/2} 2\theta\sin\theta\cos^3\theta\frac{d\theta}{\cos^2\theta} \\ &= -\mathbf{i}G2\lambda\sigma_0\int_{-\pi/2}^{\pi/2}\theta\sin 2\theta\, d\theta \\ &= -\mathbf{i}G\lambda\pi\sigma_0.\end{aligned} \tag{10-2.13}$$

▲

## 10-3. Calculation of Cogravitational Forces in Time-Independent Systems from Vector Potentials

By Eq. (2-2.6), the cogravitational force acting on a mass current distribution **J** due to an external cogravitational field **K** is

$$\mathbf{F} = \int \mathbf{J}\times\mathbf{K}\,dV. \tag{10-3.1}$$

Replacing **K** in Eq. (10-3.1) by $\nabla \times \mathbf{A}$, where **A** is the external vector potential, and splitting the integral into an integral over the surface layer and an integral over the interior of the mass current, we have

$$\mathbf{F} = \int_{S.\,layer}\mathbf{J}\times(\nabla\times\mathbf{A})\,dV + \int_{Interior}\mathbf{J}\times(\nabla\times\mathbf{A})\,dV. \tag{10-3.2}$$

As before, the integral over the surface layer can be ignored. The integral over the interior can be transformed by using vector identity (V-22) to

$$\begin{aligned}\mathbf{F} &= \int \mathbf{A}(\nabla \cdot \mathbf{J})\,dV + \int \mathbf{J}(\nabla \cdot \mathbf{A})\,dV \\ &\quad + \oint (\mathbf{A} \cdot \mathbf{J})\,d\mathbf{S} - \oint \mathbf{J}(\mathbf{A} \cdot d\mathbf{S}) \\ &\quad - \oint \mathbf{A}(\mathbf{J} \cdot d\mathbf{S}) - \int \mathbf{A} \times (\nabla \times \mathbf{J})\,dV.\end{aligned} \tag{10-3.3}$$

However, by Eq. (2-2.4), $\nabla \cdot \mathbf{J} = 0$ for time-independent systems, and, by Eq. (3-3.12), $\nabla \cdot \mathbf{A} = 0$ for time-independent systems, and, since a time-independent (steady) mass current is always parallel to its surface, also $\mathbf{J} \cdot d\mathbf{S} = 0$. Therefore, we have

$$\begin{aligned}\mathbf{F} &= \oint (\mathbf{A} \cdot \mathbf{J})\,d\mathbf{S} - \oint \mathbf{J}(\mathbf{A} \cdot d\mathbf{S}) \\ &\quad - \int \mathbf{A} \times (\nabla \times \mathbf{J})\,dV.\end{aligned} \tag{10-3.4}$$

We can also transform the second volume integral together with the second surface integral in Eq. (10-3.3) by using vector identity (V-23). We then obtain an alternative force equation

$$\begin{aligned}\mathbf{F} &= \oint (\mathbf{A} \cdot \mathbf{J})\,d\mathbf{S} - \int (\mathbf{A} \cdot \nabla)\mathbf{J}\,dV \\ &\quad - \int \mathbf{A} \times (\nabla \times \mathbf{J})\,dV.\end{aligned} \tag{10-3.5}$$

From Eq. (10-3.5) we immediately see that for constant **J** the force is simply

$$\mathbf{F} = \oint (\mathbf{A} \cdot \mathbf{J})\,d\mathbf{S}. \tag{10-3.6}$$

For a surface mass-current of density $\mathbf{J}^{(s)} = t\mathbf{J}$ per unit width, where $t$ is the thickness of the mass current sheet, the edges (rim) of the current contribute

$$\mathbf{F}_{rim} = \oint (\mathbf{A} \cdot \mathbf{J}^{(s)})\, d\mathbf{l}_{out} - \oint \mathbf{J}^{(s)}(\mathbf{A} \cdot d\mathbf{l}_{out}) \tag{10-3.7}$$

to Eq. (10-3.4) and

$$\mathbf{F}_{rim} = \oint (\mathbf{A} \cdot \mathbf{J}^{(s)})\, d\mathbf{l}_{out} \tag{10-3.8}$$

to Eqs. (10-3.5) or (10-3.6).

Although the above equations have been derived for the external potential, they remain valid if the total potential is used in them, because the self-potential cannot create a net force on a mass current.

▼

**Example 10-3.1** A long thin-walled cylinder of radius $a$ and surface mass density $\sigma_1$ rotates with angular velocity $\omega_1$ about its symmetry axis, thereby creating a surface mass-current $J_1^{(s)} = \sigma_1\omega_1 a$ in a circular direction round the symmetry axis. A larger long thin-walled cylinder of radius $b$ and surface mass density $\sigma_2$ rotates with angular velocity $\omega_2$ about its symmetry axis in the same direction, thereby creating a surface mass current $J_2^{(s)} = \sigma_2\omega_2 b$. The smaller cylinder is partially inserted into the larger cylinder coaxially with it; their common axis is the $z$ axis of cylindrical coordinates. Neglecting end effects (that is, assuming that the cogravitational field of each cylinder is confined to the interior of the cylinder), find the cogravitational force exerted by the larger cylinder upon the smaller cylinder.

We can use Eq. (10-3.8) to solve the problem. Let us assume that the smaller cylinder is to the right of the larger cylinder, that both cylinders rotate in the right-handed direction relative to the $z$ axis, and that the $z$ axis is directed to the right. The vector potential produced by the larger cylinder is then, in cylindrical coordinates,[7]

$$\mathbf{A} = -G\frac{2\pi\sigma_2\omega_2 b r}{c^2}\boldsymbol{\theta}_u, \tag{10-3.9}$$

where $\boldsymbol{\theta}_u$ is a unit vector in circular direction right-handed relative to the $z$ axis. Since the end effects are neglected and by the symmetry of the system, only the end (rim) of the smaller cylinder inside the larger cylinder contributes to the force. Since the surface mass current of the smaller cylinder is $\mathbf{J}_1^{(s)} = \sigma_1\omega_1 a\boldsymbol{\theta}_u$, the force is, by Eq. (10-3.8),

$$\begin{aligned}\mathbf{F} &= \oint\left(-G\frac{2\pi\sigma_2\omega_2 ba}{c^2}\sigma_1\omega_1 a\right)d\mathbf{l}_{out}\\ &= G\frac{4\pi^2\sigma_1\sigma_2\omega_1\omega_2 a^3 b}{c^2}\mathbf{k}.\end{aligned} \tag{10-3.10}$$

It is interesting to note that the larger cylinder repels the smaller one, which is a consequence of the general property of mass currents whereby like mass currents repel and opposite mass currents attract each other.

Note that the force represented by Eq. (10-3.10) is not the total force between the two cylinders. The total force between them is the sum of the gravitational and the cogravitational force. In this particular case (the cylinders rotate in the same direction) the cogravitational force lessens the gravitational attraction, but if the two cylinders rotated in opposite directions, then the cogravitational force would strengthen the gravitational attraction between them.

**Example 10-3.2** A spherical shell having a uniformly distributed mass of density $\rho$, inner radius $a$, and outer radius $b$ consists of two separate hemispheres. The shell rotates with angular velocity $\omega$ about its vertical symmetry axis ($z$ axis) passing at right angles to the equatorial plane separating the two hemispheres. The rotation of the shell creates a mass current density in the shell, which is, in spherical coordinates centered at the center of the shell, $\mathbf{J} = \omega \times \mathbf{r}$. Find the cogravitational force between the two hemispheres if it is known that the cogravitational vector potential inside the shell is

$$\mathbf{A} = -G\frac{2\pi\rho\omega}{15c^2}(5b^2 - 3r^2 - 2a^5/r^3)r\sin\theta\,\boldsymbol{\phi}_u, \quad (10\text{-}3.11)$$

where $\boldsymbol{\phi}_u$ is a circular unit vector in a spherical system of coordinates.[5]

Let us find the force on the upper hemisphere. To do so we can use Eq. (10-3.4). By vector identities (V-12) and (V-4), $\nabla \times \mathbf{J} = 2\rho\boldsymbol{\omega}$, and since $\boldsymbol{\omega}$ is parallel to the $z$ axis, the integral containing $\mathbf{A} \times (\nabla \times \mathbf{J})$, by symmetry, does not contribute to the net force. Thus only the surface integrals in Eq. (10-3.4) need to be considered. Since $\mathbf{A}$ is perpendicular to $d\mathbf{S}$, the second integral in Eq. (10-3.4) vanishes. Thus the force is all due to the first integral in Eq. (10-3.4).

The contribution of the flat base of the hemisphere to the force on the hemisphere is then

$$\begin{aligned}\mathbf{F}_1 &= \mathbf{k}\int_a^b G\frac{2\pi\rho\omega}{15c^2}\left(5b^2 - 3r^2 - \frac{2a^5}{r^3}\right)r\rho\omega r 2\pi r\,dr \\ &= \mathbf{k}G\frac{\pi^2\rho^2\omega^2}{15c^2}(3b^6 - 5b^2a^4 - 8ba^5 + 10a^6).\end{aligned} \quad (10\text{-}3.12)$$

Only the vertical component of $d\mathbf{S}$ makes a net contribution to the integral over the outer hemispherical surface. Hence this surface contributes

$$\begin{aligned}\mathbf{F}_2 &= -\mathbf{k}\int_0^{\pi/2} G\frac{2\pi\rho\omega}{15c^2}\left(5b^2 - 3b^2 - \frac{2a^5}{b^3}\right)b\sin\theta\rho\omega b \\ &\quad \times \sin\theta\, 2\pi b^2 \sin\theta\cos\theta\, d\theta \\ &= -\mathbf{k}G\frac{2\pi^2\rho^2\omega^2}{15c^2}(b^5 - a^5)b.\end{aligned} \quad (10\text{-}3.13)$$

The inner surface of the hemisphere contributes, similarly,

$$\mathbf{F}_3 = \mathbf{k}\int_0^{\pi/2} G\frac{2\pi\rho\omega}{3c^2}(b^2 - a^2)a\sin\theta\,\rho\omega a$$
$$\times\sin\theta\, 2\pi a^2\sin\theta\cos\theta\, d\theta \qquad (10\text{-}3.14)$$
$$= \mathbf{k}G\frac{\pi^2\rho^2\omega^2}{3c^2}(b^2 - a^2)a^4.$$

The total force is therefore

$$\mathbf{F}_1 + \mathbf{F}_2 + \mathbf{F}_3 = \mathbf{k}G\frac{\pi^2\rho^2\omega^2}{15c^2}(b^6 - 6ba^5 + 5a^6). \qquad (10\text{-}3.15)$$

As in the preceding example, this force represents a repulsion. However, it is important to remember that Eq. (10-3.15) represents only the cogravitational force between the two hemispheres and that there is also a gravitational attraction between them.

▲

## 10-4. Calculation of Cogravitational Forces in Time-Independent Systems from Scalar Potentials

Let us replace the cogravitational vector **K** in Eq. (10-3.1) by $-\nabla\varphi_c$ where $\varphi_c$ is the external cogravitational scalar potential [see Eq. (3-3.10)] . Transforming Eq. (10-3.1) as before, we have

$$\mathbf{F} = -\int_{S.\,layer}\mathbf{J}\times\nabla\varphi_c dV - \int_{Interior}\mathbf{J}\times\nabla\varphi_c dV. \qquad (10\text{-}4.1)$$

Disregarding the integral over the surface layer and using vector identity (V-11), we obtain

$$\mathbf{F} = -\int_{Interior}\mathbf{J}\times\nabla\varphi_c dV = \int_{Interior}\nabla\times(\varphi_c\mathbf{J})dV$$
$$-\int_{Interior}\varphi_c\nabla\times\mathbf{J}dV. \qquad (10\text{-}4.2)$$

The first integral on the right can be transformed by using vector identity (V-21) into a surface integral, so that the force becomes

$$\mathbf{F} = -\oint \varphi_c \mathbf{J} \times d\mathbf{S} - \int \varphi_c \nabla \times \mathbf{J}\, dV. \qquad (10\text{-}4.3)$$

For a surface current $\mathbf{J}^{(s)}$ the contribution of the rim surface is

$$\mathbf{F}_{rim} = -\oint \varphi_c \mathbf{J}^{(s)} \times d\mathbf{l}_{out}. \qquad (10\text{-}4.4)$$

Note that only the external potential can be used in the above equations because the cogravitational scalar potential is defined only for regions of space external to the source of the potential (see Section 3-3).

▼

**Example 10-4.1** A thin disk of uniform mass density $\rho$, radius $a$, and thickness $t$ rotates with angular velocity $\omega$ about its symmetry axis, which is also the $z$ axis of cylindrical coordinates. The center of the disk is at $z = 0$. At a point $z$ of the $z$ axis and perpendicular to the axis is a distant rotating ring forming a filamentary mass current $I$. The surface area of the ring is $S$. Assuming that the mass current in the ring and the rotation of the disk are right-handed relative to the $z$ axis, find the force exerted by the ring on the disk.

The rotating ring constitutes a "cogravitational dipole" and the scalar potential of the ring is, in cylindrical coordinates,[5]

$$\varphi_c = -G\frac{IS}{c^2}\left[\frac{z}{(z^2 + r^2)^{3/2}}\right], \qquad (10\text{-}4.5)$$

where $z$ is the distance from the ring and $r$ is the distance from the $z$ axis (see Section 5-2).

The rotating disk constitutes a mass current distribution $\mathbf{J} = \rho\omega r\theta_u$ for which, by vector identities (V-12) and (V-4), $\nabla \times \mathbf{J} = 2\rho\omega\mathbf{k}$. Using Eqs. (10-4.3) and (10-4.4) and taking into account that, by the symmetry of the system, only the rim of the ring contributes to the surface integral, and that the potential and the mass current density are constant at the rim, we have

$$\mathbf{F} = -\mathbf{k}G\frac{ISz}{c^2(z^2+a^2)^{3/2}}\rho\omega a t 2\pi a + \mathbf{k}\int_0^a G\frac{2ISz\rho\omega}{c^2(z^2+r^2)^{3/2}}t2\pi r dr \tag{10-4.6}$$

$$= -\mathbf{k}G\frac{4\pi IS\rho\omega t}{c^2}\left[\frac{a^2 z}{2(z^2+a^2)^{3/2}} + \frac{z}{(z^2+a^2)^{1/2}} - 1\right].$$

Note that the ring repels the disk. It is important to remember, however, that the force represented by Eq. (10-4.6) does not include the gravitational attraction between the ring and the disk.

▲

## 10-5. Calculation of Gravitational and Cogravitational Forces in Time-Dependent Systems from Potentials

The force equations derived in the preceding sections can be easily extended to time-dependent systems by using retarded gravitational and cogravitational potentials defined in Section 3-3. It is important to realize that although the potentials of time-dependent mass and mass-current distributions are retarded, the retardation does not directly affect the calculation of gravitational and cogravitational forces, because these forces are assumed to act upon masses and mass-current distribution at the time for which the forces are calculated.

However, in time-dependent systems, the gravitational field cannot be expressed in terms of a scalar potential alone. Instead, by Eq. (3-3.4), it is expressed as a combination of the retarded gravitational scalar potential and the retarded cogravitational vector potential:

$$\mathbf{g} = -\nabla\varphi - \frac{\partial \mathbf{A}}{\partial t}. \tag{3-3.4}$$

Therefore Eqs. (10-1.5), (10-1.6) and (10-1.7) acquire the

additional term

$$\mathbf{F} = -\int \rho \frac{\partial \mathbf{A}_{ret}}{\partial t} dV. \qquad (10\text{-}5.1)$$

Similarly, by Eq. (3-3.10), the cogravitational field cannot be expressed in terms of a scalar potential alone. Instead, it is expressed as a combination of the retarded cogravitational scalar potential and the retarded gravitational vector potential:

$$\mathbf{K} = -\nabla\varphi_c + \frac{\partial \mathbf{A}_{g\,ret}}{c^2 \partial t}. \qquad (3\text{-}3.10)$$

Therefore, in time-dependent systems, Eqs. (10-4.3) and (10-4.4) acquire the additional term

$$\mathbf{F} = \frac{1}{c^2}\int \mathbf{J} \times \frac{\partial \mathbf{A}_{g\,ret}}{\partial t} dV. \qquad (10\text{-}5.2)$$

As we shall presently see, Eq. (10-5.1) is very important in the generalized theory of gravitation, since it indicates the presence of an entirely new force in gravitational interactions. The properties and significance of this force, which we shall call the "gravikinetic force," will be discussed in the next chapter.

## References and Remarks for Chapter 10

**1.** All basic force equations in this chapter are derived for external, rather than for internal, gravitational and cogravitational fields. This is done for two reasons: first, only the external fields produce net forces on stationary and moving masses; second, only external fields can always be associated with scalar as well as with vector potentials.

**2.**. The electric counterpart of this equation is derived in Oleg D. Jefimenko, *Electricity and Magnetism*, 2nd ed., (Electret Scientific, Star City, 1989), pp. 210-211.

**3.** See also Oleg D. Jefimenko, "Direct calculation of electric and magnetic forces from potentials," Am. J. Phys. **58**, 625-631 (1990).
**4.** This can be easily shown by replacing the potential in Eq. (10-1.5) by a constant and again using vector identity (V-20).
**5.** In order not to dilute the presentation by excessive details, the potential is stated without derivation.
**6.** A simple way to compute this potential is to use Eq. (3-3.9).
**7.** A simple way to compute this potential is first to use Eq. (7-1.8) for finding the cogravitational field in the cylinder and then to use Eq. (3-3.3). See also Example 12-2.3.

# 11

# GRAVIKINETIC FIELD AND ITS PROPERTIES

One of the main differences between the generalized theory of gravitation and Newton's gravitational theory is that in the generalized theory of gravitation there is a special force field – the cogravitational, or Heaviside's field. The cogravitational field is produced by all moving masses, and it acts on all moving masses. In this chapter we shall learn that in the generalized theory of gravitation there is yet another force field produced by moving masses. However, in contrast with the cogravitational field, this field is produced only by masses whose velocity changes in time and, again in contrast with the cogravitational field, it acts on all masses, moving as well as stationary.

## 11-1. The Gravikinetic Field

As we know from Chapter 2, the principal gravitational field equation of the generalized theory of gravitation is

$$\mathbf{g} = -G\int\left\{\frac{[\rho]}{r^3} + \frac{1}{r^2c}\left[\frac{\partial\rho}{\partial t}\right]\right\}\mathbf{r}\,dV' + \frac{G}{c^2}\int\frac{1}{r}\left[\frac{\partial\mathbf{J}}{\partial t}\right]dV', \qquad (2\text{-}2.1)$$

where $\mathbf{J} = \rho\mathbf{v}$ is the mass current density produced by a moving mass distribution $\rho$. The first term on the right in Eq. (2-2.1) represents the retarded Newtonian gravitational field. Just like the ordinary Newtonian field, this field originates at any mass distribution $\rho$ and is responsible for the gravitational attraction. However, the last term on the right of Eq. (2-2.1) represents a gravitational field very different from the Newtonian field. As can be seen from Eq. (2-2.1), this new field is produced by a time-variable mass current $\partial\mathbf{J}/\partial t$ and it differs in two important respects from the Newtonian gravitational field: it is directed along the mass-current (more accurately, along its partial time derivative) rather than along a radius vector, and it exists only as long as the current is changing in time. Therefore the gravitational force caused by this field is also different from the ordinary gravitational (Newtonian) force. This force (designated as $\mathbf{F}_3$ in Fig. 2.2) is directed along $\partial\mathbf{J}/\partial t$ and it lasts only as long as the mass current is changing. Unlike the Newtonian gravitational force, which is always an attraction between gravitating masses, the force due to the time-variable $\mathbf{J}$ is basically a *dragging* force. If only the magnitude but not the direction of $\mathbf{J}$ changes, this force is directed parallel or antiparallel (if $\partial\mathbf{J}/\partial t$ is negative) to $\mathbf{J}$, causing a mass subjected to this force to move parallel or antiparallel to (rather than toward) the mass distribution forming the mass current. However, like the Newtonian force, the force due to the time-variable $\mathbf{J}$ acts upon all masses.

It is important to note that unlike the cogravitational field, the field produced by $\partial\mathbf{J}/\partial t$ usually is not created by masses moving with constant velocity $\mathbf{v}$ [see, however, Eq. (4-1.4)].

Since the gravitational field created by time-variable mass currents is very different from the Newtonian field and from the cogravitational field, a special name should be given to it. Taking into account that the cause of this field is a motion of masses, we may call it the *gravikinetic field*, and we may call the force which this field exerts on other masses the *gravikinetic force*. We shall designate the gravikinetic field by the vector $\mathbf{g}_k$. From Eq. (2-2.1) we thus have

$$\mathbf{g}_k = \frac{G}{c^2} \int \frac{1}{r} \left[ \frac{\partial \mathbf{J}}{\partial t} \right] dV'. \tag{11-1.1}$$

Because of the $c^2$ in the denominator in Eq. (11-1.1), the gravikinetic field cannot be particularly strong except when the mass-current responsible for it changes very fast. On the other hand, taking into account that the time scale in gravitational interactions taking place in the Universe may be very long, the ultimate effect of the gravikinetic field in such interactions may be very considerable regardless of the rate at which the mass current changes.

## 11-2. Correlation Between the Gravikinetic Field and the Cogravitational Field

If we compare Eq. (11-1.1) with the expression for the retarded cogravitational vector potential $\mathbf{A}_{ret}$ produced by a mass current $\mathbf{J}$,

$$\mathbf{A}_{ret} = - \frac{G}{c^2} \int \frac{[\mathbf{J}]}{r} dV', \tag{3-3.2}$$

we recognize that the gravikinetic field is equal to the time derivative of retarded $\mathbf{A}_{ret}$:

$$\mathbf{g}_k = - \frac{\partial \mathbf{A}_{ret}}{\partial t}. \tag{11-2.1}$$

It is interesting to note that Eq. (11-2.1) points out the possibility of a new definition and interpretation of the cogravitational vector potential. Let us integrate Eq. (11-2.1). We obtain

$$\mathbf{A}_{ret} = - \int \mathbf{g}_k dt + \text{const}. \tag{11-2.2}$$

Let us call the time integral of $\mathbf{g}_k$ the *gravikinetic impulse*. We then can say that the cogravitational vector potential created by a mass current at a point in space is equal to the negative of the gravikinetic impulse produced by this current at that point during the action of the mass current. Since the gravikinetic impulse is, in principle, a measurable quantity, we thus have an operational definition and a physical interpretation of the cogravitational vector potential.[1]

It may be useful to mention that although Eqs. (11-2.1) and (11-2.2) correlate the gravikinetic field with the cogravitational vector potential, there is no causal link between the two: the correlation merely reflects the fact that both the gravikinetic field and the cogravitational vector potential are simultaneously caused by the same mass current.[2]

A more direct (albeit not causal) relation between the gravikinetic field and the cogravitational field can be obtained as follows. Let us assume that an initially stationary mass current $\mathbf{J}(x',y',z')$ (an initially stationary rotating spherical mass, for example) moves as a whole with a constant velocity $v$ toward a stationary observer located at the origin of coordinates. The mass current is then a function of $(x'-v_x t)$, $(y'-v_y t)$, and $(z'-v_z t)$, or

$$\mathbf{J} = \mathbf{J}(x'-v_x t, y'-v_y t, z'-v_z t). \tag{11-2.3}$$

The time derivative of the current is

$$\frac{\partial \mathbf{J}}{\partial t} = -\frac{\partial \mathbf{J}}{\partial x'}v_x - \frac{\partial \mathbf{J}}{\partial y'}v_y - \frac{\partial \mathbf{J}}{\partial z'}v_z = -(\mathbf{v}\cdot\nabla')\mathbf{J}. \tag{11-2.4}$$

The gravikinetic field caused by the moving mass current is then, by Eqs. (11-1.1) and (11-2.4),

$$\mathbf{g}_k = -\frac{G}{c^2}\int\frac{[(\mathbf{v}\cdot\nabla')\mathbf{J}]}{r}dV'. \tag{11-2.5}$$

The spatial derivative appearing in Eq. (11-2.5) can be eliminated as follows. Using vector identity (V-6), which can be written as

$$\nabla'(\mathbf{v}\cdot\mathbf{J}) = (\mathbf{v}\cdot\nabla')\mathbf{J} + \mathbf{v}\times(\nabla'\times\mathbf{J}) + (\mathbf{J}\cdot\nabla')\mathbf{v} + \mathbf{J}\times(\nabla'\times\mathbf{v}), \qquad (11\text{-}2.6)$$

and taking into account that **v** is a constant vector, we obtain

$$\mathbf{g}_k = -\frac{G}{c^2}\int\frac{[\nabla'(\mathbf{v}\cdot\mathbf{J})]}{r}dV' + \frac{G}{c^2}\int\frac{[\mathbf{v}\times(\nabla'\times\mathbf{J})]}{r}dV'. \qquad (11\text{-}2.7)$$

If we compare Eq. (11-2.7) with Eq. (3-1.2) for the cogravitational field,

$$\mathbf{K} = -\frac{G}{c^2}\int\frac{[\nabla'\times\mathbf{J}]}{r}dV', \qquad (3\text{-}1.2)$$

we find that Eq. (11-2.7) can be written as

$$\mathbf{g}_k = -\frac{G}{c^2}\int\frac{[\nabla'(\mathbf{v}\cdot\mathbf{J})]}{r}dV' - \mathbf{v}\times\mathbf{K}, \qquad (11\text{-}2.8)$$

where **K** is the cogravitational field created by the moving mass current **J**.

▼

**Example 11-2.1** Show that if $\mathbf{g}_k$ is linear in time, $\mathbf{g}_k = \mathbf{a} + \mathbf{b}t$, then the retarded cogravitational vector potential in Eq. (11-2.1) can be replaced by the ordinary (unretarded) vector potential.

We shall solve this problem by using Helmholtz's theorem of vector analysis, vector identity (V-24):

$$\mathbf{V} = -\frac{1}{4\pi}\int\frac{\nabla'(\nabla'\cdot\mathbf{V}) - \nabla'\times(\nabla'\times\mathbf{V})}{r}dV'. \qquad (V\text{-}24)$$

As we know from Example 3-3.1 the divergence of the retarded cogravitational vector potential satisfies the relation

$$\nabla\cdot\mathbf{A}_{ret} = -\frac{1}{c^2}\frac{\partial\varphi_{ret}}{\partial t}, \tag{3-3.12}$$

where $\varphi$ is the retarded scalar potential of $\mathbf{g}$. Therefore, by Eqs. (11-2.1) and (3-3.12), we have

$$\nabla\cdot\mathbf{g}_k = \frac{1}{c^2}\frac{\partial^2\varphi_{ret}}{\partial t^2}. \tag{11-2.9}$$

For the curl of $\mathbf{g}_k$ we have, by Eq. (11-2.1) and by the definition of the cogravitational vector potential,

$$\nabla\times\mathbf{g}_k = -\frac{\partial\mathbf{K}}{\partial t}. \tag{11-2.10}$$

For $\nabla(\nabla\cdot\mathbf{g}_k)$ we then have, by Eq. (11-2.9), by Eq. (3-3.4), and by Eq. (11-2.1),

$$\nabla(\nabla\cdot\mathbf{g}_k) = -\frac{1}{c^2}\frac{\partial^2}{\partial t^2}(\mathbf{g} + \frac{\partial\mathbf{A}_{ret}}{\partial t}) = -\frac{1}{c^2}\frac{\partial^2}{\partial t^2}(\mathbf{g} - \mathbf{g}_k), \tag{11-2.11}$$

where $\mathbf{g}$ is the total gravitational field given by Eq. (2-2.1). For $\nabla\times(\nabla\times\mathbf{g}_k)$ we have, by Eqs. (11-2.10) and (7-1.4),

$$\nabla\times(\nabla\times\mathbf{g}_k) = \frac{4\pi G}{c^2}\frac{\partial\mathbf{J}}{\partial t} - \frac{1}{c^2}\frac{\partial^2\mathbf{g}}{\partial t^2}. \tag{11-2.12}$$

Substituting Eqs. (11-2.11) and (11-2.12) into vector identity (V-24), canceling $(1/c^2)(\partial^2\mathbf{g}/\partial t^2)$, noting that $\partial^2\mathbf{g}_k/\partial t^2 = 0$ (because $\mathbf{g}_k$ is linear in $t$), and comparing the result with Eq. (3-3.6), we finally obtain

$$\mathbf{g}_k = \frac{G}{c^2}\int\frac{1}{r}\frac{\partial\mathbf{J}}{\partial t}dV' = -\frac{\partial\mathbf{A}}{\partial t}. \tag{11-2.12}$$

▲

## References and Remarks for Chapter 11

**1.** For a related interpretation of the magnetic vector potential see Oleg D. Jefimenko, *Causality, Electromagnetic Induction, and Gravitation,* 2nd ed. (Electret Scientific, Star City, 2000) pp. 30, 31.

**2.** Nor is there, of course, a causal relation between the gravikinetic field and the cogravitational field, since they, too, are simultaneously produced by the same mass current.

# 12

# GRAVIKINETIC FORCES AND EFFECTS; GRAVITATIONAL INDUCTION

In this chapter we shall present illustrative examples on the calculation and effects of gravikinetic fields and forces. We shall compute gravikinetic forces acting on mass distributions in the vicinity of time-variable mass currents. We shall establish a connection between the change of the mechanical momentum of a mass distribution subjected to a gravikinetic field and the cogravitational vector potential associated with the time-variable mass current that produces the gravikinetic field. And we shall demonstrate how gravikinetic fields can influence the translation and rotation of masses subjected to it.

## 12-1. Gravikinetic Fields and the Mechanical Momentum

We shall now present examples on calculation of gravikinetic fields and gravikinetic forces and examples on effects of these fields and forces. We shall use examples requiring only very simple calculations. It is not the complexity of the examples that is important for our purpose. Our purpose is to provide an unambiguous demonstration of the effects and actions of gravikinetic fields; this can be best achieved with uncomplicated examples.

For simplicity, we shall limit our calculations to relatively small systems and relatively slow variations of mass currents. In such systems retardation effects are negligible (see Section 2-2), so that Eq. (11-1.1) can be written without brackets as

$$\mathbf{g}_k = -\frac{G}{c^2}\int \frac{1}{r}\frac{\partial \mathbf{J}}{\partial t}\,dV'. \tag{12-1.1}$$

If the mass current is filamentary, Eq. (12-1.1) can be written as

$$\mathbf{g}_k = -\frac{\partial I}{\partial t}\frac{G}{c^2}\int \frac{d\mathbf{l}'}{r}, \tag{12-1.2}$$

where $I$ is the mass current in the filament and $d\mathbf{l}'$ is a length element of the filament in the direction of the current. Finally, if the retardation is neglected, the gravikinetic field of a mass current $\mathbf{J}$ can be found, according to Eq. (11-2.1), from

$$\mathbf{g}_k = -\frac{\partial \mathbf{A}}{\partial t}, \tag{12-1.3}$$

where $\mathbf{A}$ is the ordinary (not retarded) cogravitational vector potential associated with $\mathbf{J}$.

When the gravikinetic force acts on a mass distribution $\rho$, it changes the mechanical momentum $\mathbf{G}_M$ of the mass distribution in accordance with

$$\Delta \mathbf{G}_M = \int \mathbf{F}\,dt = \int\int \rho\,\mathbf{g}_k\,dV dt. \tag{12-1.4}$$

If $\mathbf{g}_k$ is a function of time only, the momentum change is

$$\Delta \mathbf{G}_M = m\int \mathbf{g}_k\,dt = -\,m\Delta \mathbf{A}, \tag{12-1.5}$$

where $m$ is the total mass of the distribution, and $\Delta\mathbf{A}$ is the change in the vector potential during the time interval under consideration.

If a circular gravikinetic force acts on a mass distribution restricted to a circular motion, the angular momentum of the mass distribution changes. For a mass distribution and gravikinetic field of circular symmetry, the change in the angular momentum is

$$\Delta\mathbf{L} = \int\int \mathbf{r} \times \mathbf{g}_k dm dt = -\int \mathbf{r} \times \mathbf{A} dm, \qquad (12\text{-}1.6)$$

where $\mathbf{L}_M$ is the angular momentum and $\mathbf{r}$ is the lever arm of $\mathbf{g}_k$. As already mentioned, we are using the ordinary vector potential for simplicity; for exact calculations the retarded vector potential must be used in Eqs. (12-1.3), (12-1.4), (12-1.5), and (12-1.6).

It should be pointed out that an association between the momentum change of a body and the change of the cogravitational vector potential at the location of the body does not signify a causal relation. This association is a consequence of the fact that both a gravikinetic force and a time-variable cogravitational field (and its time-variable vector potential) are simultaneously created by a time-variable mass current.

As is known, a vector potential may contain an arbitrary additive function of zero curl ("gauge calibration"). However, only the vector potential given by Eq. (3-3.2) and by its unretarded version can be used for the calculation of the gravikinetic field.

An explanatory note is required concerning calculations of forces and torques exerted on mass distributions by gravikinetic fields. The force experienced by a mass distribution is determined, in general, by the total gravitational field given by Eq. (2-2.1), not just by the gravikinetic field, Eqs. (11-1.1), (12-1.1), or (12-1.2). Therefore a force calculated from the gravikinetic field alone may not be the true force experienced by the mass distribution under consideration. In contrast, only the gravikinetic force has an effect on the torque experienced by rings of mass and by similar objects. This is because the torque in such systems is determined by a closed line integral of the gravitational field, and only the gravikinetic field gives a non-vanishing contribution to such integrals [the first term of Eq. (2-2.1), being a function of $r$ in the direction of $\mathbf{r}$, has zero curl and therefore cannot contribute to closed line integrals; see Eqs. (7-1.19)-(7-1.23)].

## 12-2. Examples on Calculation of Gravikinetic Fields

In this section we shall present several illustrative examples on calculation of gravikinetic fields. A direct calculation of gravikinetic fields from Eq. (11-1.1) or from its unretarded versions, Eqs. (12-1.1) and (12-1.2), involves exactly the same techniques that are used for calculating cogravitational vector potentials from Eq. (3-3.2) or from its unretarded versions. These are the same techniques that have been developed for calculating the magnetic vector potential in the electromagnetic theory. Therefore we shall avoid presenting examples on direct calculation of gravikinetic fields here, since such examples would basically duplicate examples on magnetic vector potential calculations provided in most textbooks on electromagnetic theory. Instead, with one exception, we shall make use of Eq. (12-1.3), of Table 7-1 for converting electromagnetic quantities to gravitational and cogravitational quantities, and of the readily available expressions for the magnetic vector potentials.

▼

**Example 12-2.1** A straight stationary rod of length $2L$ and cross-sectional area $S$ has a uniformly distributed mass of density $\rho$ (Fig. 12.1). The rod is suddenly set in motion along its length. Find the gravikinetic field created by the rod at a distance $R$ from the rod at a point equidistant from the ends of the rod.

The electromagnetic counterpart of the moving rod is a straight wire carrying a current $I$. The magnetic vector potential for such a wire is [1]

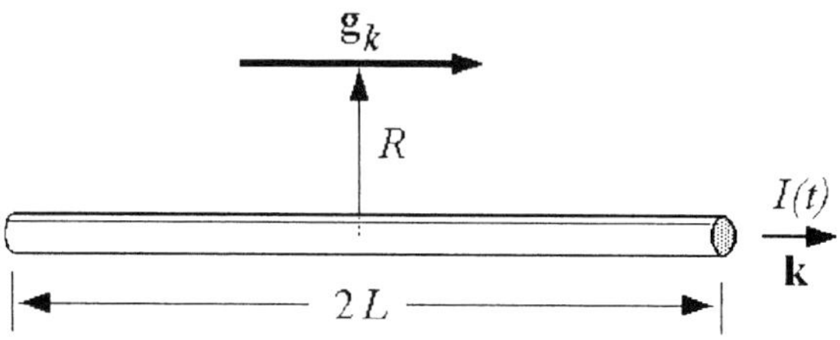


*Fig. 12.1 An accelerating rod produces a gravikinetic field.*

$$\mathbf{A} = \frac{\mu_0 I}{2\pi} \ln \frac{L + (L^2 + R^2)^{1/2}}{R} \mathbf{k}, \tag{12-2.1}$$

where **k** is a unit vector in the direction of the current $I$. By Eq. (12-1.3) and Table 7-1, the gravikinetic field of the rod is then

$$\mathbf{g}_k = \frac{\partial I}{\partial t} \frac{2G}{c^2} \ln \frac{L + (L^2 + R^2)^{1/2}}{R} \mathbf{k}, \tag{12-2.2}$$

where $I$ is now the filamentary mass current produced by the moving rod, $I = \rho v S$, ($v$ is the velocity of the rod). If the rod is long, so that $L^2 \gg R^2$, we may neglect $R^2$ in Eqs. (12-2.2) The gravikinetic field of the rod is then

$$\mathbf{g}_k = \frac{\partial I}{\partial t} \frac{2G}{c^2} \ln \frac{2L}{R} \mathbf{k}. \tag{12-2.3}$$

**Example 12-2.2** An initially stationary, thin-walled cylinder of radius $R_0$, length $2L$ and wall thickness $t$ has a uniformly distributed mass of density $\rho$ (Fig. 12.2). The cylinder is suddenly set in motion along its length. Find the gravikinetic field created by the cylinder outside and inside the cylinder.

The electromagnetic counterpart of the moving cylinder is a cylinder carrying an electric current along its length. The magnetic vector potential outside a current-carrying cylinder is the same as if the current of the cylinder were confined to the axis of the cylinder.[2] The magnetic vector potential outside the cylinder is

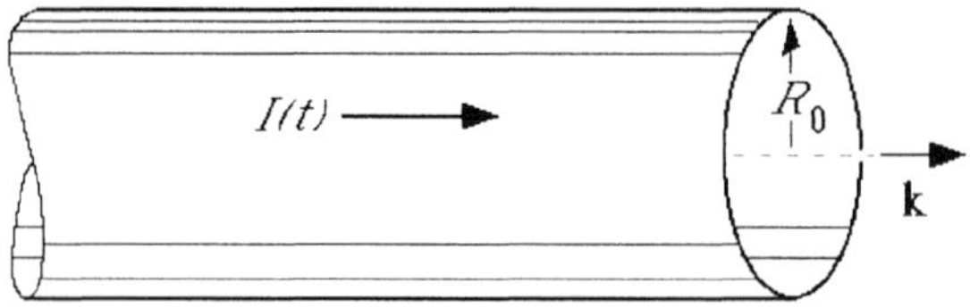


*Fig. 12.2 An accelerating cylinder creates a gravikinetic field outside and inside the cylinder.*

therefore again given by Eq. (12-2.1), and the corresponding gravikinetic field is again given by Eqs. (12-2.2) and (12-2.3). However, the mass current $I$ produced by the cylinder is $\rho v 2\pi R_0 t$.

The cogravitational vector potential inside a moving cylinder is constant and is equal to the cogravitational vector potential just outside the cylinder. (Since there is no cogravitational field inside the cylinder, this statement may appear incredulous. However, the absence of **K** inside the cylinder merely requires that **A** is the same at all points inside the cylinder. It does not require that $\mathbf{A} = 0$.) Substituting $R_0$ for $R$ in Eq. (12-2.3), we then have for the gravikinetic field inside the (long) cylinder

$$\mathbf{g}_k = \frac{\partial I}{\partial t}\frac{2G}{c^2}\ln\frac{2L}{R_0}\mathbf{k}. \tag{12-2.4}$$

**Example 12-2.3** A stationary cylinder similar to that described in Example 12-2.2 but of length $L$ is suddenly set in rotation about its symmetry axis. Neglecting end effects, find the gravikinetic field inside the cylinder.

The electromagnetic counterpart of the rotating cylinder is a cylinder carrying a circular current over its entire length. If the end effects are neglected, the magnetic field inside such cylinder is homogeneous, and the magnetic vector potential is [3]

$$\mathbf{A} = \mu_0 \frac{I}{2L} r\boldsymbol{\theta}_u, \tag{12-2.5}$$

where $r$ is the distance from the axis of the cylinder and $\boldsymbol{\theta}_u$ is the azimuthal unit vector whose direction is the same as that of the circulating current in the cylinder. The gravikinetic field is then, by Eq. (12-1.3) and Table 7-1 ,

$$\mathbf{g}_k = \frac{2\pi G}{c^2}\frac{\partial I}{\partial t}\frac{r}{L}\boldsymbol{\theta}_u, \tag{12-2.6}$$

where $I = \rho\omega R_0 Lt$ and $\omega$ is the angular velocity of the cylinder.

**Example 12-2.4** A ring of radius $b$ carries a uniformly distributed mass $m_b$ and rotates with a variable angular velocity $\omega_b$ about its symmetry axis (Fig. 12.3). Find the gravikinetic field in the plane of the ring near the center of the ring.

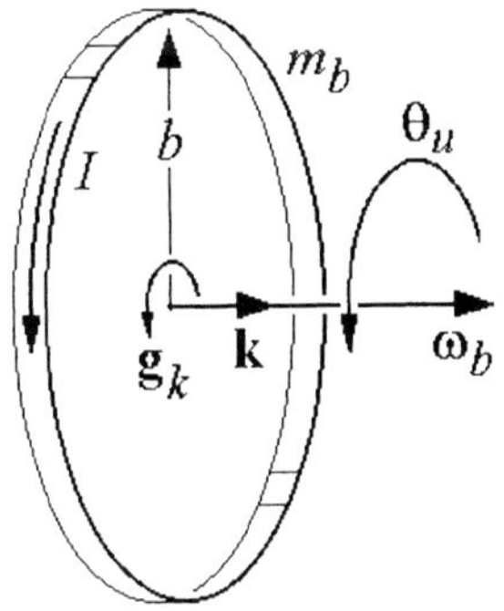


*Fig. 12.3 Calculation of the gravikinetic field near the center of a rotating ring.*

The ring constitutes a circular current $I = \omega_b m_b / 2\pi$. The cogravitational field on the axis of a similar ring was calculated in Example 2-2.1. At the center of the ring it is, by Eq. (2-2.18),

$$\mathbf{K} = -G\frac{m_b\omega_b}{c^2 b}\mathbf{k}, \tag{12-2.7}$$

where $\mathbf{k}$ is a unit vector along the axis of the ring right-handed relative to the direction of rotation.[4] Within a small region near the center of the ring this field is nearly homogeneous, so that the cogravitational vector potential in the plane of the ring is approximately[3]

$$\mathbf{A} = -G\frac{m_b \omega_b r}{2b}\boldsymbol{\theta}_u, \tag{12-2.8}$$

where $r$ is the distance from the center, and $\boldsymbol{\theta}_u$ is the azimuthal unit vector in the direction of the rotation of the ring. By Eq. (12-1.3), the gravikinetic field near the center of the ring is then approximately

$$\mathbf{g}_k = \frac{G}{c^2}\frac{\partial \omega_b}{\partial t}\frac{m_b r}{2b}\boldsymbol{\theta}_u. \tag{12-2.9}$$

**Example 12-2.5** A ring of radius $a$ carries a uniformly distributed mass $m_a$ and rotates with increasing angular velocity $\omega_a$ about its symmetry axis (Fig. 12.4). Find the gravikinetic field far from the ring in the plane of the ring.

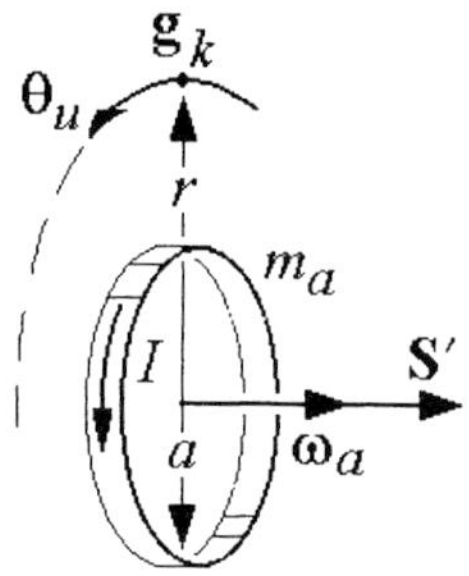


*Fig. 12.4 Calculation of the gravikinetic field at a large distance from a rotating ring.*

We shall solve this problem by direct calculation using Eq. (12-1.2). Let us first convert the line integral of Eq. (12-1.2) into a surface integral by means of vector identity (V-18). We then have [compare with Eq. (3-3.20)]

$$\mathbf{g}_k = \frac{\partial I}{\partial t}\frac{G}{c^2}\oint\frac{d\mathbf{l}'}{r} = -\frac{\partial I}{\partial t}\frac{G}{c^2}\int\frac{\mathbf{r}}{r^3}\times d\mathbf{S}', \tag{12-2.10}$$

where $d\mathbf{S}'$ is right-handed relative to $d\mathbf{l}'$ (or $I$). For field points far from the ring, $r$ may be considered constant over the entire surface area of the ring, and $r^3$ may be factored out from under the integral sign. Since we are calculating the field in the plane of the ring, $\mathbf{r}$ is perpendicular to $d\mathbf{S}'$, so that the cross product in the last integral becomes $-rdS'\boldsymbol{\theta}_u$, where the unit vector $\boldsymbol{\theta}_u$ is as shown in Fig. 12.4. Canceling $r$, replacing the integral by the area of the ring, $\pi a^2$, and replacing the current $I$ by $\omega_a m_a/2\pi$, we obtain

$$\mathbf{g}_k = \frac{\partial I}{\partial t}\frac{G\pi a^2}{4c^2r^2}\boldsymbol{\theta}_u = \frac{G}{c^2}\frac{\partial \omega_a}{\partial t}\frac{m_a a^2}{2r^2}\boldsymbol{\theta}_u . \qquad (12\text{-}2.11)$$

It is instructive to compare this result with Eq. (3-3.22). Note that $\boldsymbol{\phi}_u$ in Eq. (3-3.22) is opposite to $\boldsymbol{\theta}_u$ in Eq. (12-2.11).

▲

## 12-3. Dynamic Effects of Gravikinetic Fields; Gravitational Induction

We shall now present several examples demonstrating force effects of the gravikinetic field. For simplicity we shall use gravikinetic fields calculated in the preceding section.

The force effects that we shall show constitute the gravitational analogue of electromagnetic induction and of electromagnetic Lenz's law. As we now know, electromagnetic induction is caused by the electrokinetic field.[5] The gravikinetic field is the gravitational counterpart of the electrokinetic field, and their dynamic effects are similar, except that the gravikinetic force exerted on a mass by an increasing/decreasing gravikinetic field is parallel/antiparallel to the field, whereas the electrokinetic force exerted on a positive charge by an increasing/decreasing electrokinetic field is antiparallel/parallel to the field.

▼

**Example 12-3.1** The cylinder of Example 12-2.2 is initially at rest. A ring of mass $m_r$ and radius $R$ is placed around the cylinder

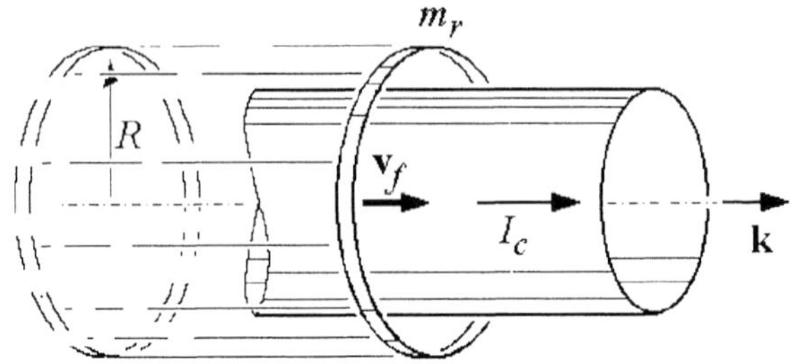


*Fig. 12.5 Accelerating cylinder drags the ring with itself.*

coaxially with it. The cylinder is then suddenly set in motion along its axis and attains a velocity $v_c$ (mass current $I_c$). The gravikinetic force causes the ring to move along (follow) the cylinder (Fig. 12.5). Assuming that no other forces act on the ring, and assuming that the ring stays near the middle of the cylinder during the time that the velocity of the cylinder changes, find the final velocity $\mathbf{v}_f$ of the ring.

According to our assumptions, the gravikinetic field through which the ring moves is a function of time only. Therefore we can use Eq. (12-1.5) for finding the final momentum and velocity of the ring. From Eqs. (12-1.5) and (12-2.3) (see Example 12-2.2), we have

$$\Delta\mathbf{G}_M = m_r\mathbf{v}_f = m_r I_c \frac{2G}{c^2}\ln\frac{2L}{R}\mathbf{k}, \tag{12-3.1}$$

so that the final velocity of the ring is

$$\mathbf{v}_f = \frac{2G}{c^2} I_c \ln\frac{2L}{R}\mathbf{k}. \tag{12-3.2}$$

Substituting $\rho v_c 2\pi R_0 t$ for $I_c$, where $\rho$ is the density of the cylinder, $R_0$ is its radius, and $t$ is its thickness, we obtain

$$\mathbf{v}_f = \frac{4G\rho v_c \pi R_0 t}{c^2}\ln\frac{2L}{R}\mathbf{k}. \tag{12-3.3}$$

The cylinder *drags* the ring so that the ring moves in the direction of the moving cylinder. It is interesting to note that the final velocity of the ring does not depend on its mass.

**Example 12-3.2** The cylinder of Example 12-2.3 is initially at rest. A small thin-walled cylinder of mass $m$ and radius $r$ is placed inside it and coaxially with it (Fig. 12.6). The larger cylinder is then suddenly set in rotation about its symmetry axis and attains a final angular velocity $\omega_0$ (mass current $I_0$). The gravikinetic force causes the small cylinder to rotate. Find the final angular velocity $\omega_f$ of the small cylinder.

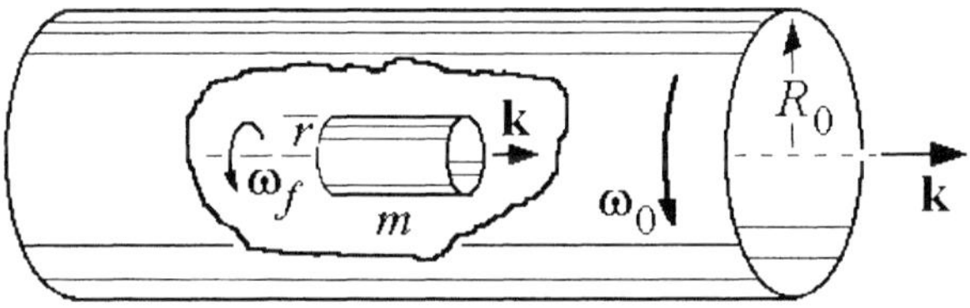


*Fig. 12.6 A small cylinder placed inside a larger cylinder rotates when the larger cylinder is set in rotation. Both cylinders rotate then in the same direction.*

Since the gravikinetic field causing the cylinder to rotate is a function of time only, we can use Eq. (12-1.5) for finding the angular velocity of the cylinder. From Eqs. (12-1.5) and (12-2.6) we have

$$\Delta \mathbf{G}_M = m\mathbf{v}_f = m\boldsymbol{\omega}_f \times \mathbf{r} = m\frac{2\pi G}{c^2}\frac{I_0 r}{L}\boldsymbol{\theta}_u, \qquad (12\text{-}3.4)$$

and since $\mathbf{r}$ is perpendicular to $\boldsymbol{\omega}_f$,

$$\boldsymbol{\omega}_f = \frac{2\pi G I_0}{c^2 L}\mathbf{k}, \qquad (12\text{-}3.5)$$

where **k** is a unit vector right-handed relative to the direction of rotation of the large cylinder. Substituting $I_0 = \rho\omega_0 R_0 Lt$, where $\rho$, $\omega_0$, $R_0$, $L$, and $t$ are the density, final angular velocity, radius, length, and thickness of the large cylinder, respectively, (see Example 12-2.3), we obtain

$$\omega_f = \frac{2\pi G\rho\omega_0 R_0 t}{c^2}\mathbf{k}. \tag{12-3.6}$$

Both cylinders rotate in the same direction. It is interesting to note that the angular velocity of the small cylinder does not depend on its mass or radius.

**Example 12-3.3** The cylinder of Example 12-2.3 initially rotates with angular velocity $\omega_0$ (mass current $I_0$) about its symmetry axis. A disk of mass $m$ and radius $a$ is placed inside the cylinder coaxially with it (Fig. 12.7). The rotation of the cylinder is then suddenly stopped. The gravikinetic force causes the disk to rotate. Find the final angular velocity $\omega_f$ of the disk.

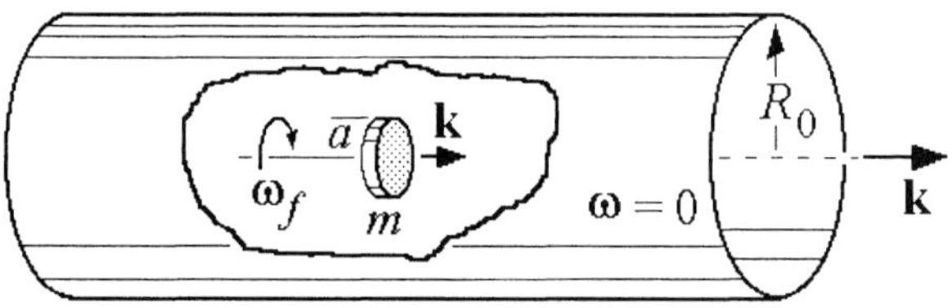


*Fig. 12.7 A small disk is placed inside a rotating cylinder. When the cylinder is stopped, the disk rotates in the direction opposite to that of the cylinder .*

The disk acquires an angular momentum $\Delta\mathbf{L}$ that can be found from Eq. (12-1.6) and Eq. (12-2.6). Let us divide the disk into elementary rings. Consider an elementary ring of radius $r$ and width

$dr$. The mass of the ring is $dm = (m/\pi a^2)2\pi r dr$. Substituting $dm$ and $-\mathbf{g}_k$ (negative sign is needed because the cylinder decelerates) from Eq. (12-2.6) in Eq. (12-1.6), integrating over $dt$, and taking into account that $\mathbf{r}$ is perpendicular to $\mathbf{g}_k$, we have

$$\Delta\mathbf{L} = -\int_0^a r\left(\frac{2G\pi I_0 r}{c^2 L}\right)\frac{m}{\pi a^2}2\pi r\,dr\mathbf{k} = -\frac{4\pi G I_0 m}{c^2 L a^2}\int_0^a r^3\,dr\mathbf{k} \qquad (12\text{-}3.7)$$

or

$$\Delta\mathbf{L} = -G\frac{\pi m a^2 I_0}{c^2 L}\mathbf{k}, \qquad (12\text{-}3.8)$$

where $\mathbf{k}$ is a unit vector right-handed relative to the direction of the rotation of the cylinder. Since the moment of inertia of the disk is $ma^2/2$, we obtain for the angular velocity

$$\omega_f = -G\frac{2\pi I_0}{c^2 L}\mathbf{k}. \qquad (12\text{-}3.9)$$

Substituting $I_0 = \rho\omega_0 R_0 L t$, where $\rho$, $\omega_0$, $R_0$, $L$, and $t$ are the density, initial angular velocity, radius, length, and thickness of the large cylinder, respectively, (see Example 12-2.3), we obtain

$$\omega_f = -G\frac{2\pi\rho\omega_0 R_0 t}{c^2}\mathbf{k}. \qquad (12\text{-}3.10)$$

The rotation of the disk is opposite to the rotation of the cylinder. Note that the angular velocity of the disk does not depend on its mass or radius.

**Example 12-3.4** The masses of the rings described in Examples 12-2.4 and 12-2.5 are $m_b$ and $m_a$, and their radii are such that $b \gg a$. The rings are placed in the same plane, and their centers coincide (Fig. 12.8). (a) Ring $b$ is given an angular acceleration $\alpha_b$. Find the angular acceleration of ring $a$ due to the gravikinetic field of ring $b$. (b) Ring $a$ is given an angular acceleration $\alpha_a$. Find the angular acceleration of ring $b$ due to the gravikinetic field of ring $a$.

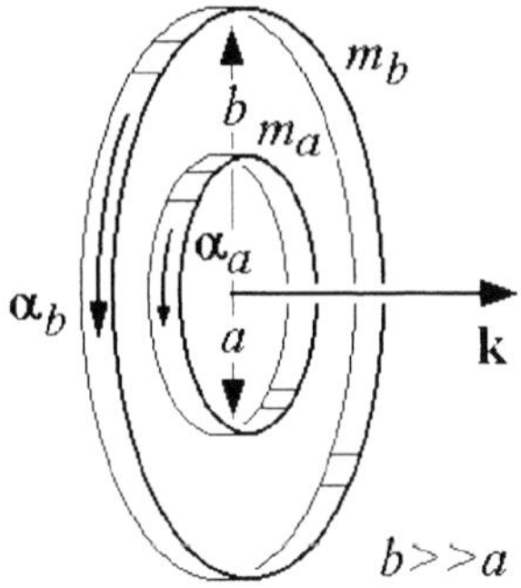


*Fig. 12.8 When one of the two mass rings is rotated, the other ring starts to rotate in the same direction.*

(a) By the definition of the gravikinetic force and by Eq. (12-2.9), the torque experienced by ring $a$ due to the gravikinetic field of ring $b$ is

$$\mathbf{T}_a = \mathbf{r} \times m_a \mathbf{g}_k = a m_a \frac{G}{c^2} \alpha_b \frac{m_b a}{2b} \mathbf{k}, \tag{12-3.11}$$

where $\mathbf{k}$ is a unit vector along the axis of the rings right-handed relative to the rotation of ring $b$. Since the moment of inertia of ring $a$ is $m_a a^2$, its angular acceleration is

$$\alpha_a = G \frac{m_b}{2bc^2} \alpha_b . \tag{12-3.12}$$

Both rings rotate in the same direction.

(b) Using Eq. (12-2.11), we find, as above in Part (a), that the angular acceleration of ring $b$ due to the gravikinetic field of ring $a$ is

$$\alpha_b = G \frac{m_a a^2}{2b^3 c^2} \alpha_a . \tag{12-3.13}$$

Once again the two rings rotate in the same direction.[6]

▲

Examples 12-3.1-12-3.4 illustrate the phenomenon of gravitational induction, whereby a changing mass current induces a secondary mass current in the neighboring bodies. The effect is similar to electromagnetic induction,[7] except that, in contrast to the latter, the direction of the induced current is the same as that of the original current if the original current increases, and is opposite to the original current if the original current decreases. Thus the sign of the "gravitational Lenz's law" is opposite to that of the electromagnetic Lenz's law.

## References and Remarks for Chapter 12

**1.** See Oleg D. Jefimenko, *Electricity and Magnetism*, 2nd ed., (Electret Scientific, Star City, 1989) p. 366.
**2.** Ref. 1, p. 367.
**3.** Ref. 1, p. 383, Problem 11.2.
**4.** Compare Ref. 1, pp. 346-347.
**5.** See Oleg. D. Jefimenko, *Causality, Electromagnetic Induction, and Gravitation*, 2nd ed., (Electret Scientific, Star City, 2000) pp. 19-66.
**6.** Similar rotation should occur when two coaxial disks are placed close to each other and one of them changes its rotational motion.
**7.** For a detailed analyses and novel interpretation of the phenomenon of electromagnetic induction see Ref. 5 and Oleg D. Jefimenko, "Presenting electromagnetic theory in accordance with the principle of causality," Eur. J. Phys. **25**, 287-296 (2004).

# II

# APPLICATIONS AND PREDICTIONS OF THE GENERALIZED THEORY OF GRAVITATION

# 13

# SOME ELEMENTARY APPLICATIONS OF THE GENERALIZED THEORY OF GRAVITATION

We shall now present several elementary illustrative examples demonstrating the use of gravitational and cogravitational equations and methods introduced in Chapters 2 to 12. As far as static gravitational systems are concerned, the generalized theory of gravitation does not add substantially to the Newtonian theory, but as we shall now see, it provides a much larger variety of methods for calculation of static gravitational systems than the Newtonian theory. The greatest impact of the generalized theory of gravitation is, however, in the domain of moving and time-dependent gravitational systems, where, as the examples presented here will show, the generalized theory of gravitation yields entirely new results not at all foreseen in the Newtonian theory.

## 13-1. Illustrative Examples on Static Gravitational Fields

As has been shown in Chapter 7, many readily available solutions of electrostatic problems can be converted to solutions of the corresponding gravitational problems by merely replacing electric quantities and constants by the corresponding gravitational quantities and constants according to Table 7-1. Several examples presented in this section make use of such conversions. The electrostatic equations used for the conversion are taken from the author's book *Electricity and Magnetism*[1]. Some readers may want to examine these equations and their derivations. For this purpose each gravitational equation appearing below and obtained by conversion is provided with the number of the page where the corresponding electrostatic equation appears in *Electricity and Magnetism* (hereafter abbreviated as EM).

▼

**Example 13-1.1** The electric field on the axis of a thin, uniformly charged disk of radius $a$ and charge $q$ at a distance $z$ from the center of the disk is

$$\mathbf{E} = \frac{q}{2\pi\varepsilon_0 a^2}\left[1 - \frac{z}{(a^2 + z^2)^{1/2}}\right]\mathbf{k}, \qquad (13\text{-}1.1)$$

where $\mathbf{k}$ is a unit vector along the axis of the disk pointing away from the disk (EM100). Using the analogy between electric and gravitational equations, find the gravitational field on the axis of a similar disk of mass $m$ (Fig. 13.1).

Replacing in Eq. (13-1.1) $\mathbf{E}$ by $\mathbf{g}$, $q$ by $m$, $\varepsilon_0$ by $-1/4\pi G$, we obtain for the gravitational field of a disk of radius $a$ and mass $m$

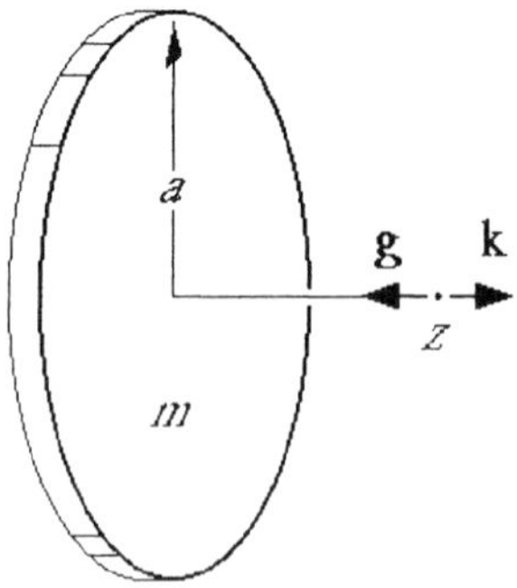


*Fig. 13.1 Calculation of the gravitational field on the axis of a disk.*

$$\mathbf{g} = -G\frac{2m}{a^2}\left[1 - \frac{z}{(a^2 + z^2)^{1/2}}\right]\mathbf{k}. \qquad (13\text{-}1.2)$$

Observe that, except for notation, Eq. (13-1.2) is the same as Eq. (3-2.16) for a disk rotating with constant angular acceleration about its symmetry axis. This may be erroneously interpreted as indicating that the axial gravitation field of a stationary disk, as well as the axial gravitation field of a rotating and accelerating disk, propagate instantaneously (with infinite speed). In reality, however, neither field propagates instantaneously: Eq. (13-1.2) and Eq. (3-2.16) hold only for the time of observation subsequent to the moment when the field signal has reached the point of observation. Note also that although the disk in this example is referred to as "stationary," it had to be created and placed in position before becoming "stationary." Prior to that time it was in the state of "being created," and thus in the state of motion, and its field propagated (and continues to propagate) with the usual speed $c$.

**Example 13-1.2** The electric force between a uniformly charged ring of charge $q'$ and radius $a$ and a thin, uniformly charged rod of charge $q$ and length $2d$ lying along the axis of the ring is

$$\mathbf{F} = \frac{qq'}{8\pi\varepsilon_0 d}\left\{\frac{1}{[a^2 + (z_0 - d)^2]^{1/2}} - \frac{1}{[a^2 + (z_0 + d)^2]^{1/2}}\right\}\mathbf{k}, \qquad (13\text{-}1.3)$$

where $z_0$ is the distance from the center of the ring to the center of the rod, and $\mathbf{k}$ is a unit vector along the axis of the ring pointing away from the ring (EM209-210, EM211-212). Find the gravitational force between a similar ring of mass $m'$ and a rod of mass $m$ (Fig. 13.2).

Substituting in Eq. (13-3.3) $m$ for $q$, $m'$ for $q'$, and $-1/4\pi G$ for $\varepsilon_0$, we obtain for the gravitational force between the ring and the rod

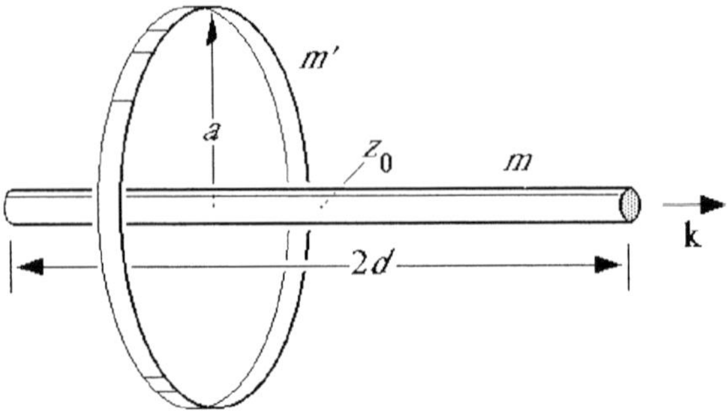


*Fig. 13.2 Calculation of the gravitational force acting on a rod inside a ring.*

$$\mathbf{F} = -G\frac{mm'}{2d}\left\{\frac{1}{[a^2 + (z_0 - d)^2]^{1/2}} - \frac{1}{[a^2 + (z_0 + d)^2]^{1/2}}\right\}\mathbf{k}. \quad (13\text{-}1.4)$$

**Example 13-1.3** The electrostatic potential inside and outside a spherical charge of uniform density $\rho$ and radius $a$ is (EM115)

$$\varphi_{\text{inside}} = \frac{q}{8\pi\varepsilon_0 a^3}(3a^2 - r^2), \qquad \varphi_{\text{outside}} = \frac{q}{4\pi\varepsilon_0 r}. \quad (13\text{-}1.5)$$

Find the gravitational potential inside and outside a similar spherical mass.

Replacing in Eq. (13-1.5) $\varepsilon_0$ by $-1/4\pi G$ and $q$ by $m$, we obtain for the potentials of a spherical mass of radius $a$

$$\varphi_{\text{inside}} = -G\frac{m}{2a^3}(3a^2 - r^2), \qquad \varphi_{\text{outside}} = -G\frac{m}{r}. \quad (13\text{-}1.6)$$

**Example 13-1.4** The electrostatic energy of a uniformly charged spherical shell of charge $q$ and radius $a$ is (EM190)

$$U = \frac{q^2}{8\pi\varepsilon_0 a}. \tag{13-1.7}$$

Find the gravitational energy of a similar shell of mass $m$.

Replacing $q$ by $m$ and $\varepsilon_0$ by $-1/4\pi G$, we obtain for the gravitational energy of a spherical shell of radius $a$ and mass $m$

$$U = -G\frac{m^2}{2a}. \tag{13-1.8}$$

**Example 13-1.5** Consider a stationary cylinder of uniform mass density $\rho$, length $2t$, and radius $a$. The axis of the cylinder is also the $z$-axis of cylindrical coordinates whose origin is at the center of the cylinder. A spherical cavity is made around an internal axial point of the cylinder at a distance $z = d$ from the center of the cylinder. A particle of mass $m$ is placed at the center of the cavity (Fig. 13.3). Find the gravitational force exerted by the cylinder on the particle.

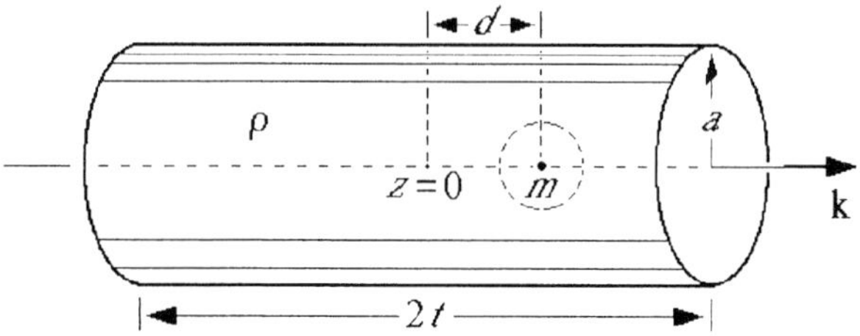


*Fig. 13.3 Calculation of the gravitational force on a point mass placed in a spherical cavity inside a cylinder.*

This problem is best solved by using Eq. (7-3.11) [since the cylinder is stationary, Eq. (7-3.11) is the same as Eq. (3-2.2)], which makes it possible to find the gravitational field of a mass by integrating over the boundary surfaces of the mass. We see by inspection that the surface of the cavity makes no contribution to the field at its center [a spherical surface produces only a radial field, all components of which meet at the center and cancel each other (compare EM105)]. Likewise, the curved surface of the cylinder makes no contribution to the field. Only the two flat surfaces of the cylinder make a contribution.

The contribution of the closest flat surface ($z>0$) of the cylinder to the field at the center of the cavity is, by Eq. (7-3.11),

$$\begin{aligned} \mathbf{g}_1 &= -G\rho\int\frac{d\mathbf{S}'}{r} = -G\rho\mathbf{k}\int_0^a\frac{2\pi R\,dR}{[R^2+(t-d)^2]^{1/2}} \\ &= -2\pi G\rho\{[a^2+(t-d)^2]^{1/2}-(t-d)\}\mathbf{k}. \end{aligned} \tag{13-1.9}$$

The contribution of the other flat surface ($z<0$) is, similarly,

$$\begin{aligned} \mathbf{g}_2 &= -G\rho\int\frac{d\mathbf{S}'}{r} = G\rho\mathbf{k}\int_0^a\frac{2\pi R\,dR}{[R^2+(t+d)^2]^{1/2}} \\ &= 2\pi G\rho\{[a^2+(t+d)^2]^{1/2}-(t+d)\}\mathbf{k}. \end{aligned} \tag{13-1.10}$$

The field at the center of the cavity is then $\mathbf{g}_1 + \mathbf{g}_2$, or

$$\mathbf{g} = 2\pi G\rho\{[a^2+(t+d)^2]^{1/2} - [a^2+(t-d)^2]^{1/2} - 2d\}\mathbf{k}. \tag{13-1.11}$$

The force on the particle of mass $m$ at the center of the cavity is therefore, by Eq. (1-1.5) or by the general force equation, Eq. (2-2.6),

$$\mathbf{F} = 2\pi Gm\rho\{[a^2+(t+d)^2]^{1/2} - [a^2+(t-d)^2]^{1/2} - 2d\}\mathbf{k}. \tag{13-1.12}$$

**Example 13-1.6** An irregular cavity has formed inside a liquid of density $\rho$ in the region where the gravitational potential is $\varphi'$. Find the buoyant force on the cavity (Fig. 13.4).

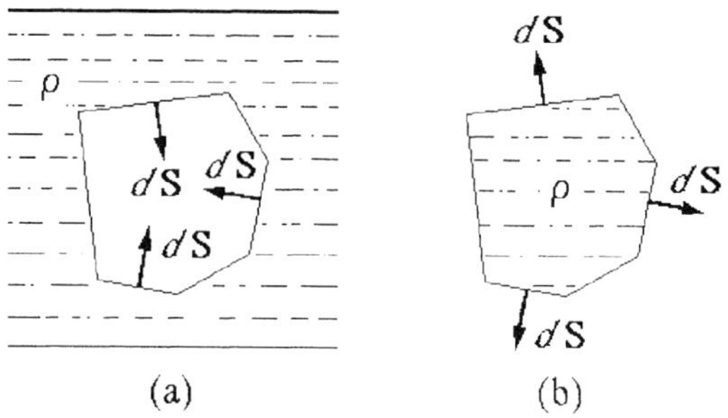


*Fig. 13.4 Calculation of the buoyant force on a cavity formed in a liquid. (a) Surface element vectors of the cavity. (b) Surface element vectors of the liquid initially contained in the cavity.*

This problem on Archimedes's principle is usually solved by means of a plausibility argument based on the consideration of the pressure inside the liquid. Here we shall provide a rigorous solution of the problem by means of Eq. (10-1.6) [or (7-3.28)]. This equation allows one to find the gravitational force on a volume bounded by a given surface. The surface element vector $d\mathbf{S}$ in this equation is directed from the mass under consideration into the empty space, regardless of whether the mass is inside or outside the bounding surface.[2] Therefore the force on the cavity and the force on the liquid initially contained in the cavity are exactly the same in magnitude, but opposite in direction. Hence, the buoyant force is equal to the weight of the liquid initially contained in the cavity.

**Example 13-1.7** A gravitational "parallel-plate capacitor" consists of two large circular plates of radius $a$ having a uniformly distributed mass $m$ (Fig. 13.5). One of the plates is in the $y,z$ plane of rectangular coordinates with its center at the origin. The second plate is at a small distance $x=d$ from the first. Using five different methods, and neglecting edge effects, find the gravitational force between the plates.[3]

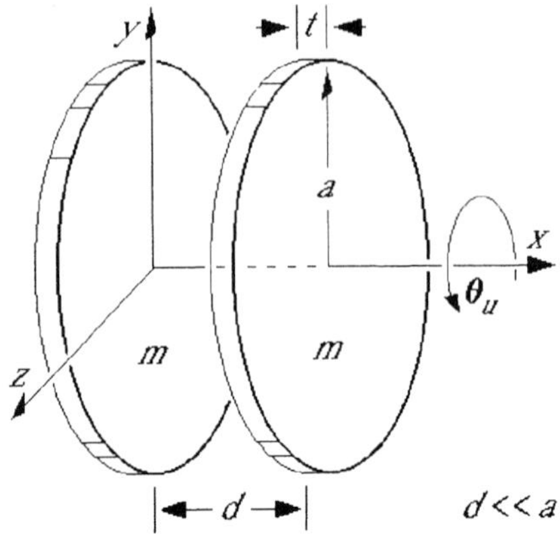


*Fig. 13.5 Calculation of the gravitational attraction between two circular plates.*

(a) *Direct calculation.* By Gauss's law, Eq. (7-3.5), and by the symmetry of the system, the gravitational field produced at $x>0$ by the first plate is

$$\mathbf{g}' = -2G\frac{m}{a^2}\mathbf{i}. \tag{13-1.13}$$

By Eq. (1-1.8) or (2-2.6), the force on the second plate is then

$$\mathbf{F} = m\mathbf{g}' = -2G\frac{m^2}{a^2}\mathbf{i}. \tag{13-1.14}$$

(b) *Force in terms of scalar potential.* The potential produced by the first plate at a distance $x$ from the origin is, by Eqs. (7-3.15) and (13-1.13)

$$\begin{aligned}\varphi' &= \int_x^0 \mathbf{g}'\cdot d\mathbf{l} + \varphi_0 = \int_x^0\left(-2G\frac{m}{a^2}\right)dx + \varphi_0 \\ &= 2G\frac{m}{a^2}x + \varphi_0,\end{aligned} \tag{13-1.15}$$

where $\varphi_0$ is a reference potential at $x=0$. Let us assume that the thickness of the second plate is $t$. The potential at the front surface and at the back surface of the plate is then

$$\varphi'_{front} = 2G\frac{m}{a^2}d + \varphi_0, \quad \varphi'_{back} = 2G\frac{m}{a^2}(d+t) + \varphi_0. \tag{13-1.16}$$

By Eq. (10-1.6) or (7-3.28), the force on the second plate is therefore [note that, by symmetry, the rim of the plate adds nothing to Eq. (10-1.6)]

$$\begin{aligned} \mathbf{F} = &- \frac{m}{\pi a^2 t}\left(2G\frac{m}{a^2}d + \varphi_0\right)(-\pi a^2\mathbf{i}) \\ &- \frac{m}{\pi a^2 t}\left[2G\frac{m}{a^2}(d+t) + \varphi_0\right]\pi a^2\mathbf{i}, \end{aligned} \tag{13-1.17}$$

or

$$\mathbf{F} = -2G\frac{m^2}{a^2}\mathbf{i}. \tag{13-1.18}$$

(c) *Force in terms of gravitational vector potential.* For $x>0$, the vector potential of the gravitational field of the first plate (homogeneous field) is [4]

$$\mathbf{A}_g = -G\frac{mr}{a^2}\boldsymbol{\theta}_u, \tag{13-1.19}$$

where $r$ is a perpendicular distance from the $x$-axis, and $\boldsymbol{\theta}_u$ is a right-handed circular unit vector around the $x$-axis [the easiest way to obtain Eq. (13-1.19) is to use Eq. (13-1.13) together with Eq. (3-3.9)]. By Eq. (10-2.5) [or (7-3.29)], the force on the second plate is then

$$\mathbf{F} = -\rho\oint\left(-G\frac{mr}{a^2}\boldsymbol{\theta}_u\right)\times d\mathbf{S}. \tag{13-1.20}$$

The surface of integration in Eq. (13-1.20) consists of the two flat surfaces and the circular rim of the second plate. By symmetry, the contributions of the two flat surfaces to the integral of Eq. (13-1.20) cancel. The only non-vanishing contribution to the integral comes from the rim of the plate. If the thickness of the plate is $t$, the surface element vector of the rim is $d\mathbf{S}=t\,d\mathbf{l}_{out}$ where $d\mathbf{l}_{out}$ is a vector representing a length element of the rim and directed radially outward from the rim. The force on the second plate is therefore [see Eq. 10-2.6]

$$\mathbf{F} = \frac{m}{\pi a^2 t}\oint G\frac{ma}{a^2}\boldsymbol{\theta}_u \times t\,d\mathbf{l}_{out}. \tag{13-1.21}$$

Simplifying, we obtain

$$\mathbf{F} = -2G\frac{m^2}{a^2}\mathbf{i}. \tag{13-1.22}$$

(d) *Force in terms of Maxwell's stress integral* (*compare EM 215-218*). The total gravitational field in the space between the two plates is zero, because there the fields of the plates have opposite directions. The total gravitational field outside the plates, to the right of the second plate, is double the field of each single plate given by Eq. (13-1.13), because there the two fields are in the same direction. We thus have

$$\mathbf{g}_{between} = 0, \qquad \mathbf{g}_{outside} = -4G\frac{m}{a^2}\mathbf{i}. \tag{13-1.23}$$

Applying Eq. (7-3.30) to a Maxwellian surface enclosing the second plate, we then obtain

$$\begin{aligned}\mathbf{F} &= \frac{1}{8\pi G}\left[0\cdot\pi a^2(-\mathbf{i}) + \left(-4G\frac{m}{\pi a^2}\right)^2\cdot\pi a^2\mathbf{i}\right]\\ &\quad - \frac{1}{4\pi G}\left[0\cdot(0\cdot\pi a^2)(-\mathbf{i}) + \left(-4G\frac{m}{\pi a^2}\right)\cdot\left(-4G\frac{m}{\pi a^2}\right)\cdot\pi a^2\mathbf{i}\right]\\ &= -2G\frac{m^2}{a^2}\mathbf{i}.\end{aligned} \tag{13-1.24}$$

(e) *Force in terms of energy*. The total field in the space between the plates and in the space outside the plates is given by

Eq. (13-1.23). According to Eq. (2-2.7) or Eq. (7-3.37), the gravitational energy density in the system is then

$$U_{v\ between} = 0,$$

$$U_{v\ outside} = -\frac{1}{8\pi G}\left(-4G\frac{m}{a^2}\right)^2 = -2G\frac{m^2}{\pi a^4}. \qquad (13\text{-}1.25)$$

Suppose now that the second plate moves through a distance $dx$. The relation between the force on the plate and the energy change associated with the displacement of the plate is

$$\mathbf{F} = -\frac{\partial U}{\partial x}\mathbf{i} = -\frac{dU}{dx}\mathbf{i}. \qquad (13\text{-}1.26)$$

The energy change associated with the displacement $dx$ is

$$dU = -\left(-2G\frac{m^2}{\pi a^4}\right)\pi a^2 dx. \qquad (13\text{-}1.27)$$

(The minus sign in front of parenthesis reflects the fact that the energy in the space between the plates is zero.) Thus the force on the second plate is

$$\mathbf{F} = -2G\frac{m^2}{a^2}\mathbf{i}. \qquad (13\text{-}1.28)$$

▲

## 13-2. Illustrative Examples on Dynamic Gravitational Fields

We shall now present illustrative examples involving nonstatic gravitational fields. These examples will depict several remarkable gravitational phenomena not revealed by the Newtonian gravitational theory. As in the preceding section, we shall start with simple conversion of electromagnetic equations to gravitational equations.

▼

**Example 13-2.1** A long beam of charged particles moves with velocity $v$ along its length. The charge density in the beam is $\rho$, the radius of the beam is $a$. The beam creates a magnetic field which, inside and outside the beam, is (EM332)

$$\mathbf{H}_{inside} = \rho\frac{\mathbf{v}\times\mathbf{r}}{2}, \qquad \mathbf{H}_{outside} = \rho a^2\frac{\mathbf{v}\times\mathbf{r}}{2r^2}, \tag{13-2.1}$$

where $\mathbf{r}$ is a radius vector directed from the axis of the beam to the point of observation. Find the cogravitational field of a similar beam of mass particles moving with velocity $v$ along its length.

Since the above expressions are for $\mathbf{H}$ rather than for $\mathbf{B}$, we must convert them to $\mathbf{B}$ by using $\mathbf{B} = \mu_0\mathbf{H}$. Then replacing $\mathbf{B}$ by $\mathbf{K}$ and $\mu_0$ by $-4\pi G/c^2$, we obtain for the cogravitational field of a beam of mass particles of density $\rho$, radius $a$, and velocity $v$

$$\mathbf{K}_{inside} = -G\frac{2\pi\rho}{c^2}\mathbf{v}\times\mathbf{r}, \qquad \mathbf{K}_{outside} = -G\frac{2\pi\rho a^2}{c^2 r^2}\mathbf{v}\times\mathbf{r}. \tag{13-2.2}$$

**Example 13-2.2** Consider a single particle of mass $m$ on the surface of the beam of mass particles described in Example 13-2.1. Find the expression for the total force (gravitational and cogravitational) acting on the particle.

Constructing a cylindrical Gaussian surface around the beam and applying Eq. (7-1.5) to this surface, we obtain for the gravitational field at the surface of the beam (compare EM89-90, EM420)

$$\mathbf{g} = -2\pi G\rho a\mathbf{r}_u, \tag{13-2.3}$$

where $\mathbf{r}_u$ is a unit vector pointing away from the axis of the beam at right angles to it. The cogravitational field at the surface of the beam is, by Eq. (13-2.2),

$$\mathbf{K} = -G\frac{2\pi\rho a}{c^2}\mathbf{v}\times\mathbf{r}_u. \tag{13-2.4}$$

The force on the mass particle (mass $m$) on the surface of the beam is then, by Eq. (2-2.6),

$$\mathbf{F} = m(\mathbf{g} + \mathbf{v}\times\mathbf{K}), \tag{13-2.5}$$

or, after substituting $\mathbf{K}$, expanding the cross product, and simplifying,

$$\mathbf{F} = -2\pi G\rho ma\left(1 - \frac{v^2}{c^2}\right)\mathbf{r}_u. \tag{13-2.6}$$

Thus the particle is always attracted to the beam, although the force of attraction is smaller than for a stationary cylinder of the same mass density and radius. The gravitational attraction always dominates over the cogravitational repulsion, so that the beam compresses as it moves. When the speed of the beam approaches $c$ (the speed of the propagation of gravitation), the force on the particle approaches zero.

**Example 13-2.3** A spherical shell of radius $R$ and uniform surface charge density $\sigma$ rotates with angular velocity $\omega$ about a diameter which is also the polar axis of spherical coordinates whose origin is at the center of the shell. The shell creates a magnetic field in the space inside and outside the shell given by (EM378)

$$\mathbf{H}_{inside} = \frac{2}{3}\sigma\omega R\mathbf{k},$$
$$\mathbf{H}_{outside} = \frac{2\sigma\omega R^4}{3r^3}\cos\theta\,\mathbf{r}_u + \frac{\sigma\omega R^4}{3r^3}\sin\theta\,\boldsymbol{\theta}_u. \tag{13-2.7}$$

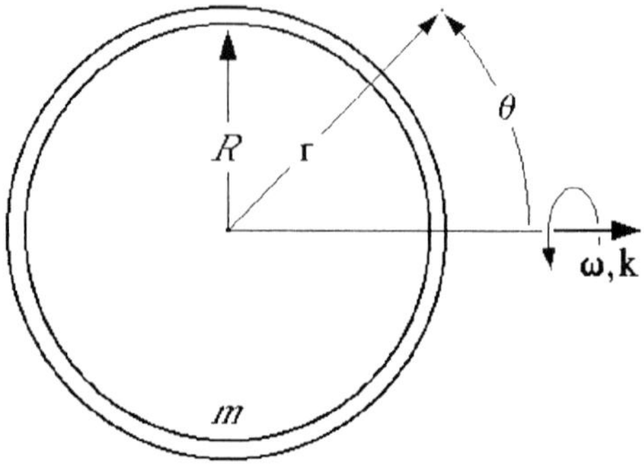


*Fig. 13.6 Calculation of the cogravitational field of a rotating spherical shell.*

Find the cogravitational field of a similar shell of uniformly distributed mass $m$ (Fig. 13.6).

Since the two equations are for the magnetic field **H** rather than for the flux density field **B**, we must first convert them to **B**, by using $\mathbf{B} = \mu_0\mathbf{H}$. Replacing then **B** by **K**, $\mu_0$ by $-4\pi G/c^2$, and $\sigma$ by $m/4\pi R^2$, we obtain for the cogravitational field of the shell of radius $R$ and mass $m$ rotating with angular velocity $\omega$

$$\mathbf{K}_{inside} = -G\frac{2m\omega}{3c^2R}\mathbf{k},$$
$$\mathbf{K}_{outside} = -G\frac{2m\omega R^2}{3c^2r^3}\cos\theta\,\mathbf{r}_u - G\frac{m\omega R^2}{3c^2r^3}\sin\theta\,\boldsymbol{\theta}_u. \tag{13-2.8}$$

The rotating spherical shell constitutes a "cogravitational dipole" (see Section 15-2) whose dipole moment is (compare EM381)

$$\mathbf{d} = -G\frac{4\pi mR^2}{3c^2}\omega. \tag{13-2.9}$$

Observe that $\omega$ in this example is right-handed relative to the $z$ axis.

As will be explained in Section 15-2, it is convenient to express the cogravitational dipole moment and the dipole field of a rotating body in terms of the angular momentum **L** of the body. The angular momentum of a spherical shell rotating about a diameter is $\mathbf{L} = (2mR^2/3)\boldsymbol{\omega}$, and therefore we have

$$\mathbf{d} = -G\frac{2\pi}{c^2}\mathbf{L} \qquad (13\text{-}2.10)$$

and

$$\mathbf{K}_{outside} = -G\frac{L}{c^2r^3}\cos\theta\,\mathbf{r}_u - G\frac{L}{2c^2r^3}\sin\theta\,\boldsymbol{\theta}_u. \qquad (13\text{-}2.11)$$

**Example 13-2.4** Show that a cogravitational dipole uniformly moving with velocity **v** appears to acquire a gravitational dipole moment

$$\mathbf{p}_{apparent} = -\frac{\mathbf{v}\times\mathbf{d}}{4\pi G}, \qquad (13\text{-}2.12)$$

where **d** is the dipole moment of the cogravitational dipole.

Consider a square frame of zero mass and length $L$ on a side. The frame supports a string of uniformly distributed particles of total mass $4m$ sliding with velocity $u \ll c$ along the sides of the frame (Fig. 13.7). Let the frame be stationary and let it be located in the $xy$-plane of rectangular coordinates with its center at the origin. Let the motion of the particles be as shown in Fig. 13.7.

The gravitational field given by the first integral of Eq. (2-2.1) can be expressed in terms of the retarded gravitational scalar potential as

$$\mathbf{g} = -G\int\left\{\frac{[\rho]}{r^2} + \frac{1}{rc}\frac{\partial[\rho]}{\partial\rho}\right\}\mathbf{r}_u\,dv' = -\nabla\varphi. \qquad (13\text{-}2.13)$$

The present-time form of the retarded gravitational scalar potential for a point mass moving with constant velocity $v$ is, by Eq. (5-6.5),

$$\varphi = -G\frac{m}{r[1-(v^2/c^2)\sin^2\theta]^{1/2}}. \qquad (5\text{-}6.5)$$

If $v \ll c$, this potential can be expressed as

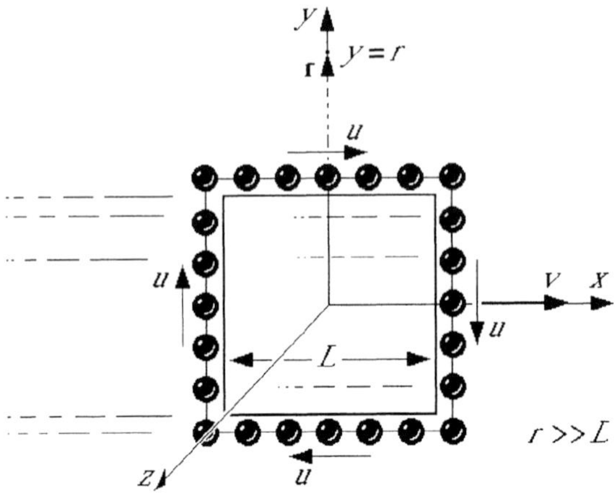


*Fig. 13.7 When a closed mass-current drifts as a whole, it appears to generate a positive as well as a negative mass. The positive (ordinary) mass appears to be generated by the particles whose resulting velocity is greater than the drift velocity v. The negative mass appears to be generated by the particles whose resulting velocity is smaller than v.*

$$\varphi = -G\frac{m}{r}\left(1 + \frac{v^2}{2c^2}\sin^2\theta\right). \tag{13-2.14}$$

Let us find the potential that the sliding particles produce at a point of the $y$ axis at a distance $r \gg L$ from the frame. Applying Eq. (13-2.14) to the particles on the horizontal sides of the frame ($\theta = \pi/2$) (since $r \gg L$, we can treat these particles as a single point mass), we have

$$\varphi_{horizontal} = -2G\frac{m}{r}\left(1 + \frac{u^2}{2c^2}\right). \tag{13-2.15}$$

Applying Eq.(13-2.14) to the particles on the vertical sides of the frame ($\theta = 0$ or $\pi$), we similarly have

$$\varphi_{\text{vertical}} = -2G\frac{m}{r}. \tag{13-2.16}$$

The total potential is therefore

$$\varphi = -4G\frac{m}{r} - G\frac{mu^2}{rc^2}. \tag{13-2.17}$$

Thus the particles appear to have acquired a mass

$$m_{apparent} = \frac{mu^2}{c^2} \tag{13-2.18}$$

as a result of their motion along the frame.

Let us now assume that the frame is moving with velocity $v$ in the positive $x$ direction. The potential due to the particles on the upper horizontal side of the frame ($y > 0$) is now

$$\begin{aligned}\varphi_{upper} &= -G\frac{m}{r}\left\{1 + \frac{(v+u)^2}{2c^2}\right\} \\ &= -G\frac{m}{r}\left\{1 + \frac{v^2}{2c^2} + \frac{uv}{c^2} + \frac{u^2}{2c^2}\right\},\end{aligned} \tag{13-2.19}$$

and the potential due to the particles on the lower horizontal side ($y < 0$) is

$$\begin{aligned}\varphi_{\text{lower}} &= -G\frac{m}{r}\left\{1 + \frac{(v-u)^2}{2c^2}\right\} \\ &= -G\frac{m}{r}\left\{1 + \frac{v^2}{2c^2} - \frac{uv}{c^2} + \frac{u^2}{2c^2}\right\}.\end{aligned} \tag{13-2.20}$$

To find the potential due to the particles on the vertical sides of the frame, we must take into account that the velocity of the particles on these sides is $(v^2+u^2)^{1/2}$, and that $\sin\theta$ for these particles is now $v/(v^2+u^2)^{1/2}$. The potential due to these particles is therefore

$$\begin{aligned}\varphi_{vertical} &= -2G\frac{m}{r}\left(1 + \frac{v^2+u^2}{2c^2}\sin^2\theta\right) \\ &= -2G\frac{m}{r}\left(1 + \frac{v^2}{2c^2}\right).\end{aligned} \tag{13-2.21}$$

Thus the total potential of the particles is now, by Eqs. (13-2.19), (13-2.20), and (13-2.21),

$$\varphi = -4G\frac{m}{r} - 2G\frac{mv^2}{rc^2} - G\frac{mu^2}{rc^2}, \tag{13-2.22}$$

so that, as a result of the motion of the frame, the particles appear to have acquired an additional mass

$$m'_{apparent} = \frac{2mv^2}{c^2}. \tag{13-2.23}$$

Observe, however, that Eqs. (13-2.19) and (13-2.20) contain inside the parentheses the terms $uv/c^2$ and $-uv/c^2$. These terms may be interpreted as representing an additional *positive* mass on the upper horizontal side of the frame and an additional *negative* mass on the lower horizontal side of the frame created by the motion of the frame. The two masses give rise to an apparent gravitational *dipole* (compare EM125,126 and EM130,131)

$$\mathbf{p}_{apparent} = \frac{muv}{c^2}L\mathbf{j} = \frac{IL^2v}{c^2}\mathbf{j}, \tag{13-2.24}$$

where $I$ is the mass current of the particles sliding along the frame which, expressed in terms of the line mass density $\lambda$, is

$$I = \lambda u = \frac{m}{L}u. \tag{13-2.25}$$

By analogy with electromagnetism, the mass particles sliding along the frame in Fig. 13.7 constitute a cogravitational dipole, whose dipole moment is (compare EM381)

$$\mathbf{d} = -\frac{4\pi G}{c^2}I\mathbf{S}, \tag{13-2.26}$$

where $\mathbf{S}$ is the surface area vector of the frame, right-handed relative to the direction of the mass current. In terms of $L$, $\lambda$, and $u$, the cogravitational dipole moment of the sliding particles is

$$\mathbf{d} = \frac{4\pi G}{c^2} IL^2\mathbf{k} = \frac{4\pi G}{c^2} \lambda u L^2 \mathbf{k}. \tag{13-2.27}$$

By Eq. (13-2.24), the apparent gravitational dipole moment can be expressed therefore as

$$\mathbf{p}_{apparent} = -\frac{\mathbf{v}\times\mathbf{d}}{4\pi G}. \tag{13-2.12}$$

**Example 13-2.5** Show that in the case of a cogravitational dipole uniformly moving with speed $v \ll c$, the gravikinetic field equation

$$\mathbf{g}_k = -\frac{G}{c^2}\int \frac{[\nabla'(\mathbf{v}\cdot\mathbf{J})]}{r} dV' - \mathbf{v}\times\mathbf{K}, \tag{11-2.8}$$

becomes

$$\mathbf{g}_k = -\mathbf{g}_{dipole} - \mathbf{v}\times\mathbf{K}, \tag{13-2.28}$$

where $\mathbf{g}_{dipole}$ is the gravitational field of the apparent gravitational dipole created by the moving cogravitational dipole and $\mathbf{K}$ is the cogravitational dipole field.

Since $v \ll c$, the retardation in Eq. (11-2.8) can be ignored. We then have

$$\mathbf{g}_k = -\frac{G}{c^2}\int \frac{\nabla'(\mathbf{v}\cdot\mathbf{J})}{r} dV' - \mathbf{v}\times\mathbf{K}. \tag{13-2.29}$$

We can transform Eq. (13-2.29) by using vector identity (V-27), obtaining

$$\mathbf{g}_k = -\frac{G}{c^2}\int \nabla\frac{\mathbf{v}\cdot\mathbf{J}}{r} dV' - \frac{G}{c^2}\int \nabla'\frac{\mathbf{v}\cdot\mathbf{J}}{r} dV' - \mathbf{v}\times\mathbf{K}. \tag{13-2.30}$$

Using vector identity (V-20), we can transform the second volume integral in Eq. (13-2.30) into a surface integral. But, because there are no mass currents at infinity, the surface integral vanishes, and so does the volume integral. Since $\nabla$ in the first integral does not operate on source-point coordinates, it can be factored out from under the integral sign. This gives

$$\mathbf{g}_k = -\frac{G}{c^2}\nabla\int\frac{\mathbf{v}\cdot\mathbf{J}}{r}dV' - \mathbf{v}\times\mathbf{K}. \qquad (13\text{-}2.31)$$

Without loss of generality, the mass current forming the cogravitational dipole can be considered filamentary. Therefore we can write

$$\mathbf{g}_k = -\frac{GI}{c^2}\nabla\oint\frac{\mathbf{v}\cdot d\mathbf{l}'}{r} - \mathbf{v}\times\mathbf{K}, \qquad (13\text{-}2.32)$$

where $d\mathbf{l}'$ is a length element vector in the direction of the mass current $I$. Factoring out $\mathbf{v}\cdot$, using vector identity (V-18), and taking into account that the linear dimensions of the cogravitational dipole are much smaller than $r$, we then have

$$\begin{aligned}\mathbf{g}_k &= \frac{GI}{c^2}\nabla\left(\mathbf{v}\cdot\int\frac{\mathbf{r}_u}{r^2}\times d\mathbf{S}'\right) - \mathbf{v}\times\mathbf{K}\\ &= \frac{GI}{c^2}\nabla\left(\mathbf{v}\cdot\frac{\mathbf{r}_u}{r^2}\times\mathbf{S}'\right) - \mathbf{v}\times\mathbf{K},\end{aligned} \qquad (13\text{-}2.33)$$

where $\mathbf{S}'$ is the surface area of the mass current loop forming the dipole. Transposing $\mathbf{v}$ and $\mathbf{r}_u$ in Eq. (13-2.33), we can write [see vector identity (V-2)]

$$\mathbf{g}_k = -\nabla\left(\frac{GI}{c^2}\frac{\mathbf{r}_u}{r^2}\cdot\mathbf{v}\times\mathbf{S}'\right) - \mathbf{v}\times\mathbf{K}. \qquad (13\text{-}2.34)$$

As was shown in Example 13-2.4, a moving cogravitational dipole generates an apparent gravitational dipole. By analogy with the electric field of an electrostatic dipole, the gravitational field of a gravitational dipole of dipole moment $\mathbf{p}$ can be written as (see

EM130)

$$\mathbf{g}_{dipole} = G\nabla \frac{\mathbf{p}_{apparent} \cdot \mathbf{r}_u}{r^2}. \tag{13-2.35}$$

Substituting $\mathbf{p}_{apparent}$ from Eq. (13-2.12) into Eq. (13-2.35) and using Eq. (13-2.26), we obtain

$$\mathbf{g}_{dipole} = \nabla\left(\frac{GI}{c^2}\frac{\mathbf{r}_u}{r^2} \cdot \mathbf{v} \times \mathbf{S}'\right). \tag{13-2.36}$$

Adding Eqs. (13-2.36) and (13-2.34), we finally obtain

$$\mathbf{g}_k = -\mathbf{g}_{dipole} - \mathbf{v} \times \mathbf{K}. \tag{13-2.28}$$

**Example 13-2.6** Consider a circular ring of radius $a$ rotating about its symmetry axis, which is parallel to the $z$ axis of rectangular coordinates (Fig. 13.8). Let the ring move with velocity $v \ll c$ along the $x$ axis, and let the center of the ring be momentarily at the origin of the coordinates. (a) Disregarding the ordinary Newtonian attraction, what is the force exerted by the ring on a point mass $m'$ located on the $z$ axis at a distance $d \gg a$ from the origin? (b) Assuming that the ring is at rest with its center at the origin, and assuming that the point mass $m'$ moves in the minus $x$ direction

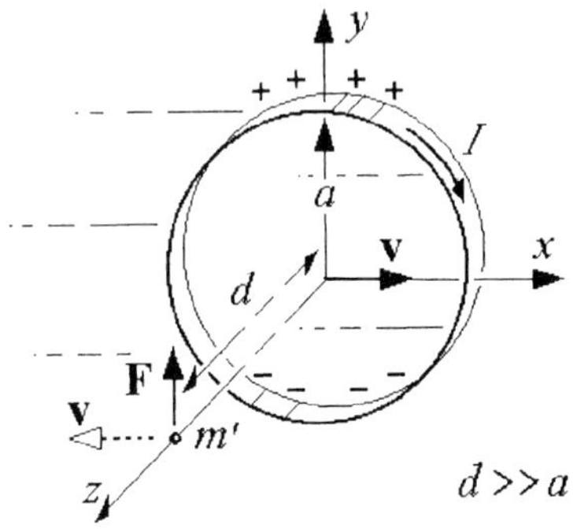


*Fig. 13.8 A rotating ring moving past a point mass exerts a force on the point mass. The plus and minus masses on the ring are fictitious: the ring appears to acquire them as a result of its motion (see Example 13-2.4).*

with velocity $v \ll c$, and ignoring the ordinary Newtonian attraction, what is the force experienced by the point mass?

(a) Since the radius of the ring is much smaller than the distance between the ring and the point mass, the mass current $I$ of the ring, viewed from the location of the point mass, constitutes a cogravitational dipole. As it was shown in Example 13-2.4, a moving cogravitational dipole appears to acquire a gravitational dipole moment, so that the total non-Newtonian gravitational field produced by the moving ring is the sum of the gravikinetic field and the gravitational dipole field. By Eq. (13-2.28), this sum is

$$\mathbf{g}_k + \mathbf{g}_{dipole} = -\mathbf{v} \times \mathbf{K}. \tag{13-2.37}$$

For the present example, since $v \ll c$, $\mathbf{K}$ in Eq. (13-2.37) is the cogravitational field produced by the ring as if it were at rest. By Eq. (2-2.17) with $x$ replaced by $d$, taking into account that $a \ll d$, and taking into account that, according to Fig. 13.8, the rotation of the ring is left-handed relative to the $z$ axis, we have

$$\mathbf{K} = G\frac{2\pi I a^2}{c^2 d^3}\mathbf{k}. \tag{13-2.38}$$

Hence, in addition to the Newtonian attraction, the force exerted by the moving rotating ring on the stationary point mass $m'$ is, by Eqs. (13-2.37) and (13-2.38),

$$\mathbf{F} = G\frac{2\pi m' v I a^2}{c^2 d^3}\mathbf{j}. \tag{13-2.39}$$

(b) If the ring is at rest, but the point mass is moving, the force acting on the point mass is given by Eq. (2-2.6). In addition to the Newtonian attraction, the point mass experiences then the force

$$\mathbf{F} = m' \mathbf{v} \times \mathbf{K} = G\frac{2\pi m' v I a^2}{c^2 d^3}\mathbf{j}, \tag{13-2.40}$$

which, under the assumed conditions (slow velocity, $\mathbf{v} = -v\mathbf{i}$ and neglected retardation), is exactly the same as the additional force experienced by the stationary point mass due to the mass current in the moving ring.

**Example 13-2.7** A rotating sphere of mass $m$ and radius $a$ moves with velocity $v \ll c$ along the $x$ axis of rectangular coordinates. The angular velocity of the sphere is $\omega$ and it is directed along the $y$ axis. A point mass $m'$ is at rest on the $z$ axis at a distance $r \gg a$ from the origin. Find the force exerted by the sphere on the point mass at the moment when the sphere passes the origin (Fig 13.9).

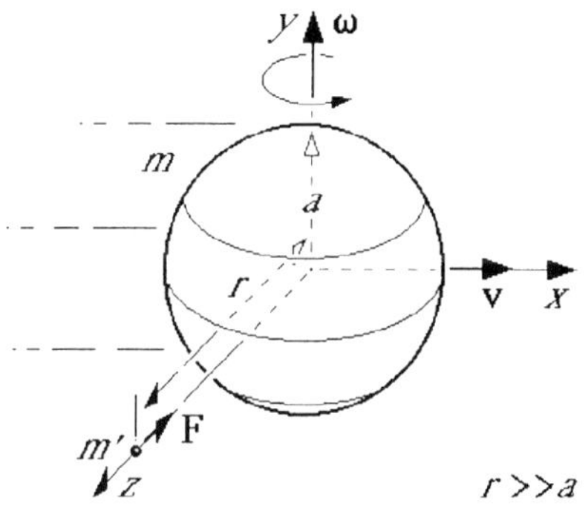


*Fig. 13.9 A rotating sphere moving past a stationary point mass exerts on the point mass not only the ordinary Newtonian gravitational force but also a velocity-dependent and a rotation-dependent force.*

The force exerted by the sphere on the point mass can be found from Eqs. (2-2.1) and (1-1.5), which yield (omitting the retardation brackets, since $v \ll c$)

$$\mathbf{F} = -m'G\int\left(\frac{\rho}{r^2}+\frac{1}{rc}\frac{\partial\rho}{\partial t}\right)\mathbf{r}_u dV' + m'\frac{G}{c^2}\int\frac{1}{r}\frac{\partial\mathbf{J}}{\partial t}dV'. \qquad (13\text{-}2.41)$$

Since the radius of the sphere is much smaller than the distance between the sphere and the point mass, the rotating sphere, from the location of the point mass, can be considered to be a point mass plus a cogravitational dipole. As was shown in Example 13-2.4, a moving cogravitational dipole appears to acquire an additional mass and creates a gravitational dipole field. By Eq. (13-2.23), a cogravitational dipole of mass $4m$ moving with velocity $v \ll c$ appears to have an additional mass $2mv^2/c^2$ [we can ignore the apparent mass given by Eq. (13-2.18), since it is insignificant unless the rotation is extremely fast]. For our sphere, whose mass is $m$, the additional mass is $mv^2/2c^2$. Designating the apparent gravitational dipole field of the sphere as $\mathbf{g}_{dipole}$, we can then express the first integral of Eq. (13-2.41) as

$$\mathbf{F}_1 = -G\frac{m'm}{r^2}\mathbf{r}_u - G\frac{m'mv^2}{2c^2r^2}\mathbf{r}_u + m'\mathbf{g}_{dipole}. \tag{13-2.42}$$

The force on the point mass is therefore

$$\mathbf{F} = -\frac{m'm}{r^2}\mathbf{r}_u - G\frac{m'mv^2}{2c^2r^2}\mathbf{r}_u + m'\mathbf{g}_{dipole} + G\frac{m'}{c^2}\int\frac{1}{r}\frac{\partial\mathbf{J}}{\partial t}dV'. \tag{13-2.43}$$

According to Eqs. (13-2.28) and (11-1.1), the last two terms in this equation are

$$\mathbf{F}_{dipole} + \mathbf{F}_k = -m'\mathbf{v}\times\mathbf{K}, \tag{13-2.44}$$

where $\mathbf{K}$ is the cogravitational field produced by the rotating sphere at the location of the point mass, when the sphere has no translational motion. This field can be found by integrating the expression for the external field of the spherical shell given in Eq. (13-2.8). The result is

$$\mathbf{K} = G\frac{m\omega a^2}{5r^3c^2}\mathbf{j}. \tag{13-2.45}$$

The total force on the point mass is therefore, by Eqs. (13-2.43), (13-2.44), and (13-2.45),

$$\mathbf{F} = -G\frac{m'm}{r^2}\mathbf{k} - G\frac{m'mv^2}{2c^2r^2}\mathbf{k} - G\frac{m'mv\omega a^2}{5r^3c^2}\mathbf{k}. \tag{13-2.46}$$

Thus the force exerted by a slowly moving, rotating sphere on a distant stationary point mass differs from the Newtonian gravitational force by the presence of two additional terms that depend on the linear and angular velocity of the sphere.

Although we have derived Eq. (13-2.46) for the moving sphere and stationary point mass, it is clear that within the accuracy of our derivations (slow velocity, neglected retardation) Eq. (13-2.46) also applies for a point mass moving relative to the sphere. Thus Eq. (13-2.46) represents a generalization of Newton's gravitational law that should preferably be used for computing planetary orbits and for similar problems of celestial mechanics (see Chapter 20).

**Example 13-2.8** Consider two point masses $m$ and $m'$. The mass $m$ is in free fall toward the ground, the mass $m'$ is at rest below $m$. At the moment when $m$ passes $m'$, $m'$ is released so that it, too, falls to the ground. (a) What is the acceleration of $m'$? (b) What is the acceleration of $m$ before and after $m'$ is released? (c) Does the acceleration of a falling body depend on its mass? (Neglect the attraction between the two masses, neglect retardation, and neglect terms of the order $v/c$ or smaller).

(a) Let us designate the acceleration of gravity vector as $\mathbf{a}$. Normally, $m'$ would then fall with the acceleration $\mathbf{a}$. However, $m'$ is subject not only to the force of gravity, but also to the force exerted upon it by the accelerating mass $m$. According to Eq. (5-4.34), this force is (within the limits of accuracy specified in the statement of the problem)

$$\begin{aligned}\mathbf{F} &= -G\frac{mm'\mathbf{r}}{r^3} - G\frac{mm'[\mathbf{r}\times(\mathbf{r}\times\mathbf{a})]}{r^3c^2} \\ &= -G\frac{mm'\mathbf{r}}{r^3} - G\frac{mm'\mathbf{r}(\mathbf{r}\cdot\mathbf{a})}{r^3c^2} + G\frac{mm'\mathbf{a}}{rc^2},\end{aligned} \tag{13-2.47}$$

where, since we neglect retardation, all the quantities are present-time quantities. The first two terms in the last expression are in the direction of the vector $\mathbf{r}$ joining the two masses and represent the attraction between $m$ and $m'$. Disregarding these terms in accordance with the statement of the problem, we are left with the gravikinetic force

$$\mathbf{F}_k = G\frac{mm'}{rc^2}\mathbf{a}, \tag{13-2.48}$$

where $r$ is the distance between the two masses. Since the force $\mathbf{F}_k$ is in the direction of the acceleration of gravity, it provides an additional acceleration to the mass $m'$. The total *initial* acceleration of $m'$ is therefore

$$\mathbf{a}_{m'\,total} = \left(1 + G\frac{m}{rc^2}\right)\mathbf{a}. \tag{13-2.49}$$

and, being in the direction of the acceleration $\mathbf{a}$ of mass $m$, represents the gravitational "drag" ("gravikinetic" force) exerted by $m$ upon $m'$.

(b) As soon as $m'$ begins to fall, it exerts an additional acceleration on $m$. Both masses now fall with an acceleration greater than $a$. The additional acceleration of $m$ enhances even further the initial acceleration of $m'$, and so on.

(c) According to the results of Parts (a) and (b), falling masses amplify the gravitational acceleration of the neighboring masses. Therefore a large mass should fall with a greater acceleration than a small mass. However, because of the $c^2$ in the denominator of Eq. (13-2.49), this effect is very small.

▲

## 13-3. Dynamic Gravitational Field Maps and Explosive Force Generated by a Fast Moving Mass

The two gravitational field maps shown in Fig. 5.4 represent the time-independent gravitational field that moves with the mass rather than the really important field that a single stationary observer would detect as the mass moves past the observer. To show the latter field, one has to construct a *dynamic gravitational field map*. Such a map depicts the gravitational field of the moving mass at a stationary point as a function of time, or, which is the same, as the function of the distance $r$ and the angle $\theta$ in the

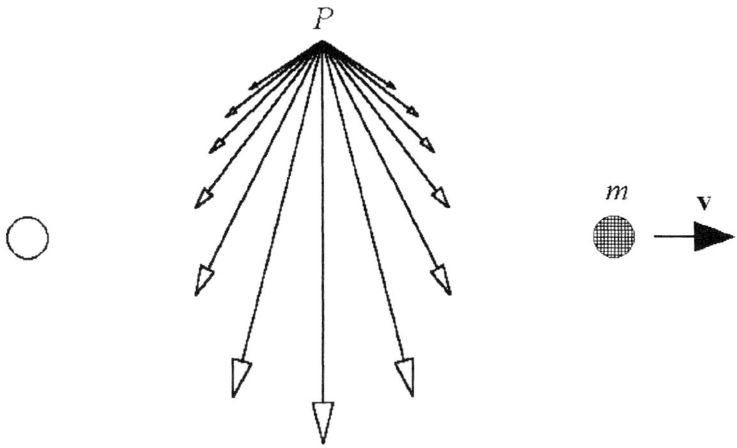


*Fig. 13.10 This dynamic map of the gravitational field of the point mass m moving with velocity* **v** *shows gravitational field vectors at the stationary point of observation P as the mass moves past P. The field vectors correspond to thirteen sequential positions of the mass. The first position is indicated by the light circle. The mass (dark circle) is at the last position. The map is drawn for v = 0.5c.*

Heaviside's Eq. (5-1.13) corresponding to the various positions of the moving mass.

A dynamic gravitational field map is shown in Fig. 13.10. The point of observation is at $P$. Thirteen different values for $r$ and $\theta$ corresponding to thirteen instantaneous sequential position occupied by the moving mass at the ends of equal time intervals $\Delta t$ were used for constructing this map. The first position of the mass is indicated by the hollow circle, the mass is at the last (thirteenth) position. The direction and strength of the gravitational field observed at $P$ is indicated by the direction and length of the arrows.

Whereas the two maps shown in Fig. 5.4 are "snapshots" of the gravitational field co-moving with the mass producing this field, the map shown in Fig. 13.10 is a "multiple exposure" map where the individual field vectors as they would be measured by the stationary observer at equal time intervals $\Delta t$ are shown all together. Of course, the entire map represents a very short event. For example, if the point $P$ is located 1 meter above the trajectory of the moving mass, the entire map represents an event that lasts only $10^{-8}$ seconds.

Closely related to the dynamic gravitational field map is the "gravitational field contour curve" representing the locus of the end points of the gravitational field vectors of a moving point mass as these vectors would be measured by the stationary observer at the point $P$. A gravitational field contour curve is strongly affected by the velocity of the mass under consideration. Three gravitational field contour curves for the same point mass moving with velocities $v = 0.01c$, $v = 0.70c$, and $v = 0.96c$, respectively, are shown in Fig. 13.10.

Dynamic gravitational field maps and the corresponding contour curves provide a new way for depicting and analyzing the gravitational field of uniformly moving point masses and reveal several important properties of this field. In particular, just by looking at Eq. (5-1.13), it may appear that a moving point mass exerts a gradually changing force on a stationary mass. However, according to Fig. 13.10, the gravitational field of a fast moving point mass, as seen by a stationary observer, is actually a momentary pulse, or burst, a sort of gravitational field explosion.

It may also appear by looking at Eq. (5-1.13) that, for a moving point mass, the gravitational field component in the direction of motion of the mass rapidly diminishes with increasing velocity of the mass and the component perpendicular to this direction rapidly increases. This appears to follow from Eq. (5-1.13) if $v \rightarrow c$ and $\theta = 0$ or $\theta = \pi/2$. However, the contour curves shown in Fig. 13.11 indicate that this assumption is only

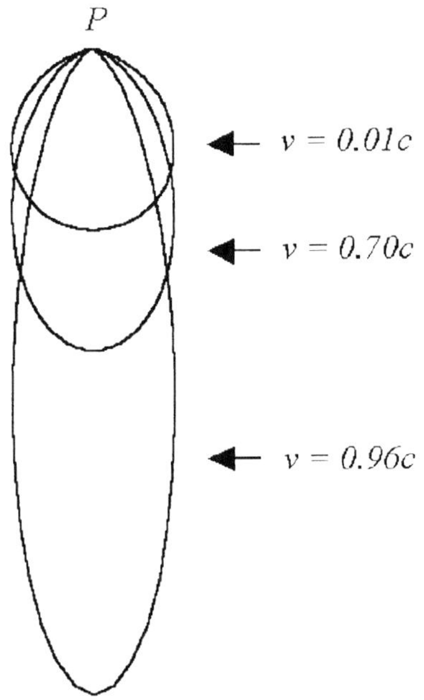


*Fig. 13.11 The lengths of the gravitational field contour curves are strongly affected by the velocity of the mass, but the widths of the curves do not noticeable depend on v. The three contour curves shown here are for the same point mass moving at velocities v = 0.01c, v = 0.70c, and v = 0.96c, as indicated.*

partially correct. Note that whereas the heights of the curves in Fig. 13.11 are strongly affected by $v$, the widths of the curves do not noticeable depend on $v$. Since the half-width of a contour curve represents the maximum value of the field component parallel to the trajectory of the moving mass, it is clear that this value is hardly affected by the speed of the mass. Of course, if $P$ is located on the trajectory of the mass (the $x$-axis), the only field component observed at $P$ is the $x$ component, and the value of this component diminishes with the distance of the charge from $P$ and with the velocity of the charge, becoming zero for $v \to c$.

Another important effect revealed by the dynamic gravitational field map shown in Fig. 13.10 and by the contour curves shown in Fig. 13.11 concerns the force exerted by a moving point mass on a stationary mass when the moving mass passes the stationary mass. As is clear from Figs. 13.10 and 13.11, this force lasts only a very short time and is essentially normal to the trajectory of the moving mass. Therefore its main effect on the stationary mass is to give a sudden thrust to the stationary mass in the direction normal to the trajectory of the moving mass. Hence the force exerted by a fast moving mass passing close to a stationary mass may have a violent explosion-like distractive effect on the stationary mass, breaking up the stationary mass by the very strong tidal forces.

Clearly, of all the graphical representations of the gravitational field of a uniformly moving point mass discussed above, the dynamic field map is by far the most important and the most informative representation.

## References and Remarks for Chapter 13

**1.** Oleg D. Jefimenko, *Electricity and Magnetism*, 2nd ed., (Electret Scientific, Star City, 1989).

**2.** Ref. 1, pp. 101-103 and 210-211.

**3.** This example involves mostly rather novel, unconventional, computations of forces. The reader is advised to consult Ref. 1, Sections 7-9 and 7-11.

**4.** The fact that the *gravitational field* can be expressed in terms of vector potential and that the *gravitational force* can be calculated from this vector potential is not well known. See Sections 3-3 and 10-2.

# 14

# TORQUE EXERTED BY A MOVING MASS ON A STATIONARY MASS

It is generally believed that a highly symmetrical mass distribution (a spherical mass of uniform mass density, for example) located in an external gravitational field experiences a force but does not experience a torque. However, according to the generalized theory of gravitation, gravitational interactions are very complex phenomena, and, as will be shown in this chapter, even a spherical mass of uniform density experiences not only a force but also a torque when located in the field of a moving mass. The torque is associated with the asymmetry of the gravitational field of the moving mass and is present even if the stationary mass is highly symmetrical, such as a sphere of uniform density. As a result of the torque, the stationary mass is set in rotation. The rotating stationary mass creates a cogravitational and a gravikinetic field that act on the moving mass, thus further contributing to the complexity of the interaction.

## 14-1. Gravitational Fields of a Point Mass Uniformly Moving Along a Straight Line and Along a Circular Orbit

We know from the derivations presented in Chapter 5 that the gravitational field **g** of a point mass *m* moving with constant velocity **v** is represented by Heaviside's formula

$$\mathbf{g} = -G\frac{m(1-v^2/c^2)}{r^3[1-(v^2/c^2)\sin^2\theta]^{3/2}}\mathbf{r}, \tag{5-1.13}$$

where $c$ is the velocity of gravitation, $\mathbf{r}$ is the present-position radius vector directed from $m$ to the point of observation, $r$ is the magnitude of $\mathbf{r}$, $v$ is the magnitude of $\mathbf{v}$, and $\theta$ is the angle between $\mathbf{r}$ and $\mathbf{v}$.

We also know from the derivations presented in Chapter 5 that the gravitational field of a point mass $m$ moving with acceleration $\dot{\mathbf{v}}$ is represented by the formula

$$\mathbf{g} = -G\frac{m}{r^3(1-\mathbf{r}\cdot\mathbf{v}/rc)^3}\left\{\left(\mathbf{r}-\frac{r\mathbf{v}}{c}\right)\left(1-\frac{v^2}{c^2}\right)+\mathbf{r}\times\left[\left(\mathbf{r}-\frac{r\mathbf{v}}{c}\right)\times\frac{\dot{\mathbf{v}}}{c^2}\right]\right\}, \tag{5-4.34}$$

where the notations are the same as in Eq. (5-1.13), except that $\dot{\mathbf{v}}$, $\mathbf{r}$ and $\mathbf{v}$ are *retarded*, that is, are evaluated for the time $t' = t - r/c$, where $t$ is the present time (the time for which $\mathbf{g}$ is evaluated). If the mass moves along a circular orbit of radius $R$, the acceleration is $\dot{\mathbf{v}} = (v^2/R^2)\mathbf{R}$, where $\mathbf{R}$ is directed from $m$ to the center of the orbit. Therefore for a mass moving along a circular orbit Eq. (5-4.34) becomes

$$\mathbf{g} = -G\frac{m}{r^3(1-\mathbf{r}\cdot\mathbf{v}/rc)^3}\left\{\left(\mathbf{r}-\frac{r\mathbf{v}}{c}\right)\left(1-\frac{v^2}{c^2}\right)+\mathbf{r}\times\left[\left(\mathbf{r}-\frac{r\mathbf{v}}{c}\right)\times\frac{v^2\mathbf{R}}{c^2R^2}\right]\right\}. \tag{14-1.1}$$

In contrast to the gravitational field of a stationary point mass, the gravitational fields represented by Eqs. (5-1.13), (5-4.34) and (14-1.1) are not radially-symmetric. As we shall presently see, it is the asymmetry of these fields that is responsible for the torque and rotation experienced by highly symmetrical mass distributions under the action of these fields. However, because of the complexity of Eqs. (5-1.13), (5-4.34) and (14-1.4), exact analytical calculations of the torque exerted on an arbitrary mass distribution by a point mass moving at an arbitrary speed is hardly possible. Therefore we shall restrict the calculations that follow to the special case of a moving point mass whose velocity is considerably smaller than the velocity of gravitation and to the special case of a stationary mass whose linear dimensions are considerably smaller than the distance of this mass from the moving mass.

## 14-2. Torque Due to a Point Mass Moving with Constant Velocity

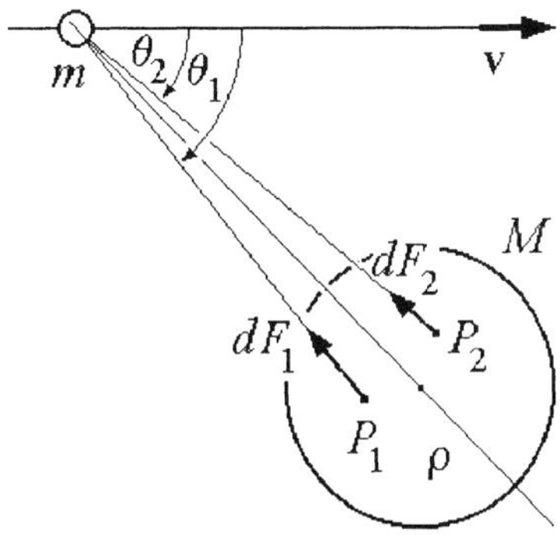


*Fig. 14.1 A point mass m moves past a stationary spherical mass M. The force $dF_1$ acting on the mass element located at $P_1$ is larger than the force $dF_2$ acting on an equal mass element located at $P_2$. Therefore the stationary mass M experiences a torque causing it to rotate.*

Let a point mass $m$ move with constant velocity **v** past a spherical mass $M$ of uniform mass density $\rho$, and let $m$ and the center of $M$ be in a plane normal to the page (Fig. 14.1). Consider two points $P_1$ and $P_2$ within $M$ located symmetrically with respect to that plane. According to Eq. (5-1.13), the force $dF_1 = g_1 dM$ exerted by $m$ on the mass element $dM$ located at $P_1$ is larger than the force $dF_2 = g_2 dM$ exerted by $m$ on the mass element $dM$ located at $P_2$ (because $\sin\theta_1$ is larger than $\sin\theta_2$). Therefore the torque with respect to the center of $M$ acting on $dM$ at $P_1$ is also larger than the oppositely directed torque with respect to the center of $M$ acting on $dM$ at $P_2$. Since the same considerations apply to all such symmetrically located points within $M$, the mass $M$, as a result of the force exerted upon it by the moving mass $m$, experiences a net torque with respect to its center and is caused to rotate about its center. In particular, for the configuration of $m$ and $M$ shown in Fig. 14.1, $M$ rotates clockwise.

As mentioned in Section 14-1, exact calculations of the torque exerted by a point mass moving at an arbitrary velocity on a stationary mass of arbitrary linear dimensions are difficult. Therefore we shall calculate the torque for the special case of a moving point mass $m$ whose velocity $\mathbf{v}$ satisfies the relation $v \ll c$, and, as the stationary mass $M$, we shall use a ring, disk and sphere whose radius $a$ satisfies the relation $a \ll r_0$, where $r_0$ is the distance between $m$ and the center of $M$.

***a. Torque on a ring of uniform density.*** Let a point mass $m$ move with constant velocity $\mathbf{v}$ in the plane of a ring of radius $a$ and cross-sectional area $S$ having a uniformly distributed mass $M$ of density $\rho$, as shown in Fig. 14.2. Let $\mathbf{v}$ satisfy the relation $v \ll c$ and let the radius vector $\mathbf{r}_0$ representing the distance from $m$ to the center of the ring satisfy the relation $a \ll r_0$. The torque $d\mathbf{T}$ with respect to the center of the ring exerted by $m$ on the mass element $\rho Sad\varphi$ contained in the shaded segment of the ring is then

$$d\mathbf{T} = \rho Sad\varphi(\mathbf{a}\times\mathbf{g}) = \mathbf{k}\rho Sa^2 g\sin\beta\, d\varphi = \mathbf{k}\rho Sa^2 g\sin(\varphi+\alpha)\,d\varphi, \tag{14-2.1}$$

where the angles $\alpha$, $\beta$, and $\varphi$ are as shown in Fig. 14.2, $\mathbf{g}$ is the gravitational field produced by $m$ at the location of the mass element, $g$ is the magnitude of $\mathbf{g}$, and $\mathbf{k}$ is a unit vector directed into the page.

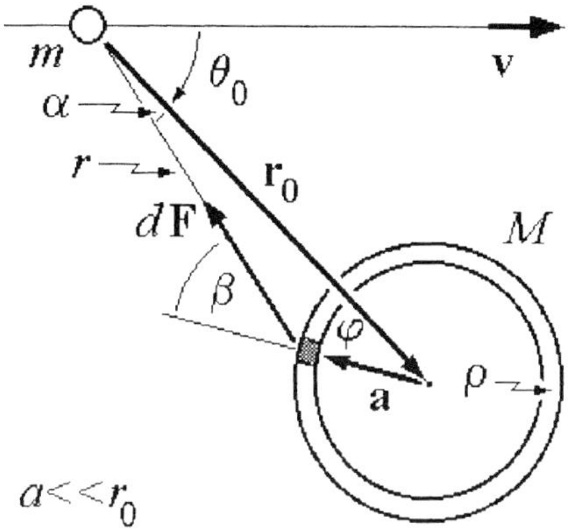


*Fig. 14.2 The torque acting on a ring of mass M is found by integrating the torque acting on the shaded segment of the ring.*

Since, by supposition, $v \ll c$, we can simplify Eq. (5-1.13) as follows:

$$\begin{aligned} g &= G\frac{m(1-v^2/c^2)}{r^2[1-(v^2/c^2)\sin^2\theta]^{3/2}} \\ &\approx G\frac{m(1-v^2/c^2)}{r^2}[1+(3/2)(v^2/c^2)\sin^2\theta] \qquad (14\text{-}2.2) \\ &\approx G\frac{m}{r^2}\left[1+\frac{v^2}{c^2}\left(\frac{3}{2}\sin^2\theta-1\right)\right]. \end{aligned}$$

Using $\theta = \theta_0 + \alpha$, where $\theta_0$ is the angle between $\mathbf{v}$ and $\mathbf{r}_0$ and taking into account that $\alpha$ is a small angle (because, by supposition, $a \ll r_0$), we have for $\sin^2\theta$

$$\begin{aligned} \sin^2\theta &= \sin^2(\theta_0+\alpha) = (\sin\theta_0\cos\alpha + \sin\alpha\cos\theta_0)^2 \approx (\sin\theta_0 + \alpha\cos\theta_0)^2 \\ &\approx \sin^2\theta_0 + 2\alpha\sin\theta_0\cos\theta_0 = \sin^2\theta_0 + \alpha\sin 2\theta_0. \qquad (14\text{-}2.3) \end{aligned}$$

Equation (14-2.2) can therefore be written as

$$g \approx G\frac{m}{r^2}\left\{1 + \frac{v^2}{c^2}\left[\frac{3}{2}\left(\sin^2\theta_0 + \alpha\sin 2\theta_0\right) - 1\right]\right\}. \tag{14-2.4}$$

We can further simplify Eq. (14-2.4) by expressing $r$ in terms of $r_0$. From Fig. 14.2 we see that

$$r^2 = a^2 + r_0^2 - 2ar_0\cos\varphi, \tag{14-2.5}$$

which, since $a \ll r_0$, can be written as

$$r^2 \approx r_0^2[1 - 2(a/r_0)\cos\varphi] \approx r_0^2. \tag{14-2.6}$$

Equation (14-2.4) becomes therefore

$$g \approx G\frac{m}{r_0^2}\left\{1 + \frac{v^2}{c^2}\left[\frac{3}{2}\left(\sin^2\theta_0 + \alpha\sin 2\theta_0\right) - 1\right]\right\}. \tag{14-2.7}$$

Now, remembering that $\alpha$ is a small angle, we simplify Eq. (14-2.1) to

$$d\mathbf{T} = \mathbf{k}\rho Sa^2 g\sin(\varphi + \alpha)\,d\varphi \approx \mathbf{k}\rho Sa^2 g(\sin\varphi + \alpha\cos\varphi)\,d\varphi. \tag{14-2.8}$$

Substituting $g$ from Eq. (14-2.7), we then have

$$d\mathbf{T} \approx G\frac{\mathbf{k}\rho Sa^2 m}{r_0^2}\left\{1 + \frac{v^2}{c^2}\left[\frac{3}{2}\left(\sin^2\theta_0 + \alpha\sin 2\theta_0\right) - 1\right]\right\}(\sin\varphi + \alpha\cos\varphi)\,d\varphi. \tag{14-2.9}$$

Finally, recognizing from Fig. 14.2 that $\alpha \approx (a \sin \varphi)/r_0$, we obtain

$$d\mathbf{T} \approx G\frac{\mathbf{k}\rho Sa^2 m}{r_0^2}\left\{1 + \frac{v^2}{c^2}\left[\frac{3}{2}\left(\sin^2\theta_0 + \frac{a\sin 2\theta_0\sin\varphi}{r_0}\right) - 1\right]\right\}\left(\sin\varphi + \frac{a\cos\varphi\sin\varphi}{r_0}\right)d\varphi. \tag{14-2.10}$$

Integrating Eq. (14-2.10) from 0 to $2\pi$, we have

$$\mathbf{T} \approx G\frac{\mathbf{k}\rho S a^2 m}{r_0^2}\int_0^{2\pi}\left\{1+\frac{v^2}{c^2}\left[\frac{3}{2}\left(\sin^2\theta_0+\frac{a\sin 2\theta_0\sin\varphi}{r_0}\right)-1\right]\right\}\left(\sin\varphi + \frac{a\cos\varphi\sin\varphi}{r_0}\right)d\varphi, \tag{14-2.11}$$

which gives for the torque acting on the entire ring

$$\mathbf{T} \approx \mathbf{k}G\frac{3\pi\rho S m a^3 v^2}{2r_0^3c^2}\sin 2\theta_0, \tag{14-2.12}$$

where we have dropped the small term with $a^2/r_0^2$.

Replacing in Eq. (4-2.12) $\rho$ by $M/2\pi aS$, we obtain the expression for the torque in terms of the mass $M$ of the ring

$$\mathbf{T} \approx \mathbf{k}G\frac{3Mma^2v^2}{4r_0^3c^2}\sin 2\theta_0. \tag{14-2.13}$$

***b. Torque on a disk of uniform density.*** We can use Eq. (14-2.12) for finding the torque acting on a small disk by considering the ring shown in Fig. 14.2 to be a differential element of the disk.

Let the thickness of the disk be $\tau$ and let its radius be $a$. Replacing $S$ in Eq. (14-2.12) by $\tau dx$, replacing $a$ by $x$, and integrating over $x$ from 0 to $a$, we obtain for the torque acting on the disk

$$\mathbf{T} \approx \mathbf{k}G\frac{3\pi\rho\tau m v^2}{2r_0^3c^2}\sin 2\theta_0\int_0^a x^3\,dx, \tag{14-2.14}$$

or

$$\mathbf{T} \approx \mathbf{k}G\frac{3\pi\rho\tau a^4 m v^2}{8r_0^3c^2}\sin 2\theta_0. \tag{14-2.15}$$

Replacing $\rho$ in Eq. (14-2.15) by $M/\pi a^2\tau$, we find the torque acting on the disk in terms of the mass $M$ of the disk

$$\mathbf{T} \approx \mathbf{k}G\frac{3Mma^2v^2}{8r_0^3c^2}\sin 2\theta_0. \qquad (14\text{-}2.16)$$

***c. Torque on a sphere of uniform density.*** Since a thin disk may be regarded as a differential element of a sphere, we can find the torque acting on a small sphere of radius $a$ by using Eq. (14-2.15). To do so, we replace in Eq. (14-2.15) $a^4$ by $(a^2 - y^2)^2$, replace $\tau$ by $dy$, and integrate over $y$ from $-a$ to $+a$. The result is

$$\mathbf{T} \approx \mathbf{k}G\frac{3\pi\rho mv^2}{8r_0^3c^2}\sin 2\theta_0\int_{-a}^{+a}(a^4 - 2a^2y^2 + y^4)dy, \qquad (14\text{-}2.17)$$

or

$$\mathbf{T} \approx \mathbf{k}G\frac{2\pi\rho ma^5v^2}{5r_0^3c^2}\sin 2\theta_0. \qquad (14\text{-}2.18)$$

Replacing $\rho$ in Eq. (14-2.18) by $3M/4\pi a^3$, we find the torque acting on the sphere in terms of the mass $M$ of the sphere

$$\mathbf{T} \approx \mathbf{k}G\frac{3Mma^2v^2}{10r_0^3c^2}\sin 2\theta_0. \qquad (14\text{-}2.19)$$

## 14-3. Torque Due to a Point Mass Moving in a Circular Orbit

We start with Eq. (14-1.1) for the gravitational field of a mass $m$ moving with uniform velocity $\mathbf{v}$ along a circular orbit of radius $R$ (Fig. 14-3). Let us find the gravitational field of $m$ at the center of the orbit. In this case $\mathbf{r} = \mathbf{R}$ and $\mathbf{r}\cdot\mathbf{v} = \mathbf{R}\cdot\mathbf{v} = 0$, so that Eq. (14-1.1) simplifies to

$$\mathbf{g} = -G\frac{m}{R^3}\left\{\mathbf{R}\left(1 - \frac{v^2}{c^2}\right) - \mathbf{v}\frac{R}{c}\right\}. \qquad (14\text{-}3.1)$$

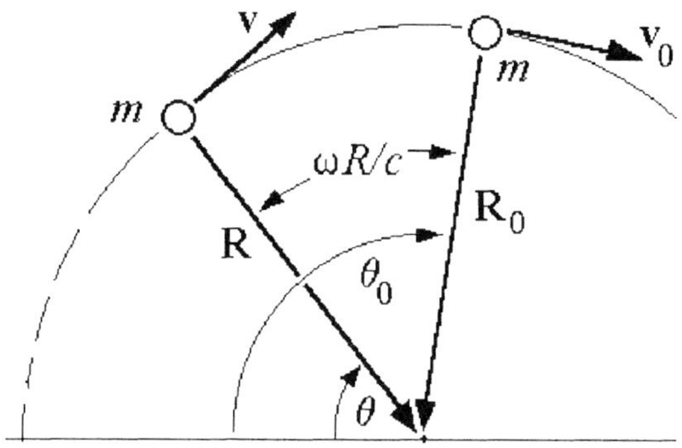


*Fig. 14.3 Correlation between the retarded quantities $\theta$,* **R**, *and* **v** *and the present-time quantities $\theta_0$,* $\mathbf{R}_0$, *and* $\mathbf{v}_0$ *for a point mass m moving along a circular orbit.*

Equation (14-3.1) expresses the gravitational field in terms of the retarded position vector and retarded velocity vector of the mass. We shall now convert Eq. (14-3.1) to present-time quantities by resolving the retarded position vector **R** and the retarded velocity vector **v** into their components along the present position vector $\mathbf{R}_0$ and the present velocity vector $\mathbf{v}_0$ (the radius $R$ of the orbit is, of course, not affected by retardation and need not be converted). Since the angle between the present position vector and the retarded position vector is $\theta_0 - \theta = \omega R/c = v/c$, where $\omega$ is the angular velocity of the mass, we obtain for the two components of **g**

$$g_{R_0} = -G\frac{m}{R^3}\left\{\left(1-\frac{v^2}{c^2}\right)R\cos(v/c) + \frac{Rv}{c}\sin(v/c)\right\}, \qquad (14\text{-}3.2)$$

$$g_{v_0} = -G\frac{m}{R^3}\left\{\left(1-\frac{v^2}{c^2}\right)R\sin(v/c) - \frac{Rv}{c}\cos(v/c)\right\}, \qquad (14\text{-}3.3)$$

and for the total field

$$\mathbf{g} = -G\frac{m}{R^3}\left\{\left[\left(1-\frac{v^2}{c^2}\right)\cos(v/c) + \frac{v}{c}\sin(v/c)\right]\mathbf{R}_0 + \left[\left(1-\frac{v^2}{c^2}\right)\frac{R}{v}\sin(v/c) - \frac{R}{c}\cos(v/c)\right]\mathbf{v}_0\right\}. \tag{14-3.4}$$

By supposition, $v \ll c$. Therefore, in the calculations that follow, we shall neglect terms smaller than $(v/c)^3$. Expanding $\sin(v/c)$ and $\cos(v/c)$ in Eq. (14-3.4) into power series of $v/c$ and dropping terms containing $v/c$ to powers higher than 3, we obtain

$$\mathbf{g} = -G\frac{m}{R^3}\left\{\left(1-\frac{v^2}{2c^2}\right)\mathbf{R}_0 - \frac{2Rv^2}{3c^3}\mathbf{v}_0\right\}. \tag{14-3.5}$$

The $\mathbf{R}_0$ component of Eq. (14-3.5) is radially symmetric and therefore cannot contribute to the torque on a highly symmetrical mass $M$ at the center of the orbit. Therefore, in the calculations that follow we only need to consider the $\mathbf{v}_0$ component of $\mathbf{g}$

$$\mathbf{g}_{v_0} = G\frac{2mv^2}{3R^2c^3}\mathbf{v}_0. \tag{14-3.6}$$

Consider points $P_1$ and $P_2$ inside a stationary spherical mass $M$. Let $P_1$ and $P_2$ be located symmetrically relative to the line normal to line connecting the center of $M$ with the point mass $m$ orbiting around $M$ (Fig. 14.4). Using Eq. (14-3.6), assuming that the density of $M$ is uniform and assuming that the radius of $M$ is much smaller than $R$, we find that the force $dF_1 = g_1 dM$ with which $m$ acts in the direction parallel to the velocity vector $\mathbf{v}$ on the mass element $dM$ at $P_1$ is larger than the force $dF_2 = g_2 dM$ with which $m$ acts in the same direction on the mass element $dM$ at $P_2$ (because the distance from $m$ to $P_1$ is smaller than the distance from $m$ to $P_2$). Since the same considerations apply to all such symmetrically located points within $M$, the mass $M$ experiences a torque and is caused to rotate about its center.

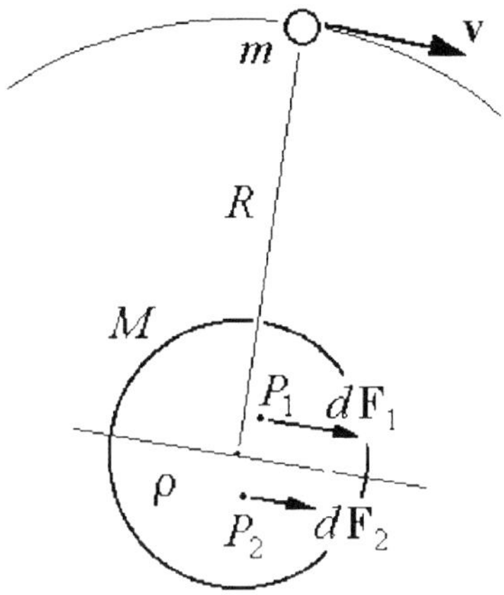


*Fig. 14.4 A point mass m moves in a circular orbit about a spherical mass M of uniform density ρ. The force $dF_1$ acting on the mass element located at $P_1$ is larger than the force $dF_2$ acting on an equal mass element located at $P_2$. Therefore the mass M experiences a torque causing it to rotate.*

***a. Torque on a ring of uniform density.*** Let $m$ rotate about a ring of uniform mass density $\rho$ and total mass $M$ whose center is at the center of the orbit and whose plane coincides with the plane of the orbit of $m$ (Fig. 14.5). Let $\mathbf{v}$ satisfy the relation $v \ll c$ and let the radius of the ring $a$ satisfy the relation $a \ll R$. Although we have derived Eq. (14-3.6) for the center of the orbit, it is approximately valid for points close to the center, and since the radius of our ring is much smaller than the radius of the orbit, we can use Eq. (14-3.6) for finding an approximate expression for the torque exerted by $m$ on the ring.

For the torque $d\mathbf{T}$ with respect to the center of the ring exerted by $m$ on the mass element $\rho Sad\varphi$ contained in the shaded segment of the ring we then have

$$d\mathbf{T} = \rho Sad\varphi(\mathbf{a} \times \mathbf{g}_{v_0}) = \mathbf{k}\rho Sa^2 g_{v_0} \sin\beta\, d\varphi = \mathbf{k}\rho Sa^2 g_{v_0} \cos\varphi\, d\varphi, \tag{14-3.7}$$

where the angles $\beta$ and $\varphi$ are as shown in Fig. 14.5, $g_{v_0}$ is the magnitude of $\mathbf{g}_{v_0}$ at the location of the shaded segment, and $\mathbf{k}$ is a unit vector directed into the page. According to Eq. (14-3.6),

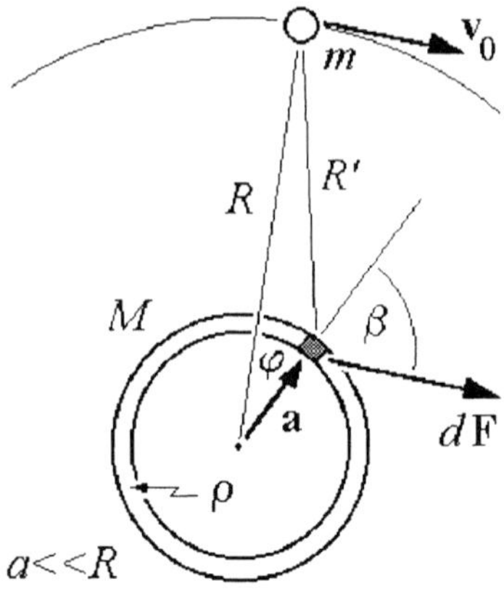


*Fig. 14.5 The torque acting on the ring of mass M located at the center of the orbit of m is found by integrating the torque acting on the shaded segment of the ring. Note that the force acting on the shaded sector is in the direction of the velocity vector* $\mathbf{v}_0$ *of m.*

$$g_{v_0} = G\frac{2mv^3}{3R'^2c^3}, \tag{14-3.8}$$

where $R'$ is the distance between $m$ and the shaded segment of the ring and $v$ is the magnitude of the velocity vector $\mathbf{v}_0$ (which, of course, is the same as the magnitude of the velocity vector $\mathbf{v}$). From Fig. 14.5 we see that, since $a \ll R$, the distance from $m$ to the shaded segment element of the ring is approximately

$$R' \approx R - a\cos\varphi, \tag{14-3.9}$$

and therefore

$$\frac{1}{R'^2} \approx \frac{1}{(R - a\cos\varphi)^2} \approx \frac{1}{R^2}\left(1 + \frac{2a}{R}\cos\varphi\right). \tag{14-3.10}$$

Substituting Eqs. (14-3.8) and (14-3.10) into Eq. (14-3.7), we have

$$d\mathbf{T} \approx G\mathbf{k}\rho S a^2\frac{2mv^3}{3R^2c^3}\left(1 + \frac{2a}{R}\cos\varphi\right)\cos\varphi\, d\varphi. \tag{4-3.11}$$

Integrating Eq. (14-3.11) from 0 to $2\pi$, we obtain for the torque acting on the ring

$$\mathbf{T} \approx G\mathbf{k}\rho S a^3 \frac{2\pi m v^3}{3R^3c^3}. \tag{14-3.12}$$

Replacing $\rho$ in Eq. (14-3.12) by $M/2\pi aS$, we obtain the expression for the torque in terms of the mass $M$ of the ring

$$\mathbf{T} \approx G\mathbf{k}\frac{Mma^2v^3}{3R^3c^3}. \tag{14-3.13}$$

***b. Torque on a disk of uniform density.*** We can use Eq. (14-3.12) for finding the torque acting on a small disk by considering the ring shown in Fig. 14.5 to be a differential element of the disk.

Let the thickness of the disk be $\tau$ and let its radius be $a$. Replacing $S$ in Eq. (14-3.12) by $\tau dx$, replacing $a$ by $x$, and integrating over $x$ from 0 to $a$, we obtain for the torque acting on the disk

$$\mathbf{T} \approx G\mathbf{k}\frac{2\pi\rho\tau m v^3}{3R^3c^3}\int_0^a a^3 dx, \tag{14-3.14}$$

or

$$\mathbf{T} \approx G\mathbf{k}\frac{\pi\rho\tau a^4 m v^3}{6R^3c^3}. \tag{14-3.15}$$

Replacing $\rho$ in Eq. (14-3.15) by $M/\pi a^2\tau$, we find the torque acting on the disk in terms of the mass $M$ of the disk

$$\mathbf{T} \approx G\mathbf{k}\frac{Mma^2v^3}{6R^3c^3}. \tag{14-3.16}$$

***c. Torque on a sphere of uniform density.*** Since a thin disk may be regarded as a differential element of a sphere, we can find the torque acting on a small sphere of radius $a$ by using Eq. (14-3.15). To do so, we replace $a^4$ in Eq. (14-3.15) by $(a^2 - y^2)^2$, replace $\tau$ by $dy$, and integrate over $y$ from $-a$ to $+a$. The result is

$$\mathbf{T} \approx G\mathbf{k}\frac{\pi\rho m v^3}{6R^3c^3}\int_{-a}^{+a}(a^4 - 2a^2y^2 + y^4)dy, \qquad (14\text{-}3.17)$$

or

$$\mathbf{T} \approx G\mathbf{k}\frac{8\pi\rho m a^5 v^3}{45R^3c^3}. \qquad (14\text{-}3.18)$$

Replacing $\rho$ in Eq. (4-3.18) by $3M/4\pi a^3$, we find the torque acting on the sphere in terms of the mass $M$ of the sphere

$$\mathbf{T} \approx G\mathbf{k}\frac{2Mma^2v^3}{15R^3c^3}. \qquad (14\text{-}3.19)$$

## 14-4. The Differential Rotation of the Sun

The calculations presented above were for the case of a solid stationary mass. A very interesting and important effect should exist, however, if the stationary mass is in a gaseous or liquid state. In this case the rotation of the stationary mass under the action of a moving mass will not occur with one single angular velocity. In particular, the equatorial regions of the stationary mass (regions close to the plane of the orbit of the moving mass) will rotate faster than the region closer to the poles, because the force and the torque exerted by the moving mass on the equatorial regions are larger than the force and torque exerted on the polar regions. For the same reason, the outer regions of the mass will rotate faster than the interior regions.

It is very likely that the non-uniform rotation of the Sun is a manifestation of this effect. As is known, the equatorial regions of the Sun rotate faster than the rest of the Sun. The regions of the Sun near its equator rotate once every 25 days. The Sun's rotation rate decreases with increasing latitude, so that its rotation rate is slowest near its poles. At its poles the Sun rotates once every 36 days.

Until now the differential rotation of the Sun was one of the great astronomical puzzles. However, the generalized theory of gravitation provides a clear and convincing explanation of this effect: according to the generalized theory of gravitation, the faster rotation of the equatorial regions of the Sun is a consequence of the torque exerted on the Sun by the planets orbiting around it. Since all of the planets in the Solar system revolve around the Sun in the same direction, the torque that they exert on the Sun is also in the same direction, and, as a result, the Sun, too, rotates in the same direction.

The explanation of the heretofore unexplained differential rotation of the Sun is a very important result supporting the validity of the generalized theory of gravitation.

## 14-5. Discussion

As we have seen, a moving mass does not merely attract or repel a stationary mass, but also exerts a torque on it and thus causes it to rotate even if the stationary mass is highly symmetric and has a uniform mass density. The direction of rotation depends on the direction of velocity of the moving mass.

In particular, when a point mass, starting from infinity, moves with constant speed along a straight line past a spherical mass, the point mass, as it comes closer to the spherical mass, exerts a torque on the spherical mass causing it to rotate so that the part of the spherical mass nearest to the point mass moves in the direction along which the point mass is moving. But then, as the point mass moves away from the spherical mass, the direction of the torque is reversed and the spherical mass tends to rotate so that its part nearest to the point mass moves in the direction opposite to the direction along which the point mass is moving.

According to Eq. (14-2.19), the torque is greatest at $\theta_0 = \pi/4$ and $\theta_0 = 3\pi/4$, and it is zero at $\theta_0 = 0$, $\theta_0 = \pi/2$, and $\theta_0 = \pi$.

If a point mass moves along a circular orbit around a spherical mass located at the center of the orbit, the torque exerted by the point mass on the spherical mass is always in the same direction and causes the spherical mass to rotate in the same sense in which the point mass revolves.

Clearly, the dynamics of the interaction between a moving point mass and a stationary mass distribution is much more complicated than previously believed. The torque exerted by the moving mass on the stationary mass and the subsequent rotation of the stationary mass are only the initial stages of a very complex sequence of events. When the stationary mass rotates, it creates a cogravitational field. In the case of a point mass moving along a straight line,[1] the torque acting on the stationary mass is a function of time and therefore the angular velocity of the stationary mass is also a function of time. Therefore the mass current formed by the stationary (now rotating) mass is time dependent and, hence, creates a gravikinetic field. The cogravitational field and the gravikinetic field of the stationary (now rotating) mass acts in turn on the moving point mass and affects its motion unless the motion is somehow controlled by external means. This is quite different from the simple attraction between a moving point mass and a stationary mass according to Newton's theory.

The interaction between an orbiting point mass and a spherical mass at the center of the orbit is even more complex. In principle such a system can be closed and need not depend on external forces for its stability. However, because of the torque acting on the central mass, the stability of the system is not at all certain. First, because, by Eq. (14-3.19), the torque exerted by the orbiting point mass is always present, the angular velocity of the central mass constantly accelerates. Therefore the cogravitational field resulting from the rotation is also always present and so is the gravikinetic field. Clearly, under these conditions the orbiting mass cannot move with constant speed, and the radius of the orbit cannot remain the same unless there exists some additional mechanism that keeps the speed and the radius constant. Furthermore, there is a problem with the conservation of angular momentum. In a closed system, the sum of the mechanical angular momentum and the field angular momentum must remain the same at all times. This means that the cogravitational field of the rotating mass, the cogravitational field of the moving point mass, and the gravitational and gravikinetic fields of the moving and of the stationary (now rotating) mass at all times maintain a very precise balance.

In summary then, the interaction between a moving mass and a stationary mass is an exceedingly complex phenomenon, the details of which are yet to be determined. However, it is quite clear that by assuming that the gravitational interaction between a moving and a stationary mass is merely a Newtonian attraction, one cannot obtain correct solutions of the problems involving moving and stationary masses (this is particularly important in connection with Mercury's perihelion anomaly; see Chapter 20).

Of course, because of the factor $v^2/c^2$ in Eq. (14-2.19) and $v^3/c^3$ in Eq. (14-3.19), the torque exerted by a moving mass on a stationary mass is usually very small. However, taking into account that the time scale in cosmic systems, and in our Solar system in particular, is extremely long, the cumulative effect of the gravitational torque in stellar and planetary systems may be very significant.[2]

## References and Remarks for Chapter 14

**1.** Because of the interaction between the two masses such a motion is in general impossible unless the point mass is by some means constrained to maintain its speed and trajectory. However, if in Eq.

(14-2.19) M $\ll$ $m$, then neither speed nor trajectory of $m$ will be significantly affected by the force exerted on $m$ by $M$.

**2.** Rotational effect similar to those described in this chapter can also occur in electromagnetic systems. See Oleg D. Jefimenko, "Torque exerted by a moving electric charge on a stationary electric charge distribution," J. Phys. A: Math. Gen. **35**, 5305-5314 (2002).

# 15

# MORE ABOUT ORBITAL MOTION AND ROTATION

In this chapter we shall investigate in greater detail gravitational and cogravitational fields and interactions of bodies in the state of orbital and rotational motion. We shall find that the interaction between such bodies is even more complex than as explained in the preceding chapter. We shall find that rotating bodies experience additional forces and experience torques under the action of external cogravitational fields. We shall find that the cogravitational field of a rotating central body affects the periods of revolution of the planets or satellites orbiting the central body.

## 15-1. Gravitational and Cogravitational Fields Produced by a Mass Moving Along a Circular Orbit

As we know from Sections 14-1 and 14-3, a point mass $m$ moving with velocity $v$ along a circular orbit of radius $R$ creates at the center of the orbit a gravitational field

$$\mathbf{g} = -G\frac{m}{R^3}\left\{\mathbf{R}\left(1-\frac{v^2}{c^2}\right)-\mathbf{v}\frac{R}{c}\right\}, \qquad (14\text{-}3.1)$$

where $\mathbf{R}$ and $\mathbf{v}$ are the retarded position radius vector and the retarded velocity vector. Expressed in terms of the present position radius vector $\mathbf{R}_0$ and the present-time velocity vector $\mathbf{v}_0$ with terms smaller than $(v/c)^3$ neglected, this field is

$$\mathbf{g} = -G\frac{m}{R^3}\left\{\left(1 - \frac{v^2}{2c^2}\right)\mathbf{R}_0 - \frac{2Rv^2}{3c^3}\mathbf{v}_0\right\}. \qquad (14\text{-}3.5)$$

Let us now find the cogravitational field produced by $m$ at the center of the orbit. Applying Eq. (4-2.10) to the gravitational field given by Eq. (14-3.1), we obtain

$$\mathbf{K} = -G\frac{m}{R^4 c}\mathbf{R}\times\left\{\mathbf{R}\left(1 - \frac{v^2}{c^2}\right) - \mathbf{v}\frac{R}{c}\right\}, \qquad (15\text{-}1.1)$$

or, since $\mathbf{R} \times \mathbf{R} = 0$,

$$\mathbf{K} = -G\frac{m}{c^2R^3}[\mathbf{v}\times\mathbf{R}]. \qquad (15\text{-}1.2)$$

Although $\mathbf{v}$ and $\mathbf{R}$ in Eq. (15-1.2) are retarded, their cross product is not affected by conversion to the present velocity vector and present position vector of the mass, because the cross product is the same for all points of the orbit. Therefore the cogravitational field given by Eq. (15-1.2) is exactly as expected from the gravitational equivalent of the electromagnetic Biot-Savart law, Eq. (7-3.19) (note, however, that the cogravitational field is always counterclockwise relative to the angular velocity vector of the orbiting mass).

Equations (14-3.1) and (14-3.5) for the gravitational field are quite unexpected. Intuitively, one would expect the field to be the Newtonian field (possibly with a factor) directed to the center of the orbit. Contrary to expectations, the true gravitational field of a point mass moving with constant speed in a circular orbit is very different from the Newtonian field. First, the field has a component parallel to the instantaneous velocity vector, and thus is *not* directed to the center of the orbit. Second, the field is not proportional to $1/r^2$.

As far as the Solar system is concerned, it is clear from the derivations presented here and throughout the book that Newton's gravitational law cannot be used as a rigorous basis for planetary dynamics. The problem is that, even if the gravitational field of the Sun is exactly a Newtonian field, so that the gravitational force exerted by the Sun on planets is the ordinary $1/r^2$ force, the gravitational force exerted by the planets on the Sun is, by Eqs. (14-3.4) and (14-3.5), neither radial nor proportional to $1/r^2$ [the fact that Eqs. (14-3.4) and (14-3.5) have been obtained for a circular rather than for an elliptical orbit cannot possibly change the essence of the information provided by Eqs. (14-3.4) and (14-3.5)]. Therefore the dynamics of the Solar system based on Newton's gravitational law can at best be only approximately correct, although the corrections associated with the velocity and acceleration of the planets are clearly very small.

The most remarkable property of the gravitational field of an orbiting mass represented by Eq. (14-3.5) is the presence of the field component in the direction of the instantaneous velocity vector of the mass, so that the field is not even radial. In our Solar system, this new component of the gravitational field may have important consequences both on the motion of the Sun and on the motion of planets. Although the field given by Eq. (14-3.5) is for the center of the orbit, this field should be approximately correct within a certain region of space around the center of the orbit. As far as the Sun is concerned, the new component of the gravitational field exerts then a torque on the Sun and causes it to rotate in the direction of the orbital velocity of the planet (see Section 14-4). Outer planets should produce a similar effect on the motion of the inner planets, causing an acceleration (and deceleration) of their orbital velocities and, what is most important, causing a secular motion of the large axes of the orbits of the inner planets in the direction of the orbital velocity of the outer planets (see Chapter 20).

The cogravitational field created by planets also has an effect on the dynamics of the solar system. Inner planets moving in the cogravitational field created by the outer planets experience a cogravitational force in accordance with Eq. (2-2.6) and an additional cogravitational "dipole" force associated with the rotation of the force-experiencing planet (see Section 15-3).

## 15-2. Cogravitational Field Produced by a Rotating Body; Association with Angular Momentum

In Example 3-3.2 we found that a ring of mass $m$ and radius $a$ rotating with angular velocity $\omega$ produces a cogravitational field which at distances $r \gg a$ from the ring is

$$\mathbf{K} = -\frac{Gma^2\omega}{2c^2r^3}(2\cos\theta\,\mathbf{r}_u + \sin\theta\,\boldsymbol{\theta}_u), \tag{3-3.23}$$

where $\theta$ is the angle between the angular velocity vector $\omega$ ($x$ axis in Fig. 3.2) and $\mathbf{r}$. Let us rewrite Eq. (3-3.23) as

$$\mathbf{K} = \frac{d}{2\pi r^3}\cos\theta\,\mathbf{r}_u + \frac{d}{4\pi r^3}\sin\theta\,\boldsymbol{\theta}_u, \tag{15-2.1}$$

where $d$ is the magnitude of the vector

$$\mathbf{d} = -G\frac{2\pi ma^2}{c^2}\omega \tag{15-2.2}$$

(the direction of $\mathbf{d}$ is opposite to the angular velocity vector $\omega$).

In electromagnetic theory, an equation analogous to Eq. (15-2.1) represents the magnetic field of a "magnetic dipole", and the equation analogous to Eq. (15-2.2) represents the "dipole moment" of the magnetic dipole. By analogy, we shall call the field represented by Eq. (15-2.1) and by all equations exhibiting the same dependence of $\mathbf{K}$ on $r$ and $\theta$ the "cogravitational dipole" field, and we shall call the coefficient $d$ appearing in these equations the "cogravitational dipole moment."

There is a simple correlation between the cogravitational dipole moment and the angular momentum of all axially-symmetric rotating bodies. Comparing the dipole moment of the rotating ring given by Eq. (15-2.2) with the equation for the angular momentum of the same ring

$$\mathbf{L} = ma^2\omega, \tag{15-2.3}$$

we see that the cogravitational dipole moment of the ring can be expressed as

$$\mathbf{d} = -G\frac{2\pi}{c^2}\mathbf{L} \tag{15-2.4}$$

and that the dipole field of the ring can therefore be expressed as

$$\mathbf{K} = -G\frac{L}{c^2r^3}\cos\theta\,\mathbf{r}_u - G\frac{L}{2c^2r^3}\sin\theta\,\boldsymbol{\theta}_u, \tag{15-2.5}$$

where $\theta$ is the angle between the angular momentum vector **L** and **r**. Observe that in our definition of the cogravitational dipole moment, vectors **d** and **L** are opposite to each other.

In vector notation, Eq. (15-2.1) can be written as

$$\mathbf{K} = \frac{1}{4\pi r^5}[3(\mathbf{d}\cdot\mathbf{r})\mathbf{r} - r^2\mathbf{d}]. \tag{15-2.6}$$

and Eq. (15-2.5) can then be written as

$$\mathbf{K} = -G\frac{1}{2c^2r^5}[3(\mathbf{L}\cdot\mathbf{r})\mathbf{r} - r^2\mathbf{L}]. \tag{15-2.7}$$

It is interesting to note that Eqs. (15-2.4), (15-2.5) and (15-2.7) hold for all axially-symmetric bodies with axially-symmetric mass density (that is, density that is a function of distance from the symmetry axis and a function of distance along the axis only). This follows from the fact that all such bodies can be considered as consisting of elementary rings for each of which Eqs. (15-2.4), (15-2.5) and (15-2.7) are valid, and therefore Eqs. (15-2.4), (15-2.5) and (15-2.7) are valid also for the entire body. Thus, for example, Eqs. (15-2.4), (15-2.5) and (15-2.7) hold for the spherical shell discussed in Example 13-2.3 and for the sphere discussed in Example 13-2.7.

## 15-3. Cogravitational Force and Torque Experienced by a Rotating Body

Like all mass distributions, a cogravitational dipole experiences a gravitational force under the action of any gravitational field. However, since a cogravitational dipole is a moving (spinning) mass distribution, and therefore constitutes a mass current, it also experiences a cogravitational force, but only when it is located in an inhomogeneous external cogravitational field. In a homogeneous cogravitational field it does not experience a net cogravitational force because, due to its symmetry, the resultant of the cogravitational forces acting on all its elements is equal to zero.

A cogravitational dipole also experiences a torque when located in an external cogravitational field, regardless whether the field is homogeneous or inhomogeneous. The torque causes the axis of the dipole to precess about the direction of the cogravitational field.

The force experienced by a cogravitational dipole (in addition to the ordinary Newtonian attraction) is, by Eq. (7-3.32),

$$\mathbf{F} = -\frac{c^2}{4\pi G}(\mathbf{d}\cdot\nabla)\mathbf{K}', \qquad (7\text{-}3.32)$$

and the torque is, by Eq. (7-3.33),

$$\mathbf{T} = -\frac{c^2}{4\pi G}\mathbf{d}\times\mathbf{K}'. \qquad (7\text{-}3.33)$$

The primes in these equations indicate that the cogravitational field acting on the dipole is due to external sources.

As a spinning body a cogravitational dipole constitutes a gyroscope and precesses when a torque is applied to it. The angular velocity of precession of a cogravitational dipole is[1]

$$\Omega = \frac{T}{L\sin\theta} \tag{15-3.1}$$

where $T$ is the torque acting on the dipole, $L$ is the magnitude of the angular momentum of the dipole and $\theta$ $(0<\theta\leq\pi/2)$ is the angle between **d** and **K**′. The direction of $\Omega$ is determined by ascertaining that the vectors **Ω**, **L** and **T**, in the order stated, form a right-handed system ("**ΩLT**" rule).

[**Note:** Eq. (15-3.1) holds rigorously only if $\Omega$ is much smaller than the rotational angular velocity of the body forming the cogravitational dipole moment; this happens, for example, if $L$ is relatively large and $T$ is relatively small.]

For an axially-symmetric force-experiencing or torque-experiencing body it is convenient to express Eqs. (7-3.32) and (7-3.33) in terms of the angular momentum of the body. By Eqs. (15-2.4), (7-3.32) and (7-3.33), we then have

$$\mathbf{F} = \frac{1}{2}(\mathbf{L}\cdot\nabla)\mathbf{K}' \tag{15-3.2}$$

and

$$\mathbf{T} = \frac{1}{2}\mathbf{L}\times\mathbf{K}'. \tag{15-3.3}$$

For practical applications Eq. (15-3.2) should preferably be expressed in scalar form. Expanding Eq. (15-3.2), we obtain

$$\begin{aligned} F_x &= \frac{1}{2}\left(L_x\frac{\partial K_x'}{\partial x} + L_y\frac{\partial K_x'}{\partial y} + L_z\frac{\partial K_x'}{\partial z}\right) \\ F_y &= \frac{1}{2}\left(L_x\frac{\partial K_y'}{\partial x} + L_y\frac{\partial K_y'}{\partial y} + L_z\frac{\partial K_y'}{\partial z}\right) \\ F_z &= \frac{1}{2}\left(L_x\frac{\partial K_z'}{\partial x} + L_y\frac{\partial K_z'}{\partial y} + L_z\frac{\partial K_z'}{\partial z}\right) \end{aligned} \tag{15-3.4}$$

There are several important consequences of Eqs. (7-3.32), (7-3.33) and (15-3.1)-(15.3.4). First, because all rotating bodies have a cogravitational dipole moment and therefore experience a force in an inhomogeneous cogravitational field, the weight of a rotating body (attraction to the central body) depends on the rotation of the body if the central body is also rotating. In our Solar system this effect is too small to be measured at this time, but it has major theoretical implications. Second, because rotating bodies experience a torque and execute a precessional motion under the action of a cogravitational field, it is possible to measure the cogravitational field by observing the precession of a test body located in this field. (The existence of the cogravitational field of the Earth will probably be established with the help of Gravity Probe B, a satellite in a polar orbit carrying gyroscopes for detecting and measuring the cogravitational effect of the rotating Earth)[2].

▼

**Example 15-3.1** Two spherical bodies rotate about their diameters. The angular momentum of the first body is $\mathbf{L}_1$, that of the second is $\mathbf{L}_2$. The first body is at the origin of rectangular coordinates. The second body is on the $x$ axis at a distance $x$ from the first. The radius of the second body is much smaller than $x$. Find the cogravitational force and torque experienced by the second body under the action of the first body if the angular momenta of the two bodies are directed along the $x$ axis.

Since the angular momenta of the two bodies are along the $x$ axis, the only components of their angular momenta are $L_{x1}$ and $L_{x2}$, respectively.

The cogravitational field produced by the first body at the location of the second body is, by Eq. (15-2.5),

$$\mathbf{K}_{x1} = -G\frac{L_{x1}}{c^2x^3}\mathbf{i}. \tag{15-3.5}$$

Differentiating Eq. (15-3.5) with respect to $x$, we have

$$\frac{\partial K_{x1}}{\partial x} = G\frac{3L_{x1}}{c^2x^4}. \qquad (15\text{-}3.6)$$

Substituting Eq. (15-3.6) into the first Eq. (15-3.4) and taking into account that the angular momentum of the second body has only the $x$ component, we obtain for the cogravitational force acting on the second body

$$F_x = G\frac{3L_{x2}L_{x1}}{2c^2x^4}. \qquad (15\text{-}3.7)$$

Thus the first body repels the second body along the line joining them.

By Eqs. (15-2.5) and (15-3.3), the two bodies do not exert a torque on each other because the angular momenta of the two bodies are parallel.

**Example 15-3.2** Two spherical bodies rotate about their diameters. The angular momentum of the first body is $\mathbf{L}_1$, that of the second body is $\mathbf{L}_2$. The first body is at the origin of rectangular coordinates and its angular momentum is directed along the $x$ axis. The second body is on the $x$ axis at a distance $x$ from the first and its angular momentum is in the $y$ direction. The radii of the two bodies are much smaller than $x$. Find the cogravitational force exerted by the two bodies upon each other, the torque exerted by the two bodies upon each other and the angular velocity of precession of each body resulting from the torque experienced by the body.

Since the angular momentum of the first body is along the $x$ axis the cogravitational field produced by the first body at the location of the second body is, by Eq. (15-2.5),

$$\mathbf{K}_1 = -G\frac{L_1}{c^2x^3}\mathbf{i}. \qquad (15\text{-}3.8)$$

Denoting the angular momentum of the second body as $L_{y2}$ and examining Eq. (15-3.4), it appears that the only force that the second body could experience under the action of the first body is due to the second term of the first Eq. (15-3.4),

$$F_x = \frac{1}{2} L_y \frac{\partial K_x'}{\partial y}, \tag{15-3.9}$$

because this is the only term in Eq. (15-3.4) with $x$ and $y$ subscripts matching the subscripts of $L_2$ and $K_1$ in our system of the two bodies. However, looking at Eq. (15-3.8), it is clear that the derivative of $K_{x1}$ with respect to $y$ is zero, so that it appears that the second body experiences no force at all. But a more careful examination of the system reveals that there is a force on the second body after all, because the cogravitational field of the first body is not confined to the $x$ axis. Indeed, if, by using Eq. (15-2.7), we express $\mathbf{K}_1$ as a function of the distance $r$ from the location of the first body (origin of coordinates) in the $xy$ plane, we find that

$$\mathbf{K}_1 = -G\frac{1}{2c^2r^5}[3(L_1x)\mathbf{r} - \mathbf{L}_1r^2] = -G\frac{1}{2c^2r^5}[3(L_1x)(x\mathbf{i} + y\mathbf{j}) - L_1r^2\mathbf{i}]. \tag{15-3.10}$$

Thus there are two components of $\mathbf{K}_1$ in the $xy$ plane: the $x$ component and the $y$ component. Therefore, by Eq. (15-3.4), the second body, whose angular momentum is in the $y$ direction, can experience a force if the derivative of at least one of the two components of $\mathbf{K}_1$ with respect to $y$ does not vanish on the $x$ axis. Examining Eq. (15-3.10) we recognize that the derivative of the $x$ component of $\mathbf{K}_1$ does vanish on the $x$ axis, but the derivative of the $y$ component does not.

According to Eq. (15-3.10), the $y$ component of $\mathbf{K}_1$ is

$$K_{y1} = -G\frac{3L_1xy}{2c^2r^5}. \tag{15-3.11}$$

or, expressing $r$ in the denominator of Eq. (5-3.11) in terms of $x$ and $y$,

$$K_{y1} = -G\frac{3L_1xy}{2c^2(x^2+y^2)^{5/2}}. \tag{15-3.12}$$

Differentiating Eq. (15-3.12) with respect to $y$, we obtain

$$\frac{\partial K_{y1}}{\partial y} = -G\frac{3L_1x[(x^2+y^2)^{5/2} - 5(x^2+y^2)^{3/2}y^2]}{2c^2(x^2+y^2)^5}, \tag{15-3.13}$$

which on the $x$ axis becomes

$$\frac{\partial K_{y1}}{\partial y} = -G\frac{3L_1}{2c^2x^4}. \tag{15-3.14}$$

Substituting Eq. (15-3.14) into the second Eq. (5-3.4), we obtain for the cogravitational force acting on the second body

$$F_y = -G\frac{3L_2L_1}{4c^2x^4}. \tag{15-3.15}$$

Thus the first body pushes the second body downward in the negative $y$ direction. By the symmetry of the system it is clear that the second body pushes the first body with the same force upward.

According to Eq. (15-3.3), the torque experienced by the second body under the action of the first is

$$\mathbf{T}_2 = \frac{1}{2}\mathbf{L}_2 \times \mathbf{K}_1. \tag{15-3.16}$$

Since $\mathbf{L}_2$ is in the $y$ direction and $\mathbf{K}_1$ is in the negative $x$ direction, the torque is

$$\mathbf{T}_2 = \frac{1}{2}L_2K_1\mathbf{k} \tag{15-3.17}$$

or, with Eq. (15-3.8),

$$\mathbf{T}_2 = G\frac{L_1L_2}{2c^2x^3}\mathbf{k}. \tag{15-3.18}$$

field. Because of the rapid motion of an electron along its orbit, the orbit can be treated as a current-carrying rigid circle, and the orbit can then be treated as a magnetic dipole. The external magnetic field exerts a torque on this dipole and causes its axis (the symmetry axis of the orbit) to precess about the direction of the field.[3]

Although planets and satellites do not move very fast, the time scale in the Solar system is extremely long. Therefore the orbit of a planet or satellite can likewise be treated as a rotating rigid circle forming a mass current $m/t$, where $m$ is the mass of the planet or satellite and $t$ is the period of revolution of the planet or satellite under consideration. If the orbital angular velocity of the planet (satellite) is $\omega$ and the radius of the orbit is $r$, the angular momentum of the rigid orbit is, by Eq. (15-2.3),

$$\mathbf{L}_o = mr^2\omega, \tag{15-5.1}$$

and, by Eq. (15-2.4), the dipole moment of the orbit is

$$\mathbf{d}_o = -G\frac{2\pi}{c^2}\mathbf{L}. \tag{15-5.2}$$

In the presence of an external cogravitational field $\mathbf{K}'$, the orbit then experiences a torque, which, by Eq. (15-3.3), is

$$\mathbf{T}_o = \frac{1}{2}\mathbf{L}_o \times \mathbf{K}', \tag{15-5.3}$$

and, by Eq. (15-3.1), the orbit precesses about the direction of $\mathbf{K}'$ with angular velocity

$$\Omega_o = \frac{T_o}{L_o \sin\theta}, \tag{15-5.4}$$

where $\theta$ is the angle between $\mathbf{L}_o$ and $\mathbf{K}'$. Since $\omega = 2\pi/t$, and since $T_0 = (1/2)|\mathbf{L}_o \times \mathbf{K}'| = (1/2)L_oK'\sin\theta$, the precession angular velocity of the orbit is

$$\Omega_o = \frac{K'}{2} \tag{15-5.5}$$

regardless of the angle between the direction of the cogravitational field and the axis of the orbit.

The cogravitational field **K**′ responsible for the precession of the orbit of a satellite or planet can be produced by the rotating central body, by other rotating bodies, by other orbiting satellites or planets, and by the rotating galaxy. Thus, for example, if a satellite revolves around a rotating central body, and the cogravitational field created by the central body at the points of the satellite's orbit is **K**′, the orbit of the satellite will precess in accordance with Eq. (15-5.5). For a satellite in direct orbit, the precession will be, by the **ΩLT** rule, in the retrograde direction.

## References and Remarks for Chapter 15

**1.** See, for example, A. P. French, *Newtonian Mechanics* (Norton, New York, 1971) pp. 680, 681.
**2.** Actually, the purpose of Gravity Probe B is to measure the "gravimagnetic field" of the Earth. The gravimagnetic field is a concept of the general relativity theory associated with rotating bodies and differs from the cogravitational field of a rotating body in the generalized theory of gravitation by a factor of 4 (it is presumably four times larger than the cogravitational field; see Chapter 20 for details). Qualitatively, however, the two fields are the same.
**3.** See, for example, R. A. Becker, *Introduction to Theoretical Mechanics* (McGraw-Hill, New York, 1954) pp. 304, 305.

# 16

# TRANSFORMATION OF ENERGY AND MOMENTUM IN GRAVITATIONAL AND COGRAVITATIONAL INTERACTIONS

As we know from Chapters 2 and 8, gravitational and cogravitational fields are repositories of energy and momentum. A very important aspect of gravitational and cogravitational interactions is the exchange of energy and momentum between gravitational and cogravitational fields and bodies located in these fields. In this chapter we shall analyze the mechanism of this exchange. We shall learn how this mechanism affects the kinetic energy of bodies moving in the presence of gravitational fields. And we shall discover the actual physical nature of gravitational and cogravitational forces.

## 16-1. Energy Exchange Between a Gravitational Field and Bodies Moving in It

One of the most eloquent examples of the effectiveness of the generalized theory of gravitation is its ability to explain the details of the process responsible for the variation of kinetic energy of bodies moving under the action of gravitational fields. As is known, the motion of stellar bodies and the fall of bodies under the action of the gravitational field is associated with conversion of potential energy into kinetic energy and vice versa. In particular, when a body is falling under the action of the gravitational field of the Earth, its potential energy diminishes and its kinetic energy increases. But how, exactly, does this come about? How is this energy exchange actually accomplished? In the past this phenomenon was simply interpreted as a result of the energy conservation, but the process, or mechanisms, of the energy exchange remained unknown. As we shall now see, the generalized theory of gravitation explains this heretofore hidden process with perfect clarity.

Let a body of mass $m$ fall under the action of the Earth's gravitational field $\mathbf{g}$ (Fig. 16.1). Note that the magnitude of $\mathbf{g}$ is equal to the acceleration of gravity $g$. Let the velocity of the body at the moment of observation be $\mathbf{v}$. Like all moving masses, the falling body creates around itself a cogravitational field $\mathbf{K}$ left-handed relative to the velocity vector of the body. Therefore, according to Eq. (2-2.9) (gravitational Poynting vector equation)

$$\mathbf{P} = \frac{c^2}{4\pi G}\mathbf{K}\times\mathbf{g}, \tag{2-2.9}$$

there is a flow of gravitational energy $U_{\mathrm{gr}}$ at the surface of the falling body directed into the body. The rate at which the gravitational energy enters the body is

$$\frac{dU_{gr}}{dt} = \frac{c^2}{4\pi G}\oint(\mathbf{K}\times\mathbf{g})\cdot d\mathbf{S}_{in} = \frac{c^2}{4\pi G}\oint(\mathbf{g}\times\mathbf{K})\cdot d\mathbf{S}, \tag{16-1.1}$$

where $d\mathbf{S}_{in}$ is a surface element vector of the falling body directed into the body, and $d\mathbf{S}$ is a surface element vector directed, as usually accepted in vector analysis, from the body into the surrounding space; the integration is over the entire surface of the falling body. Transposing in the integrand the cross and the dot

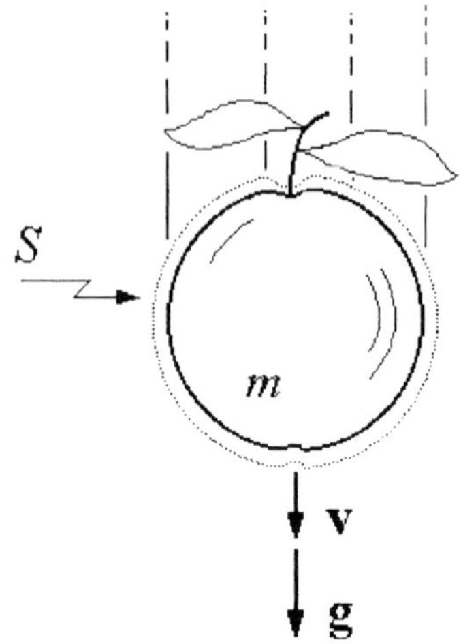


*Fig. 16.1 The generalized theory of gravitation provides a clear explanation of the mechanism of energy exchange involved in gravitational interactions: the increase of the kinetic energy of a body moving under the action of a gravitational field occurs as a consequence of the influx* of *gravitational field energy into the body via the gravitational Poynting vector.*

and factoring out the constant vector **g** together with the dot from under the integral sign, we have

$$\frac{dU_{gr}}{dt} = \frac{c^2}{4\pi G}\oint \mathbf{g}\cdot(\mathbf{K}\times d\mathbf{S}) = \frac{c^2}{4\pi G}\mathbf{g}\cdot\oint \mathbf{K}\times d\mathbf{S}. \quad (16\text{-}1.2)$$

Converting now the last surface integral into the volume integral by using vector identity (V-21), we obtain

$$\frac{dU_{gr}}{dt} = -\frac{c^2}{4\pi G}\mathbf{g}\cdot\int \nabla\times\mathbf{K}\,dV. \quad (16\text{-}1.3)$$

By Eq. (7-1.4), since **g** is not a function of time,

$$\nabla\times\mathbf{K} = -\frac{4\pi G}{c^2}\rho\mathbf{v}. \quad (16\text{-}1.4)$$

Therefore Eq. (16-1.3) reduces to

$$\frac{dU_{gr}}{dt} = \mathbf{g} \cdot \int \rho \mathbf{v} dV. \tag{16-1.5}$$

Factoring out the constant vector **v** from under the integral sign, we obtain

$$\frac{dU_{gr}}{dt} = \mathbf{g} \cdot \mathbf{v} \int \rho dV. \tag{16-1.6}$$

Thus, since **g** and **v** are parallel, and since the last integral in Eq. (16-1.6) represents the mass of the falling body, we find that when the body is falling, there is an influx of the gravitational field energy (potential energy) into the body at the rate

$$\frac{dU_{gr}}{dt} = \mathbf{g} \cdot \mathbf{v} m = mvg. \tag{16-1.7}$$

Let us now consider the kinetic energy. The kinetic energy of a falling body increases at the rate

$$\frac{dU_{kin}}{dt} = \frac{d}{dt}\left(\frac{mv^2}{2}\right) = mv\frac{dv}{dt} = mvg, \tag{16-1.8}$$

where g is the acceleration of the falling body. However, as was mentioned above, $g$ in Eq. (16-1.7) is the same acceleration, and therefore the rate at which the kinetic energy of the falling body increases is equal to the rate of influx of the gravitational field energy into the body.[1]

Thus the generalized theory of gravitation provides a clear explanation of the mechanism of the energy exchange involved in gravitational interactions: the increase of the kinetic energy of the body moving under the action of a gravitational field occurs as a consequence of the gravitational field energy influx into the body via the gravitational Poynting vector. Essentially the same considerations apply to the case when a body moves against the gravitational field, in which case its kinetic energy diminishes due to an outflow of energy from the body into the surrounding space again via the gravitational Poynting vector.

The simplicity of the above calculations tends to hide the utmost significance of the obtained results. The fact is that no gravitational theory can be considered definitive if it cannot provide a clear explanation of the mechanism of conversion of "gravitational potential energy" into the kinetic energy of falling bodies. Therefore, in spite of their simplicity, the above calculations constitute an exceptionally important proof of the validity of the generalized theory of gravitation and, at the same time, reveal the true nature of the "gravitational potential energy."

## 16-2. The Physical Nature of Gravitational and Cogravitational Forces

Gravitational interaction of celestial bodies is a very mysterious phenomenon. It is traditionally attributed (without any further explanation) to the action of forces of "universal gravitation." But where are the threads, the ropes, the chains or the springs that pull celestial bodies one to the other? How does the Earth "know" that it needs to revolve around the Sun? How does it "feel" where the Sun is located? As far as we know, there exists no material connection between celestial bodies. But if there is no material connection, does it not mean that gravitational interactions are not a manifestation of the action of forces, but a manifestation of the existence of some heretofore overlooked agent or mechanism? The generalized theory of gravitation answers this question with perfect clarity.

According to the generalized theory of gravitation, gravitational and cogravitational fields are repositories of not only energy but also of momentum $\mathbf{G}$ [see Eq. (2-2.11)], and a direct exchange (transformation) of momentum can occur between a gravitational-cogravitational field and a body (mass) located in this field. As a result of such an exchange, a momentum is transferred from the field to the body (or vice versa), increasing (or decreasing) the mechanical momentum $\mathbf{G}_M$ of the body. The momentum exchange takes place in accordance with Eq. (2-2.12),

$$\frac{d\mathbf{G}_M}{dt} = -\frac{1}{4\pi G}\int \frac{\partial}{\partial t}(\mathbf{K}\times\mathbf{g})\,dV + \frac{1}{4\pi G}\left[\frac{1}{2}\oint(\mathbf{g}^2+c^2\mathbf{K}^2)d\mathbf{S} - \oint \mathbf{g}(\mathbf{g}\cdot d\mathbf{S}) - c^2\oint \mathbf{K}(\mathbf{K}\cdot d\mathbf{S})\right], \tag{2-2.12}$$

If $d\mathbf{G}_M/dt$ in this equation is positive, a momentum is transferred from the field to the body; if $d\mathbf{G}_M/dt$ is negative, a momentum is transferred from the body to the field (for simplicity, we shall consider in the discussion that follows the case of the positive $d\mathbf{G}_M/dt$ only).

The first integral (the volume integral) in Eq. (2-2.12) is evaluated over the region of the gravitational-cogravitational field containing the body under consideration and represents the rate of change of the gravitational-cogravitational field momentum in this region. The remaining integrals (surface integrals) are evaluated over the boundary surface of the region over which the first integral is evaluated and represent the flux of gravitational-cogravitational field momentum through this surface.

Thus, the increase of the mechanical momentum of the body occurs at the expense of the field momentum lost by the region in which the body is located, as well as at the expense of the field momentum entering the region from the surrounding space. The total momentum of the field and the body always remains the same (compare Chapter 8). It should be noted that the transfer of gravitational-cogravitational field momentum into mechanical momentum (and vice versa) is closely connected with the transfer of field energy into kinetic energy of the body (and vice versa) discussed in Section 16.1 (clearly, there cannot be an exchange of mechanical energy without a simultaneous exchange of momentum, because both the kinetic energy of a body and the momentum of the body depend on the velocity of the body).

Since the effect of a force cannot be distinguished from that of a change of mechanical momentum, and since force is a much more familiar concept than momentum, we naturally see "force actions" in gravitational and cogravitational interactions, although what happens in reality is a straightforward momentum exchange (transformation) between the gravitational-cogravitational field and the body (mass) located in this field. Thus we must conclude that "force" in gravitational systems is a convenient and important mathematical device, but not the physical effect, entity, or agent responsible for gravitational interactions.

It is important to note that although forces (in the conventional sense of the word "force") do not play a role in gravitational interactions, this does not at all diminish the practical significance of Eq. (2-2.6) and of all other force equations agreeing with Eq. (2-2.6). In Example 16-2.2 we shall show that Eq. (2-2.6) is a direct consequence of Eq. (2-2.12). Therefore the results obtained with the help of force equations are indistinguishable from those obtained from Eq. (2-2.12), but force equations are much simpler and much more convenient to use than Eq. (2-2.12).[2]

Let us now illustrate the details of gravitational momentum transfer into mechanical momentum by means of the following example.

▼

**Example 16-2.1** A long cylinder of mass $m$, length $l$ and radius $a$, with $l \gg a$, falls along its axis under the action of the Earth's gravitational field $\mathbf{g}_e$ (Fig. 16.2). The velocity of the cylinder at the time of observation is $\mathbf{v}$. Analyze momentum transfer in the system.

Let us find the gravitational self-field created by the cylinder outside itself. Since $l \gg a$, we can neglect the end effects of the cylinder. The gravitational field of the cylinder is then, according to Gauss's theorem of vector analysis [see Eq. (7-1.5); see also EM89, 90],

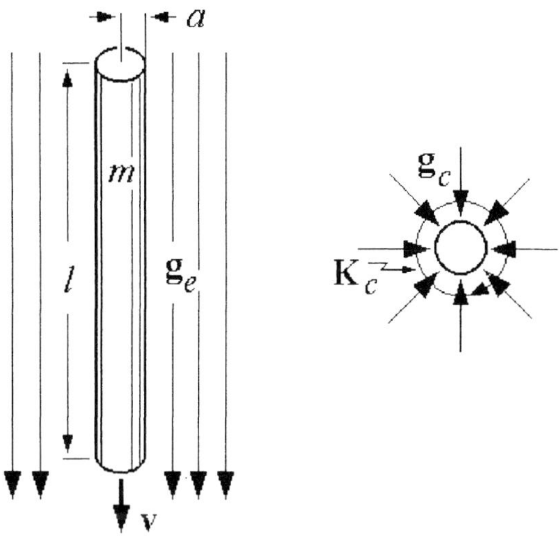


*Fig. 16.2 A cylinder of mass m falls under the action of the Earth's gravitational field* $\mathbf{g}_e$. *On the right is the end view of the cylinder showing its gravitational self-field* $\mathbf{g}_c$ *and its cogravitational field self-field* $\mathbf{K}_c$.

$$\mathbf{g}_c = -G\frac{2m}{lr}\mathbf{r}_u, \tag{16-2.1}$$

where $\mathbf{r}_u$ is a unit vector at right angles to the axis of the cylinder directed from the axis into the surrounding space.

The total gravitational field outside the cylinder is the sum of the cylinder's self-field $\mathbf{g}_c$ and of the external field $\mathbf{g}_e$ in which the cylinder moves:

$$\mathbf{g} = \mathbf{g}_c + \mathbf{g}_e = -G\frac{2m}{lr}\mathbf{r}_u + \mathbf{g}_e. \tag{16-2.2}$$

The falling cylinder also creates a cogravitational field $\mathbf{K}_c$. As we shall see, the magnitude of this field does not matter in the present case – only its direction is important. According to Eq. (7-1.8), $\mathbf{K}_c$ is a circular field directed clockwise as seen from below the cylinder (as was explained in Section 7-1, the cogravitational field is left-handed relative to the mass current by which it is created). Since there is no other cogravitational field in the system under consideration, $\mathbf{K}_c$ is the total cogravitational field $\mathbf{K}$ of the system.

Let us now construct a cylindrical surface enclosing the cylinder just outside the cylinder, and let us apply the first integral of Eq. (2-2.12) to the enclosed volume and apply the remaining integrals to the surface enclosing the cylinder. Since the gravitational and cogravitational fields inside the cylinder are not functions of time, and since we neglect the end effects of the cylinder, the first integral in Eq. (2-2.12) (volume integral) vanishes, and Eq. (2-2.12) reduces to

$$\frac{d\mathbf{G}_M}{dt} = \frac{1}{4\pi G}\left[\frac{1}{2}\oint(\mathbf{g}^2 + c^2\mathbf{K}^2)d\mathbf{S} - \oint \mathbf{g}(\mathbf{g}\cdot d\mathbf{S}) - c^2\oint \mathbf{K}(\mathbf{K}\cdot d\mathbf{S})\right]. \tag{16-2.3}$$

The first integral in this equation vanishes by symmetry [to every $d\mathbf{S}$ at a point of the cylindrical surface there corresponds an equal but opposite $d\mathbf{S}$ at a diametrically opposite point, while $\mathbf{g}^2$ and $\mathbf{K}^2$ are the same at both points; and on the two flat ends of the cylinder $d\mathbf{S}$'s are also in opposite directions, while $\mathbf{g}^2$ and $\mathbf{K}^2$ are the same at both ends]. The last integral vanishes because on the cylindrical surface $\mathbf{K}$ is perpendicular to $d\mathbf{S}$, so that $\mathbf{K}\cdot d\mathbf{S} = 0$, and on the two flat ends of the cylinder $d\mathbf{S}$'s are in opposite directions, while $\mathbf{K}$ is the same at both ends. Thus only the second integral survives in Eq. (16-2.3) so that

$$\frac{d\mathbf{G}_M}{dt} = -\frac{1}{4\pi G}\oint \mathbf{g}(\mathbf{g}\cdot d\mathbf{S}). \tag{16-2.4}$$

Substituting Eq. (16-2.2) into Eq. (16-2.4) and taking into account that at the surface of the cylinder $r = a$, we obtain

$$\frac{d\mathbf{G}_M}{dt} = -\frac{1}{4\pi G}\oint\left(-G\frac{2m}{la}\mathbf{r}_u + \mathbf{g}_e\right)\left[\left(-G\frac{2m}{la}\mathbf{r}_u + \mathbf{g}_e\right)\cdot d\mathbf{S}\right]. \quad (16\text{-}2.5)$$

On the cylindrical surface, $\mathbf{g}_e$ is perpendicular to $d\mathbf{S}$, so that $\mathbf{g}_e \cdot d\mathbf{S} = 0$, and on the flat ends of the cylinder $d\mathbf{S}$'s are in opposite directions, while $\mathbf{g}_e$ is the same at both ends. Hence Eq. (16-2.5) reduces to

$$\frac{d\mathbf{G}_M}{dt} = \frac{1}{4\pi G}\oint\left(-G\frac{2m}{la}\mathbf{r}_u + \mathbf{g}_e\right)G\frac{2m}{la}\mathbf{r}_u \cdot d\mathbf{S}. \quad (16\text{-}2.6)$$

Factoring out the constants and taking into account that $\mathbf{r}_u$ is parallel to $d\mathbf{S}$ on the cylindrical surface (so that $\mathbf{r}_u \cdot d\mathbf{S} = dS$) and perpendicular to $d\mathbf{S}$ on the flat ends (so that the flat ends make no contribution to the integral), we obtain

$$\frac{d\mathbf{G}_M}{dt} = \frac{m}{2\pi la}\int\left(-G\frac{2m}{la}\mathbf{r}_u + \mathbf{g}_e\right)dS, \quad (16\text{-}2.7)$$

where the integration is now over the cylindrical surface. Since to every $\mathbf{r}_u$ at a point of the cylindrical surface there corresponds an equal but opposite $\mathbf{r}_u$ at a diametrically opposite point, the first term in the integrand makes no contribution to the integral, and we have

$$\frac{d\mathbf{G}_M}{dt} = \frac{m}{2\pi la}\int \mathbf{g}_e dS. \quad (16\text{-}2.8)$$

Factoring out $\mathbf{g}_e$ and integrating, we obtain the final result:

$$\frac{d\mathbf{G}_M}{dt} = \frac{m}{2\pi la}\mathbf{g}_e\int dS = \frac{m2\pi al}{2\pi la}\mathbf{g}_e = m\mathbf{g}_e. \quad (16\text{-}2.9)$$

And so we have arrived at the well-known expression $m\mathbf{g}_e$ which is customarily considered to represent the force with which the Earth's gravitational field acts on falling bodies. However, according to our calculations, $m\mathbf{g}_e$ is not a force, but the rate of change of the momentum of $m$ (mass of the cylinder). Thus our calculations have clearly shown that what appears to us to be the force acting on a mass in a gravitational field is in reality the invisible process of direct momentum transfer from the gravitational (or cogravitational) field to the body located in this field.

**Example 16-2.2** Show that the basic gravitational and cogravitational force equation, Eq. (2-2.6), is a direct consequence of the momentum equation, Eq. (2-2.12).

Let us modify vector identity (V-22) by setting in it $\mathbf{B} = \mathbf{A}$. We then obtain a new vector identity which we shall call (V-22m):

$$\frac{1}{2}\oint A^2 d\mathbf{S} - \oint \mathbf{A}(\mathbf{A}\cdot d\mathbf{S}) = \int [\mathbf{A}\times(\nabla\times\mathbf{A}) - \mathbf{A}(\nabla\cdot\mathbf{A})]dV, \tag{V-22m}$$

where $\mathbf{A}$ is an arbitrary vector field. Applying now vector identity (V-22m) to the surface integrals in Eq. (2-2.12). We obtain

$$\frac{d\mathbf{G}_M}{dt} = \frac{1}{4\pi G}\int \frac{\partial}{\partial t}(\mathbf{g}\times\mathbf{K})dV$$
$$-\frac{1}{4\pi G}\int [(\nabla\cdot\mathbf{g})\mathbf{g} + c^2(\nabla\cdot\mathbf{K})\mathbf{K} - \mathbf{g}\times(\nabla\times\mathbf{g}) - c^2\mathbf{K}\times(\nabla\times\mathbf{K})]\,dV. \tag{16-2.10}$$

Now, by Eqs. (7-1.1)-(7-1.4),

$$\nabla\cdot\mathbf{g} = -4\pi G\rho, \qquad \nabla\cdot\mathbf{K} = 0,$$
$$\nabla\times\mathbf{g} = -\frac{\partial\mathbf{K}}{\partial t}, \qquad \nabla\times\mathbf{K} = -\frac{4\pi G}{c^2}\mathbf{J} + \frac{1}{c^2}\frac{\partial\mathbf{g}}{\partial t}. \tag{6-2.11}$$

Substituting these expressions into Eq. (16-2.10), we have

$$\frac{d\mathbf{G}_M}{dt} = \frac{1}{4\pi G}\int \frac{\partial}{\partial t}(\mathbf{g}\times\mathbf{K})dV \qquad (16\text{-}2.12)$$
$$- \frac{1}{4\pi G}\int\left[-4\pi G\rho\,\mathbf{g} + \mathbf{g}\times\frac{\partial\mathbf{K}}{\partial t} - \mathbf{K}\times\left(-4\pi G\mathbf{J} + \frac{\partial\mathbf{g}}{\partial t}\right)\right]dV.$$

Since

$$\frac{\partial}{\partial t}(\mathbf{g}\times\mathbf{K}) = \frac{\partial\mathbf{g}}{\partial t}\times\mathbf{K} + \mathbf{g}\times\frac{\partial\mathbf{K}}{\partial t} \qquad (16\text{-}2.13)$$

and

$$\frac{\partial\mathbf{g}}{\partial t}\times\mathbf{K} = -\mathbf{K}\times\frac{\partial\mathbf{g}}{\partial t}, \qquad (16\text{-}2.14)$$

the expressions containing the time derivatives cancel, and we are left with

$$\frac{d\mathbf{G}_M}{dt} = \int(\rho\mathbf{g} - \mathbf{K}\times\mathbf{J})dV = \int(\rho\mathbf{g} + \mathbf{J}\times\mathbf{K})dV = \int\rho(\mathbf{g} + \mathbf{v}\times\mathbf{K})dV, \qquad (16\text{-}2.15)$$

which is the same as Eq. (2-2.6), except that instead of the usual force on the left side of the equation we have the rate of change of the mechanical momentum $\mathbf{G}_M$ of the mass caused by the action of the fields $\mathbf{g}$ and $\mathbf{K}$.

In connection with the above derivation, it may be noted that for time-independent systems Eq. (2-2.12) reduces to Eqs. (7-3.30) and (7-3.36), which are the gravitational and cogravitational equivalents of electromagnetic Maxwell's stress integrals[3] (with the rate of change of mechanical momentum in place of the usual force):

$$\frac{d\mathbf{G}_M}{dt} = \frac{1}{8\pi G}\oint(\mathbf{g}^2 + c^2\mathbf{K}^2)d\mathbf{S} - \frac{1}{4\pi G}\oint\mathbf{g}(\mathbf{g}\cdot d\mathbf{S}) - \frac{c^2}{4\pi G}\oint\mathbf{K}(\mathbf{K}\cdot d\mathbf{S}). \qquad (16\text{-}2.14)$$

▲

## References and Remarks for Chapter 16

**1.** A less general case of the gravitational and kinetic energy exchange was previously considered in D. Bedford and P. Krumm, "The gravitational Poynting vector and energy transfer," Am. J. Phys. 55, 362-363 (1987); see also Oleg D. Jefimenko, *Causality, Electromagnetic Induction and Gravitation,* 2nd ed., (Electret Scientific, Star City, 2000) pp. 125-127.
**2.** For a similar but much more extensive and detailed analysis of the physical nature of electromagnetic forces see Oleg D. Jefimenko *Electromagnetic Retardation and Theory of Relativity*, 2nd ed., (Electret Scientific, Star City, 2004) pp. 302-330.
**3.** See Oleg D. Jefimenko, *Electricity and Magnetism*, 2nd ed., (Electret Scientific, Star City, 1989) pp. 215-216 and 446-447.

# 17

# PHYSICAL LINK BETWEEN GRAVITATIONAL AND ELECTROMAGNETIC FIELDS

Gravitational fields attract photons. As a result, light beams are deflected by gravitational fields from the paths that they would have in field-free space. On the other hand, according to Maxwellian electromagnetic theory, light consists of electromagnetic waves, and the propagation of all electromagnetic waves is governed by the laws of electromagnetism. Hence, at least to some extent, gravitation and electromagnetism are physically linked together. In this chapter we shall explore some aspects of this link.

## 17-1. Coupling of Gravitational and Electromagnetic Fields

Consider a light ray propagating with velocity $c$ in a region free from a gravitational field. Let it strike at a grazing incidence a thin transparent boundary between the field-free region and a region where the gravitational field is $\mathbf{g}$ (Fig. 17.1) (such a configuration of gravitational fields can be created with the help of "gravitational parallel-plate capacitors" discussed in Example 13-1.7). Upon entering the region where $\mathbf{g} \neq 0$, the ray is deflected, because the photons in the ray are attracted by the

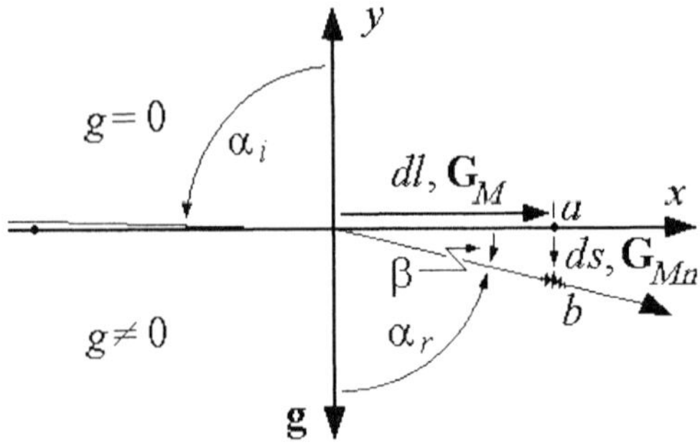


*Fig. 17.1 A light ray propagating with velocity c in a region free from gravitational field strikes at a grazing incidence a boundary between the field-free region and a region where the gravitational field is* **g**. *Upon entering the region where* **g** $\neq$ 0, *the ray is deflected, because the photons in the ray are attracted by the gravitational field.*

gravitational field. Let us find the angle $\beta$ between the deflected ray and the boundary, assuming that the ray is only slightly deflected from its original path.

Acting on an individual photon in the ray, the gravitational field creates a component of the photon's mechanical momentum normal to the boundary, $\mathbf{G}_{Mn}$, in accordance with the formula

$$\frac{d\mathbf{G}_{Mn}}{dt} = m\mathbf{g}, \qquad (17\text{-}1.1)$$

where $m$ is the photon's mass. Immediately after crossing the boundary, the trajectory of the photon is still a straight line and its velocity is still essentially $c$.[1] Therefore we can replace $dt$ in Eq. (17-1.1) by $dl/c$, where $dl$ is the distance travelled by the photon during the time $dt$. We then obtain from Eq. (17-1.1) and Fig. 17.1

$$d\mathbf{G}_{Mn} = -\mathbf{j}\frac{mg}{c}dl. \tag{17-1.2}$$

As is known, the mechanical momentum of the photon in a field-free space is $G_M = mc$. Since the velocity of the photon immediately after crossing the boundary is still essentially $c$, we can assume that the tangential (with respect to the boundary) component of the photon's mechanical momentum over the distance $dl$ is the same as in the original ray, that is $\mathbf{G}_M = mc\mathbf{i}$. Consequently, taking into account that, by supposition, the deflection of the ray is small and using Eq. (17-1.2), we have for the angle $\beta$ between the deflected light ray and the boundary

$$\sin\beta \approx \tan\beta = \frac{dG_{Mn}}{G_M} = \frac{mgdl}{cmc} = \frac{gdl}{c^2}. \tag{17-1.3}$$

Furthermore, according to Fig. 17.1, $dl = ds/\tan\beta \approx ds/\sin\beta$, so that, using Eq. (17-1.3), we can write

$$\sin^2\beta \approx \frac{gds}{c^2} \tag{17-1.4}$$

or,

$$\cos^2\beta \approx 1 - \frac{gds}{c^2}. \tag{17-1.5}$$

We can simplify Eq. (17-1.5) by noticing that, by Eq. (7-3.15), $gds$ is the difference of gravitational scalar potentials $\varphi_a$ and $\varphi_b$ between the points $a$ and $b$ shown in Fig. 17.1. Setting $\varphi_a = 0$ (that is, using the potential at the boundary as the reference potential), we then have $gds = -\varphi_b$, or, simply, $gds = -\varphi$. Eq. (17-1.5) then becomes

$$\cos^2\beta \approx 1 + \frac{\varphi}{c^2} \tag{17-1.6}$$

(observe that $\varphi$ is negative).

Now, by Snell's law,[2]

$$\frac{\sin\alpha_i}{\sin\alpha_r} = \frac{\sin\alpha_i}{\cos\beta} = n\,, \tag{17-1.7}$$

where $n$ is the index of refraction in the space where the refracted ray is located. Since in the present case $\alpha_i = \pi/2$, the index of refraction in the space below the boundary is, by Eq. (17-1.7),

$$n \approx \frac{1}{\cos\beta}\,. \tag{17-1.8}$$

or, with Eq. (17-1.6),

$$n \approx \frac{1}{(1 + \varphi/c^2)^{1/2}}\,. \tag{17-1.9}$$

Taking into account that $n = c/v$, where $v$ is the velocity of the refracted ray,[2] our calculations show that the velocity of light in a region of space where the gravitational potential is $\varphi$ is

$$v \approx c(1 + \varphi/c^2)^{1/2} \approx c(1 + \varphi/2c^2)\,, \tag{17-1.10}$$

and, since $\varphi$ is negative, is smaller that the velocity of light in a field-free region.[3]

According to the Maxwellian electromagnetic theory, the index of refraction of a medium is determined by the relative permittivity $\varepsilon$ and the relative permeability $\mu$ of the medium in accordance with the formula[2]

$$n = (\varepsilon\mu)^{1/2}\,. \tag{17-1.11}$$

Therefore we must conclude that the gravitational field, as far as electromagnetic phenomena are concerned, constitutes a medium whose $\varepsilon$ and $\mu$ are, by Eqs. (7-1.9) and (17-1.11)

$$\varepsilon\mu \approx \frac{1}{1 + \varphi/c^2} \approx 1 - \varphi/c^2\,. \tag{17-1.11}$$

The consequences of this result may be very considerable. All electromagnetic forces and interactions depend on $\varepsilon$ and $\mu$. Larger $\varepsilon$ results in smaller electric fields and smaller electric forces, larger $\mu$ results in larger magnetic fields and stronger magnetic forces. Insofar as the interatomic forces are basically electric forces, atomic energy levels should shift towards lower energy levels by gravitational fields and therefore the radiation by atoms and molecules located in a gravitational field should be shifted towards the "red" end of the spectrum. For the same reason, electric and atomic clocks should slow down in the presence of gravitational fields (see also Section 17-3). Many other similar effects are clearly possible. Of course, all such effects are very small except, possibly, in very strong gravitational fields. It should be noted, however, that our calculations apply to relatively weak gravitational fields.

## 17-2. The Bending of Light Under the Action of a Gravitational Field

Consider a light beam passing across a gravitational field. The gravitational field attracts the photons in the beam and deflects the beam from its original direction. Let the beam pass near a large mass, such as the Sun. The gravitational field of the Sun attracts the photons in the beam and causes the beam to bend. To determine the exact shape of the beam and its deflection from the original direction one needs to treat the Sun's gravitational field as a spherical lens whose index of refraction depends on the distance (potential) from the Sun in accordance with Eq. (17-1.9) (if the Sun's field is "weak"). This is a difficult problem and we shall not attempt to solve it here. There is, however, a simpler way to obtain an approximate solution of the problem. We do not expect the beam to be bent very much and therefore we can assume that the angle $\alpha$ (see Fig. 17.2) between the final and the original direction of the beam is small. Let us find this angle.

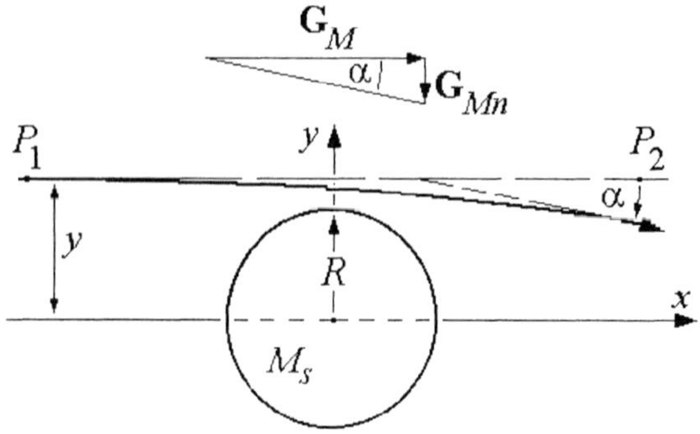


*Fig. 17.2 Light ray passing close to the Sun is bent by the Sun's gravitational field.*

Acting on an individual photon in the beam, the component of the Sun's gravitational field normal to the beam creates a component of the photon's mechanical momentum normal to the beam, $\mathbf{G}_{Mn}$, in accordance with the formula

$$\frac{d\mathbf{G}_{Mn}}{dt} = m\mathbf{g}_n, \tag{17-2.1}$$

where $m$ is the photon's mass and $\mathbf{g}_n$ is the normal component of the Sun's gravitational field. Let the mass of the Sun be $M_s$, let the Sun be at the origin of rectangular coordinates and let the light beam be originally parallel to the $x$ axis at a distance $y$ above the axis. The component of the Sun's gravitational field normal to the beam is then, by Eq. (7-3.10) or (7-3.9), (as is known, the external gravitational field of a uniformly distributed spherical mass is the same as if the entire mass was at the center of the sphere)

$$\mathbf{g}_n = \mathbf{g}_y = -G\frac{M_s}{r^3}y\mathbf{j}, \tag{17-2.2}$$

where $r$ is the distance between the sun and the photon under consideration. The momentum component $\mathbf{G}_{Mn}$ acquired by the photon during its flight past the Sun is therefore

$$\mathbf{G}_{Mn} = -\mathbf{j}\int G\frac{mM_s}{r^3}y\,dt. \tag{17-2.3}$$

Assuming that the speed of the photon remains essentially constant and equal to $c$ during its flight past the Sun, we can replace $dt$ in Eq. (17-2.3) by $dx/c$. Expressing then $r$ in terms of $x$ and $y$, we can write Eq. (17-2.3) as

$$\mathbf{G}_{Mn} = -\mathbf{j}\frac{GmM_s}{c}\int_{-P_1}^{P_2}\frac{y}{(x^2+y^2)^{3/2}}dx, \tag{17-2.4}$$

where $-P_1$ and $P_2$ are the starting and the ending $x$ coordinate of the photon's trajectory, respectively. Since the mass of photons is small, the light beam does not bend very much, and its distance, $y$, from the $x$ axis remains essential constant. Integrating Eq. (17-2.4) while keeping $y$ constant, we obtain

$$\mathbf{G}_{Mn} = -\mathbf{j}\frac{GmM_s}{c}\frac{xy}{y^2(x^2+y^2)^{1/2}}\Bigg|_{-P_1}^{P_2} \tag{17-2.5}$$

and, assuming that $P_1 = P_2 \gg y$,

$$\mathbf{G}_{Mn} = -\mathbf{j}\frac{2GmM_s}{cy}. \tag{17-2.6}$$

As is known, the mechanical momentum of the photon in a field-free space is $G_M = mc$ and the direction of the momentum is the same as the direction of the photon's velocity. Assuming that the point $P_1$ is essentially in a field-free space, the original mechanical momentum of the photon under consideration is $\mathbf{G}_M = mc\mathbf{i}$. By symmetry, the $x$ component of the photon's velocity at the point $P_2$ is the same as at the point $P_1$. Therefore its mechanical momentum along the $x$ axis at the point $P_2$ is again $\mathbf{G}_M = mc\mathbf{i}$. Consequently, the angle between the final and the original direction of the light beam passing the Sun (see Fig. 17.2) is given by the formula

$$\sin\alpha \approx \frac{G_{Mn}}{G_M} = \frac{2GmM_s}{mccy}, \tag{17-2.7}$$

or, taking into account that, by our suppositions, $G_{Mn} \ll G_M$

$$\alpha \approx \frac{2GM_s}{c^2y}. \tag{17-2.8}$$

For light grazing the surface of the Sun, Eq. (17-2.8) becomes

$$\alpha \approx \frac{2GM_s}{c^2R}, \tag{17-2.9}$$

where $R$ is the radius of the Sun.

It should be noted that the actual deflection of a light beam by the Sun is larger than as shown by Eq. (17-2.9) because the space in the vicinity of the Sum is not a vacuum. The Sun is actually surrounded by a material medium – solar corona, zodiacal cloud, solar wind, etc., all of which make the index of refraction in the space near the Sun larger than one. Also, one should remember that our solution is only approximately correct because, as mentioned at the beginning of this section, to find the actual deflection of the beam, the Sun's gravitational field should be treated as a complex spherical lens.

## 17-3. Gravitational Shift of Spectral Lines

Consider a body moving in a gravitational field. According to Eqs. (16-1.7) and (16-1.8), as a result of the energy exchange between the gravitational field and the body, the kinetic energy of the body increases or decreases at the rate

$$\frac{dU_{kin}}{dt} = \pm m\mathbf{g} \cdot \mathbf{v}, \qquad (17\text{-}3.1)$$

where "+" applies when the body moves in the direction of the gravitational field, and "−" applies when the body moves in the opposite direction. Let us express $\mathbf{v}$ as $ds/dt$, where $ds$ is the distance traveled by the body during the time interval $dt$. From Eq. (17-3.1) we then have

$$\frac{dU_{kin}}{dt} = \pm m\mathbf{g} \cdot \frac{ds}{dt} \qquad (17\text{-}3.2)$$

or

$$dU_{kin} = \pm m\mathbf{g} \cdot d\mathbf{s}. \qquad (17\text{-}3.3)$$

By Eq. (17-3.3) (omitting the subscript *kin* for simplicity), if the body travels from a point $r_1$ to a point $r_2$, its kinetic energy changes by

$$U_2 - U_1 = \pm m \int_{r_1}^{r_2} \mathbf{g} \cdot d\mathbf{s}, \qquad (17\text{-}3.4)$$

where $U_2$ and $U_1$ denote the kinetic energy of the body at points $r_2$ and $r_1$, respectively. By Eq. (7-3.15), equation (17-3.4) can be written as

$$U_2 - U_1 = m(\varphi_1 - \varphi_2), \qquad (17\text{-}3.5)$$

where $\varphi_1$ and $\varphi_2$ are the gravitational potentials at points $r_1$ and $r_2$, respectively.

Let us now assume that the body under consideration is a photon and let us assume that it moves from a region of stronger gravitational field (point $r_1$, smaller $\varphi$ - remember that in general the gravitational potential is a negative quantity) to a region of weaker gravitational field (point $r_2$, larger $\varphi$). As is known, the mass of a photon is $m = E/c^2$ and its frequency is $\nu = E/h$, where $E$ is the energy of the photon, $c$ is the velocity of light and $h$ is Planck's constant. In this case, according to Eq. (17-3.5), as the photon moves from $r_1$ to $r_2$, its energy, and therefore its frequency, diminishes. From Eq. (17-3.5) we then have for the change of the photon's frequency ("frequency shift")

$$\nu_2 - \nu_1 = \frac{U_2 - U_1}{h} = \frac{E}{hc^2}(\varphi_1 - \varphi_2), \tag{17-3.6}$$

where $\nu_2$ and $\nu_1$ are the frequencies of the photon at points $r_2$ and $r_1$, respectively, and $E$ is the average energy of the photon during its motion from $r_1$ to $r_2$.

Let us further assume that the frequency shift is small compared with $\nu_1$, so that photon's mass, $m = E/c^2 = h\nu/c^2$ remains essentially the same as it was at the point $r_1$. In this case we can replace in Eq. (17-3.6) the average energy of the photon by $h\nu_1$, obtaining

$$\nu_2 - \nu_1 = \frac{\nu_1}{c^2}(\varphi_1 - \varphi_2). \tag{17-3.7}$$

For the ratio of the frequency shift to the initial frequency we then have

$$\frac{\nu_2 - \nu_1}{\nu_1} = \frac{\varphi_1 - \varphi_2}{c^2}. \tag{17-3.8}$$

If the source of the photon is within a gravitational field and the observer is at a field-free point ($\varphi_2 = 0$), the shift is toward smaller frequencies ("red shift" of the spectral lines in the spectrum of the source). If the source of the photon is at a field-free point ($\varphi_1 = 0$) and the observer is in a gravitational field, the shift is toward larger frequencies ("violet shift" of the spectral lines in the spectrum of the source).

In particular, the gravitational red shift for light emitted by atoms on the surface of the Sun, as observed far from the Sun, is, according to Eqs. (17-3.8) and (7-3.12) [or (7-3.13)],

$$\frac{\Delta\nu}{\nu_1} = \frac{GM_s}{Rc^2}. \qquad (17\text{-}3.9)$$

If the energy of the photon experiences a large change under the action of the gravitational field, the calculations of the frequency shift are difficult. Clearly, however, the essential result of the calculations will not differ significantly for the results presented above.

It may be noted that Eq. (17-3.9) applies also to the frequency shift of atomic clocks and, assuming that the frequency shift is small, can be written as

$$\frac{\Delta\nu}{\nu} = \frac{GM}{Rc^2}, \qquad (17\text{-}3.10)$$

where $M$ the mass of the spherical body, $R$ is the distance from its center, and $\nu$ is the frequency of the clock in a field-free space.

## References and Remarks for Chapter 17

**1.** Actually, the velocity of the photon is now $v = c/n$, where $n$ is the index of refraction in the region below the boundary. Our calculations and the derivations that follow are for "weak" gravitational fields, in which $n \approx 1$.

**2.** See, for example, Oleg D. Jefimenko, *Electricity and Magnetism*, 2nd ed., (Electret Scientific, Star City, 1989) pp. 535-539.

**3.** Although Eq. (17-1.10) has been derived specifically for the velocity of light, it clearly applies to any electromagnetic radiation, including radio and radar signals.

# 18

# GRAVITATIONAL AND COGRAVITATIONAL WAVES

In the broadest sense of the word, a wave is any "disturbance" that propagates in space. In the very narrow sense, a wave is a "disturbance" of sinusoidal shape. Gravitational and cogravitational fields can exist as waves. In this chapter we shall derive equations describing basic properties of gravitational and cogravitational waves and we shall learn how such waves can be generated.

## 18-1. The Existence of Gravitational and Cogravitational Waves

As we know, the similarity of Eqs. (7-1.1)-(7-1.4) with Maxwell's electromagnetic equations suggests that many electromagnetic phenomena have their gravitational and cogravitational counterparts. In particular it may be expected that there should exist gravitational and cogravitational waves similar to the electromagnetic waves. We shall now show by direct calculations how Eqs. (7-1.1)-(7-1.4) predict the existence of gravitational and cogravitational waves.

We start with Eq. (7-1.3)

$$\nabla \times \mathbf{g} = -\frac{\partial \mathbf{K}}{\partial t}. \qquad (7\text{-}1.3)$$

Taking the curl of this equation, we obtain

$$\nabla \times \nabla \times \mathbf{g} = -\frac{\partial}{\partial t}\nabla \times \mathbf{K}. \tag{18-1.1}$$

Substituting Eq. (7-1.4),

$$\nabla \times \mathbf{K} = -\frac{4\pi G}{c^2}\mathbf{J} + \frac{1}{c^2}\frac{\partial \mathbf{g}}{\partial t} \tag{7-1.4}$$

into Eq. (18-1.1), we obtain

$$\nabla \times \nabla \times \mathbf{g} = \frac{4\pi G}{c^2}\frac{\partial \mathbf{J}}{\partial t} - \frac{1}{c^2}\frac{\partial^2 \mathbf{g}}{\partial t^2}, \tag{18-1.2}$$

or

$$\nabla \times \nabla \times \mathbf{g} + \frac{1}{c^2}\frac{\partial^2 \mathbf{g}}{\partial t^2} = \frac{4\pi G}{c^2}\frac{\partial \mathbf{J}}{\partial t}. \tag{18-1.3}$$

Similarly, taking the curl of Eq. (7-1.4), we have

$$\nabla \times \nabla \times \mathbf{K} = -\frac{4\pi G}{c^2}\nabla \times \mathbf{J} + \frac{1}{c^2}\frac{\partial \nabla \times \mathbf{g}}{\partial t}, \tag{18-1.4}$$

and, substituting Eq. (7-1.3) into Eq. (18-1.4), we obtain

$$\nabla \times \nabla \times \mathbf{K} = -\frac{4\pi G}{c^2}\nabla \times \mathbf{J} - \frac{1}{c^2}\frac{\partial^2 \mathbf{K}}{\partial t^2}, \tag{18-1.5}$$

or

$$\nabla \times \nabla \times \mathbf{K} + \frac{1}{c^2}\frac{\partial^2 \mathbf{K}}{\partial t^2} = -\frac{4\pi G}{c^2}\nabla \times \mathbf{J}. \tag{18-1.6}$$

Equations (18-1.3) and (18-1.6) are mathematical expressions for waves propagating in space with velocity $c$. In the present case they represent waves carrying with themselves the gravitational field $\mathbf{g}$ and the cogravitational field $\mathbf{K}$, respectively.

If in the region under consideration there are no mass currents, Eqs. (18-1.3) and (18-1.6) reduce to

$$\nabla \times \nabla \times \mathbf{g} + \frac{1}{c^2}\frac{\partial^2 \mathbf{g}}{\partial t^2} = 0 \tag{18-1.7}$$

and

$$\nabla \times \nabla \times \mathbf{K} + \frac{1}{c^2}\frac{\partial^2 \mathbf{K}}{\partial t^2} = 0. \qquad (18\text{-}1.8)$$

Now, by vector identity (V-16)

$$\nabla \times \nabla \times \mathbf{g} = \nabla(\nabla \cdot \mathbf{g}) - \nabla^2 \mathbf{g} \qquad (18\text{-}1.9)$$

and

$$\nabla \times \nabla \times \mathbf{K} = \nabla(\nabla \cdot \mathbf{K}) - \nabla^2 \mathbf{K}. \qquad (18\text{-}1.10)$$

Furthermore, by Eq. (7-1.2), $\nabla \cdot \mathbf{K} = 0$, and, by Eq. (7-1.1), in a mass-free region of space also $\nabla \cdot \mathbf{g} = 0$. Therefore in a region of space where there are no masses and no mass currents, Eqs. (18-1.3) and (18-1.6) become the more familiar "wave equations"

$$\nabla^2 \mathbf{g} - \frac{1}{c^2}\frac{\partial^2 \mathbf{g}}{\partial t^2} = 0 \qquad (18\text{-}1.11)$$

and

$$\nabla^2 \mathbf{K} - \frac{1}{c^2}\frac{\partial^2 \mathbf{K}}{\partial t^2} = 0. \qquad (18\text{-}1.12)$$

To demonstrate that Eqs. (18-1.11) and (18-1.12) do indeed represent waves, let us assume that the fields **g** and **K** in Eqs. (18-1.11) and (18-1.12) depend only on one coordinate in a rectangular system of coordinates (the wavefront of such waves is a plane and therefore such waves are called plane waves). Let us assume that **g** and **K** in Eqs. (18-1.11) and (18-1.12) depend only on the $z$ coordinate. In this case $\partial \mathbf{g}/\partial x = \partial \mathbf{g}/\partial y = 0$ and $\partial \mathbf{K}/\partial x = \partial \mathbf{K}/\partial y = 0$, so that Eqs. (18-1.11) and (18-1.12) become

$$\frac{\partial^2 \mathbf{g}}{\partial z^2} - \frac{1}{c^2}\frac{\partial^2 \mathbf{g}}{\partial t^2} = 0 \qquad (18\text{-}1.13)$$

and

$$\frac{\partial^2 \mathbf{K}}{\partial z^2} - \frac{1}{c^2}\frac{\partial^2 \mathbf{K}}{\partial t^2} = 0. \qquad (18\text{-}1.14)$$

The solution of each of these equations is the sum of the two arbitrary vector functions

$$\mathbf{f}_1(t - z/c) + \mathbf{f}_2(t + z/c) \tag{18-1.15}$$

as can be verified by direct substitution.

The function $\mathbf{f}_1(t - z/c)$ is an expression for a wave propagating with speed $c$ in the positive direction of the $z$ axis. This can be seen from the fact that, since the argument of the function is $t - z/c$, the value of the function at the time $t_0$ and point $z_0$ is the same as at a later time $t_1$ and a further point $z_1$, provided that

$$(t_1 - z_1/c) = (t_0 - z_0/c). \tag{18-1.16}$$

Solving Eq. (18-1.16) for $c$, we have

$$c = \frac{z_1 - z_0}{t_1 - t_0}. \tag{18-1.17}$$

But $z_1 - z_0$ is the distance traveled by the wave during the time $t_1 - t_0$. Therefore $c$ is the velocity with which the wave propagates.

Similarly, the function $\mathbf{f}_1(t + z/c)$ is a mathematical expression for a wave propagating with velocity $c$ in the negative direction of the $z$ axis.

Although we usually associate waves with a sinusoidal shape, in physics a wave is actually a "disturbance" of any shape that propagates in space in accordance with Eq. (18-1.15) or in accordance of a similar equation not restricted to the $z$ axis only.

## 18-2. Direction of Gravitational and Cogravitational Field Vectors in Plane Waves

For a plane wave which is a function of $z$ and $t$ only, all partial derivatives with respect to $x$ and $y$ vanish. Since, by Eq. (7-1.2),

$$\nabla \cdot \mathbf{K} = \frac{\partial K_x}{\partial x} + \frac{\partial K_y}{\partial y} + \frac{\partial K_z}{\partial z} = 0, \qquad (18\text{-}2.1)$$

we then have $\partial K_z/\partial z = 0$. Therefore the cogravitational field in a plane wave has no varying component along the direction of the propagation of the wave. Similarly, by Eq. (7-1.1), in a mass-free space (we are considering mass-free space),

$$\nabla \cdot \mathbf{g} = \frac{\partial g_x}{\partial x} + \frac{\partial g_y}{\partial y} + \frac{\partial g_z}{\partial z} = 0 \qquad (18\text{-}2.2)$$

and therefore $\partial g_z/\partial z = 0$, so that the gravitational field in a plane wave has no varying component along the direction of the propagation. This means that the wave is transverse – that is, the gravitational and cogravitational field vectors of the wave are perpendicular to the direction of the propagation of the wave.

We can obtain a more complete picture of the orientation of field vectors in a plane wave as follows. According to the preceding section, the field vectors in a plane wave propagating in the positive $z$ direction are functions of $t - z/c$. By vector identity (V-13), we then have

$$\nabla \times \mathbf{g} = \nabla(t - z/c) \times \frac{\partial \mathbf{g}}{\partial(t - z/c)} = -\frac{\mathbf{k}}{c} \times \frac{\partial \mathbf{g}}{\partial(t - z/c)}, \qquad (18\text{-}2.3)$$

where $\mathbf{k}$ is a unit vector in the $z$ direction. Also, we have

$$\frac{\partial \mathbf{K}}{\partial t} = \frac{\partial \mathbf{K}}{\partial(t - z/c)}. \qquad (18\text{-}2.4)$$

But according to Eq. (7-1.3),

$$\nabla \times \mathbf{g} = -\frac{\partial \mathbf{K}}{\partial t}. \qquad (7\text{-}1.3)$$

Therefore

$$-\frac{\mathbf{k}}{c} \times \frac{\partial \mathbf{g}}{\partial(t - z/c)} = -\frac{\partial \mathbf{K}}{\partial(t - z/c)}. \qquad (18\text{-}2.5)$$

Integrating this equation, we obtain

$$\mathbf{K} = \frac{1}{c}\mathbf{k} \times \mathbf{g}. \tag{18-2.6}$$

Since the wave is transverse, so that, $\mathbf{k} \perp \mathbf{g}$, Eq. (18-2.6) shows that, in a plane wave, the cogravitational field vector, the unit vector in the direction of propagation, and the gravitational field vector are mutually perpendicular and form a right-handed system in the order stated.

## 18-3. Energy Relations in Plane Gravitational and Cogravitational Waves

By the energy law Eq. (2-2.7), the energy density in a plane gravitational-cogravitational wave propagating through free space is

$$U_v = -\frac{1}{8\pi G}\mathbf{g}^2 - \frac{1}{8\pi G}c^2\mathbf{K}^2. \tag{18-3.1}$$

By Eq. (18-2.6), considering only the magnitudes of $\mathbf{K}$ and $\mathbf{g}$,

$$c^2\mathbf{K}^2 = \mathbf{g}^2 \tag{18-3.2}$$

Therefore

$$\frac{1}{8\pi G}\mathbf{g}^2 = \frac{1}{8\pi G}c^2\mathbf{K}^2, \tag{18-3.3}$$

so that the energy of the wave is divided equally between the gravitational and the cogravitaional fields.

Let us now calculate the gravitational Poynting's vector associated with a plane gravitational-cogravitational wave. Using Eq. (2-2.9) and Eq. (18-2.6), we have

$$\mathbf{P} = \frac{c^2}{4\pi G}\mathbf{K}\times\mathbf{g} = \frac{c}{4\pi G}(\mathbf{k}\times\mathbf{g})\times\mathbf{g}. \tag{18-3.4}$$

Transposing in the last expression $\mathbf{k} \times \mathbf{g}$ and $\mathbf{g}$ and using vector identity (V-3), we have

$$\mathbf{P} = \frac{c}{4\pi G}\mathbf{g}(\mathbf{k}\cdot\mathbf{g}) - \frac{c}{4\pi G}\mathbf{k}(\mathbf{g}\cdot\mathbf{g}), \qquad (18\text{-}3.5)$$

and since **k** is perpendicular to **g**, we obtain

$$\mathbf{P} = -\frac{c}{4\pi G}\mathbf{g}^2\mathbf{k}. \qquad (18\text{-}3.6)$$

or, with Eq. (18-3.3),

$$\mathbf{P} = -\frac{c^3}{4\pi G}\mathbf{K}^2\mathbf{k}. \qquad (18\text{-}3.7)$$

Dividing Eqs. (18-3.6) and (18-3.7) by 2, adding them, using Eq. (18-3.1), and using $c\mathbf{k} = \mathbf{c}$, we can express **P** as

$$\mathbf{P} = U_v\mathbf{c}, \qquad (18\text{-}3.8)$$

where $U_v$ is the energy density, and **c** is the velocity vector of the wave. Thus gravitational-cogravitational field energy is propagated by the gravitational-cogravitational wave with speed $c$ in the direction in which the wave propagates.

Note, however, that $U_v$ is negative. An important consequence of the negative $U_v$ in gravitational-cogravitational waves is that, in contrast to the electromagnetic waves, a gravitational-cogravitational wave striking a body pulls the body toward the wave, that is, exerts a negative rather than a positive pressure on the body.[1]

## 18-4. Generation of Sinusoidal Gravitational-Cogravitational Waves

Consider a ring whose mass is distributed along its circumference with uniform line density $\lambda$ (Fig. 18.1). Let it oscillate about its symmetry axis with angular velocity $\omega$ according to the formula

$$\alpha = \alpha_0 \sin\omega t, \qquad (18\text{-}4.1)$$

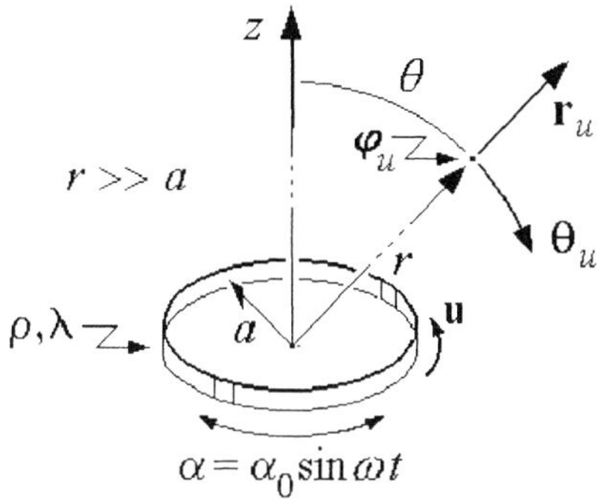


*Fig. 18.1 Gravitational-cogravitational waves can be generated by an oscillating ring. The unit vector $\varphi_u$ is directed into the page.*

where $\alpha_0$ is the amplitude of the oscillations. The linear velocity **u** of the ring is given by the formula

$$\mathbf{u} = a\frac{d\alpha}{dt}\varphi_u = a\alpha_0\omega(\cos\omega t)\varphi_u, \tag{18-4.2}$$

where $a$ is the radius of the ring and $\varphi_u$ is the circular (azimuthal) unit vector directed along the velocity vector **u** of the ring.

The gravitational field of the ring is given by the Eq. (2-2.1),

$$\mathbf{g} = -G\int\left\{\frac{[\rho]}{r^3} + \frac{1}{r^2c}\left[\frac{\partial\rho}{\partial t}\right]\right\}\mathbf{r}\,dV' + \frac{G}{c^2}\int\frac{1}{r}\left[\frac{\partial(\rho\mathbf{u})}{\partial t}\right]dV', \tag{2-2.1}$$

Since the mass density in the ring is constant, the first integral in Eq. (2-2.1) makes no contribution to the emission of waves. Therefore, in the calculations that follow, we shall only be concerned with the last integral of Eq. (2-2.1). Substituting Eq. (18-4.2) into this integral and factoring out the constants, we obtain

$$\mathbf{g} = \varphi_u\frac{a\alpha_0\omega G}{c^2}\int\frac{1}{r}\left[\frac{\partial(\rho\cos\omega t)}{\partial t}\right]dV' = -\varphi_u\frac{a\alpha_0\omega^2 G}{c^2}\int\frac{[\rho\sin\omega t]}{r}dV'. \tag{18-4.3}$$

Just as in Eq. (2-2.1), the brackets in Eq. (18-4.3) signify retardation and indicate that the present time $t$ in the expressions between the brackets must be replaced by the past time, the "retarded time" $t' = t - r/c$. Substituting $t'$ in Eq. (18-4.3) and removing the brackets we obtain

$$\mathbf{g} = -\varphi_u \frac{a\alpha_0\omega^2 G}{c^2} \int \frac{\rho \sin\omega(t - r/c)}{r} dV' . \qquad (18\text{-}4.4)$$

Note that the product $\rho dV'$ is equal to the mass element $dm$ contained in the volume element $dV'$ of the ring. However, the same mass element can be represented by the product $\lambda dl'$, where $\lambda$ is the linear density of the mass of the ring, $\lambda = m/2\pi a$, and where $dl'$ is an element of the length (circumference) of the ring. Replacing $\rho dV'$ in Eq. (18-4.4) by $\lambda dl'$, factoring out $\lambda$ (which is a constant) from under the integral sign, placing $\varphi_u$ under the integral sign, and replacing the product $\varphi_u dl'$ by the length element vector $d\mathbf{l}'$ of the ring, we obtain

$$\mathbf{g} = -\frac{a\alpha_0\omega^2\lambda G}{c^2} \oint \frac{\sin\omega(t - r/c)}{r} d\mathbf{l}' . \qquad (18\text{-}4.5)$$

where the integration is over the circumference of the ring.

Using vector identity (V-18), we transform the contour integral in Eq. (18-4.5) into the surface integral:

$$\begin{aligned} \mathbf{g} &= -\frac{a\alpha_0\omega^2\lambda G}{c^2} \oint \frac{\sin\omega(t - r/c)}{r} d\mathbf{l}' \\ &= -\frac{a\alpha_0\omega^2\lambda G}{c^2} \int d\mathbf{S}' \times \nabla \frac{\sin\omega(t - r/c)}{r} , \end{aligned} \qquad (18\text{-}4.6)$$

where the integration is over the surface area of the ring. Computing the gradient in the integrand, we get

$$\nabla \frac{\sin \omega(t - r/c)}{r} = -\left\{ \frac{\omega}{rc} \cos \omega(t - r/c) + \frac{1}{r^2} \sin \omega(t - r/c) \right\} \mathbf{r}_u, \tag{18-4.7}$$

where $\mathbf{r}_u$ is a unit vector directed from an element of the surface area of the ring to the point where the gravitational field $\mathbf{g}$ is being observed.

When considering emissions of waves, one is usually interested in waves far from the emitter. In this case we can neglect the term with $1/r^2$ in Eq. (18-4.7). Substituting the remaining expression into Eq. (18-4.6) and transposing $d\mathbf{S}'$ and $\mathbf{r}_u$, we obtain

$$\mathbf{g} = -\frac{a\alpha_0\omega^3\lambda G}{c^3} \int \frac{\cos \omega(t - r/c)}{r} \mathbf{r}_u \times d\mathbf{S}'. \tag{18-4.8}$$

Far from the emitter $r \gg a$ and therefore the distance from all the points of the ring to the point of observation is practically the same. Hence, in this case, the integral in Eq. (18-4.8) reduces simply to the product of the integrand and the area of the ring $\pi a^2$. Replacing $\lambda$ by $m/2\pi a$, taking into account that the magnitude of the vector product $\mathbf{r}_u \times d\mathbf{S}'$ is $\pi a^2 \sin\theta$ (see Fig. 18.1), and taking into account that the direction of this vector product is against the azimuthal unit vector $\boldsymbol{\varphi}_u$, we finally obtain for the gravitational field g emitted by the ring:

$$\mathbf{g} = G\frac{m\alpha_0 a^2\omega^3 \cos \omega(t - r/c)}{2rc^3} (\sin\theta)\, \boldsymbol{\varphi}_u. \tag{18-4.9}$$

A similar expression can be derived for the cogravitational field $\mathbf{K}$ emitted by the ring (for brevity we shall give it here without derivation):

$$\mathbf{K} = -G\frac{m\alpha_0 a^2\omega^3 \cos \omega(t - r/c)}{2rc^4} (\sin\theta)\, \boldsymbol{\theta}_u, \tag{18-4.10}$$

where $\boldsymbol{\theta}_u$ is the polar unit vector.

Equations (18-4.9 and (18-4.10) represent spherical sinusoidal waves propagating in space with velocity $c$ in the direction of increasing $r$.

With the help of the gravitational Poynting vector we can calculate the radiation power $W$ of the gravitational emission of the oscillating ring.[2] For the average power emitted during one complete oscillation of the ring we obtain

$$W_{av} = -G\frac{m^2\alpha_0^2 a^4\omega^6}{12c^5}. \qquad (18\text{-}4.11)$$

An important characteristic of Eq. (18-4.11) is the negative sign in front of it. It indicates that the gravitational waves carry negative energy, and that the oscillating ring absorbs energy from the surrounding space.

It may be noted that the magnitude of the gravitational and cogravitational radiation fields is much too small to be measured in a laboratory at this time. For example, according to Eqs. (18-4.9)-(18-4.11), a ring of mass 1 kg and radius 1 m oscillating with an amplitude of 1 rad and circular frequency 1 $\sec^{-1}$ would produce at a distance of 10 m from itself a gravitational and cogravitational wave of amplitude $g \approx 1.2 \cdot 10^{-36}$ m/sec$^2$, $K \approx 4 \cdot 10^{-45}$ sec$^{-1}$, and average power flow of $W_{av} \approx -2.3 \cdot 10^{-54}$ watts.

## References and Remarks for Chapter 18

**1.** The calculations of the negative pressure by gravitational-cogravitational waves are similar to the corresponding calculations of the positive pressure by electromagnetic waves. For the latter calculations see Oleg D. Jefimenko, *Electricity and Magnetism*, 2nd ed., (Electret Scientific, Star City, 1989) pp. 513-514.

**2.** See also Oleg. D. Jefimenko, *Causality, Electromagnetic Induction, and Gravitation*, 2nd ed., (Electret Scientific, Star City,

2000) pp. 132-133. The calculations are similar to the corresponding calculations of the power emitted by a magnetic dipole antenna. See Ref. 1, pp. 562-565.

# 19

# GRAVITATION AND ANTIGRAVITATION

According to Einstein's mass-energy relation, any energy has a certain mass. But mass is the source of gravitation. Therefore any energy, including gravitational energy, should be a source of gravitation. In this chapter we shall complete our development and generalization of Newtonian gravitational theory by investigating how gravitational fields are affected by the gravitational energy contained in them. For simplicity, we shall discuss time-independent fields only.[1]

## 19-1. Gravitational Energy as a Source of Gravitation

As explained in Section 1-1, the basic equations of Newton's theory of gravitation are, in modern formulation, the field laws

$$\nabla \times \mathbf{g} = 0, \tag{1-1.3}$$

and

$$\nabla \cdot \mathbf{g} = -4\pi G\rho. \tag{1-1.4}$$

As stated in Section 2-2 and elaborated in Chapter 8, gravitational fields are repositories of energy. By Eq. (2-2.7), the density of gravitational field energy contained in a gravitational field is (for simplicity of presentation we shall not consider the cogravitational field energy in the discussion that follows)

$$U_v = -\frac{\mathbf{g}^2}{8\pi G}. \tag{19-1.1}$$

However, according to Einstein's mass-energy relation

$$U = mc^2, \tag{19-1.2}$$

any energy has a mass given by $m = U/c^2$, where $c$ is the velocity of light. Hence we must conclude that the gravitational energy density given by Eq. (19-1.1) has a mass density

$$\rho_g = -\frac{\mathbf{g}^2}{8\pi Gc^2}. \tag{19-1.3}$$

But then the source of gravitation should be not just the ordinary mass density $\rho$ but the sum of $\rho$ and $\rho_g$, in which case Eq. (1-1.4) should be replaced by[2,3]

$$\nabla \cdot \mathbf{g} = -4\pi G\rho + \frac{\mathbf{g}^2}{2c^2}. \tag{19-1.4}$$

From now on, we shall assume that the divergence law of gravitation is given not by Eq. (1-1.4) but by Eq. (19-1.4) instead. Observe that the last term in this equation contains the total gravitational field **g**. This means that the equation takes into account the effect of the gravitational energy upon itself.

Note that the mass density of the gravitational field, $\rho_g$, is negative. Thus Eq. (19-1.4) indicates that there may exist not only ordinary attractive gravitational fields but also repulsive, or antigravitational, fields. It also indicates that the field outside a uniform spherical mass distribution depends not only on the magnitude of this distribution but also on its internal field, so that such a mass distribution cannot be replaced by an equal point mass at its center, as can be done in the conventional Newtonian theory. Finally, Eq. (19-1.4) gives us at least a partial explanation for the behavior of gravitational field lines in gravitational fields.

It is well known that the electric field lines in electrostatic fields always have a beginning (on positive charges) and an end (on negative charges). But, according to Eq. (1-1.4), the gravitational field lines have no beginning, they just end on mass elements. A clue to the mystery of their beginning is now given by Eq. (19-1.4): at least some gravitational field lines begin on $\rho_g$ in the space around and within mass distributions.

The basic field equations of Newtonian gravitational theory, Eqs. (1-1.3) and (1-1.4), are usually solved by means of the gravitational potential $\varphi$, defined by

$$\mathbf{g} = -\nabla\varphi. \tag{1-1.9}$$

By combining Eqs. (1-1.9) and (1-1.4), one obtains

$$\nabla^2\varphi = 4\pi G\rho, \tag{1-1.10}$$

which, subject to appropriate boundary conditions, can be solved for a variety of mass distributions. The field **g** can then be found from $\varphi$ by means of Eq. (1-1.9). In the Newtonian theory one can also use integral methods for finding **g**. Any of the following expressions can be used for finding **g** given by Eqs. (1-1.3) and (1-1.4) [4]

$$\mathbf{g} = -G\int\frac{\rho}{r^2}\mathbf{r}_u dV', \tag{1-1.7}$$

$$\mathbf{g} = G\int\frac{\nabla'\rho}{r}dV', \tag{19-1.5}$$

or

$$\varphi = -G\int\frac{\rho}{r}dV'. \tag{1-1.11}$$

Unfortunately, none of the above techniques or equations can be employed for finding gravitational fields given by Eqs. (1-1.3) and (19-1.4), since Eq. (19-1.4) is nonlinear in **g**. Thus, when the effect of the gravitational energy on the gravitational field is taken into account, one cannot in general find the gravitational field **g** from a given mass distribution $\rho$. There is, however, a way out of this difficulty: one can postulate a certain field **g** satisfying Eq. (1-1.3) and then from Eq. (19-1.4) one can find the mass distribution

$$\rho = -\frac{\nabla\cdot\mathbf{g}}{4\pi G} + \frac{\mathbf{g}^2}{8\pi Gc^2} \tag{19-1.6}$$

producing this field. Examples of such calculations are given in the next section. One can also obtain approximate solutions for nonlinear gravitational fields by assuming that gravitational energy is entirely due to the true mass, thus ignoring the effect of the gravitational energy upon itself (the true mass is the mass as such, excluding the associated gravitational energy mass).

## 19-2. Examples of Nonlinear Gravitational Fields

We shall now present illustrative examples demonstrating basic properties of nonlinear gravitational fields. All fields in these examples are spherically symmetric and are in a radial direction. Hence they automatically have a zero curl and thus satisfy Eq. (1-1.3).[5] Of course, even if a field satisfies Eqs. (1-1.3) and (19-1.4), it still may be physically meaningless. Therefore we shall restrict our choice of fields to those that satisfy the following validity conditions:

(a) the energy of the field must be finite,
(b) the field must be finite at $r=0$,
(c) the true mass density $\rho$ must be either positive or zero,
(d) the field must be everywhere continuous.

▼

**Example 19-2.1** Find the mass distribution producing the field

$$\mathbf{g} = -G\frac{m}{a^3}\mathbf{r} \quad \text{for} \quad r \le a, \tag{19-2.1}$$

and

$$\mathbf{g} = -G\frac{m}{r^2}\mathbf{r}_u \quad \text{for} \quad r > a. \tag{19-2.2}$$

Note that in the conventional Newtonian theory this field is produced by a sphere of radius $a$, mass $m$, and uniform density $\rho = 3m/4\pi a^3$.

Substituting Eqs. (19-2.1) and (19-2.2) into Eq. (19-1.6) and differentiating, we obtain

$$\rho = \frac{m}{4\pi a^3}\left(3 + G\frac{mr^2}{2c^2a^3}\right) \quad \text{for} \quad r \le a, \tag{19-2.3}$$

and

$$\rho = G\frac{m^2}{8\pi c^2 r^4} \quad \text{for} \quad r > a. \tag{19-2.4}$$

An important consequence of this solution is that a $1/r^2$ field is produced not by a sphere, but by a mass distribution extending all the way to infinity (although the greatest mass density is within the sphere; see Fig. 19.1). Another important consequence is that $m$ in Eqs. (19-2.1) and (19-2.2) is not the true mass of the sphere. The true mass of the sphere, obtained by integrating Eq. (19-2.3), is

$$m_0 = m\left(1 + G\frac{m}{10c^2a}\right), \tag{19-2.5}$$

so that

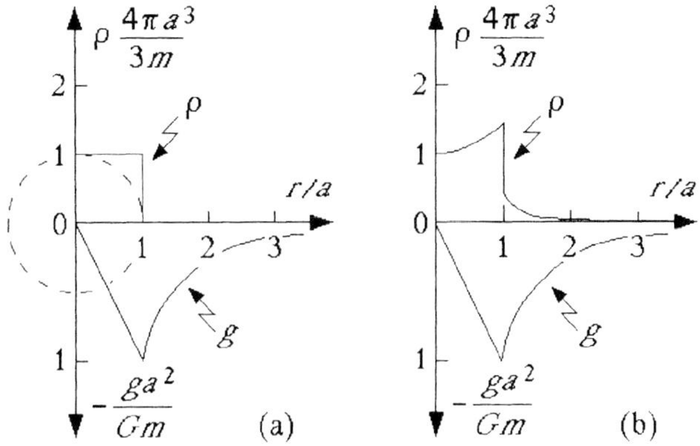


*Fig. 19.1 (a) According to Newton's theory, the gravitational field shown in this figure is produced by a mass of uniform density confined to a sphere. (b) According to the nonlinear theory of gravitation, the same field is produced by a mass of variable density occupying all space. (The scale for the field is twice as large as the scale for the mass density.)*

$$m = \frac{5c^2a}{G}\left[\left(1 + G\frac{2m_0}{5c^2a}\right)^{1/2} - 1\right]. \tag{19-2.6}$$

If $2Gm_0/5c^2a \ll 1$, which is usually the case, Eq. (19-2.6) can be written as

$$m \approx m_0\left(1 - G\frac{m_0}{10c^2a}\right). \tag{19-2.7}$$

The true mass external to the sphere, obtained by integrating Eq. (19-2.4), is

$$m_e = G\frac{m^2}{2c^2a}, \tag{19-2.8}$$

and the total true mass as seen from $r=\infty$, $m_t=m_0+m_e$, is

$$m_t = m\left(1 + G\frac{3m}{5c^2a}\right). \tag{19-2.9}$$

**Example 19-2.2** Find the mass distribution that produces the field given by Eq. (19-2.1) for $r<a$ and produces the field

$$\mathbf{g} = -G\frac{2c^2\mu}{2c^2r^2 - G\mu r}\mathbf{r}_u \qquad \text{for} \qquad r > a, \tag{19-2.10}$$

where

$$\mu = \frac{2c^2ma}{2c^2a + Gm}. \tag{19-2.11}$$

Note that this expression for $\mu$ makes $\mathbf{g}$ continuous at $r=a$.

Since the field for $r \leq a$ is the same as in Example 19-2.1, the mass density $\rho$ for $r \leq a$ is the same as that given by Eq. (19-2.3). Substituting Eq. (19-2.10) into Eq. (19-1.6) and differentiating, we obtain for the mass density in the remaining space

$$\rho = 0, \qquad r > a. \tag{19-2.12}$$

Thus the field under consideration is produced by a mass confined to a sphere of radius $a$.

Let us investigate this field in some detail. If in Eq. (19-2.10) $G\mu/2c^2r \ll 1$, then the equation can be written as

$$\mathbf{g} = -G\frac{\mu}{(1 - G\mu/2c^2r)r^2}\mathbf{r}_u \approx -G\frac{\mu(1 + G\mu/2c^2r)}{r^2}\mathbf{r}_u$$

or

$$\mathbf{g} \approx -G\frac{\mu}{r^2}\mathbf{r}_u - G^2\frac{\mu^2}{2c^2r^3}\mathbf{r}_u. \tag{19-2.13}$$

The first term in this equation is the simple Newtonian gravitational field of a sphere. However, the mass $\mu$ in this equation is not the true, or "naked," mass of the sphere. To find the true mass, we need to solve Eq. (19-2.11) for $m$ and then substitute the result into Eq. (19-2.5). This gives for the true mass of the sphere

$$m_0 = \frac{\mu}{1 - G\mu/2c^2a}\left[1 + G\frac{\mu}{10c^2a(1 - G\mu/2c^2a)}\right]. \tag{19-2.14}$$

The true mass $m_0$ is larger than $\mu$, which was to be expected, since $\mu$ is the sum of the true mass and the negative mass of the gravitational energy. Of course, the mass responsible for the observable gravitational field outside the sphere is not $m_0$ but $\mu$.

Let us now assume that the sphere producing the field under consideration is the Sun, and let us change the designation of the mass in Eq. (19-2.13) from $\mu$ to the more familiar $m$. We then have for the gravitational field of the Sun

$$\mathbf{g} \approx -G\frac{m}{r^2}\mathbf{r}_u - G^2\frac{m^2}{2c^2r^3}\mathbf{r}_u. \tag{19-2.15}$$

Consider now a planet in an orbit around the Sun. Let us designate the mass of the planet as $m'$. The gravitational force acting on the planet is then $m'$ multiplied by the right side of Eq. (19-2.15). For a nearly circular orbit, the gravitational force acting on the planet is equal to the centripetal force applied to the planet: $Gmm'/r^2 = m'v^2/r$, where $v$ is the velocity of the planet. Introducing $v$ into Eq. (19-2.15), we therefore can write for the force exerted by the Sun on the planet

$$\mathbf{F} \approx -G\frac{mm'}{r^2}\mathbf{r}_u - G\frac{mm'v^2}{2c^2r^2}\mathbf{r}_u. \tag{19-2.16}$$

As we shall see in the next chapter, the last term in this equation has a very special significance for the dynamics of our Solar system: it causes a perihelion advance of planetary orbits and it affects the accuracy of our determination of planetary masses and of the mass of the Sun.

**Example 19-2.3** Find the mass distribution producing the field

$$\mathbf{g} = -G\frac{m}{a^2}(2e^{1-r/a} - 1)\mathbf{r}_u \quad \text{for} \quad a \le r \le 2a. \qquad (19\text{-}2.17)$$

Note that this field becomes *antigravitational* for $r > (\ln 2 + 1)a$.

Substituting Eq. (19-2.17) into Eq. (19-1.6) and differentiating, we obtain

$$\rho = \frac{m}{4\pi a^3}\left[2(r/a-1)e^{1-r/a} + \frac{2a}{r}(2e^{1-r/a}-1) + G\frac{m}{2c^2 a}(2e^{1-r/a}-1)^2\right].$$

(19-2.18)

▲

## 19-3. Properties of Gravitational Fields in Free Space

The most interesting aspect of the effect of gravitational energy on gravitational fields is the possibility of the existence of mass distributions creating antigravitational fields in free space. Naturally, if such mass distributions are to be stable under gravitational forces alone, the internal gravitational field of the mass distributions must be attractive everywhere within the distributions. The question arises therefore: can there exist a mass distribution producing an attractive field at all points within itself, but a repulsive field outside?

To answer this question, we shall consider the most general expression for a spherically symmetric field,

$$\mathbf{g} = Af(r)\mathbf{r}_u, \qquad (19\text{-}3.1)$$

where $A$ is a constant and $f(r)$ is any function of $r$, and shall determine $f(r)$ for $\rho = 0$.

Substituting Eq. (19-3.1) into Eq. (19-1.6) and setting $\rho = 0$, we have

$$0 = -\frac{\nabla \cdot \mathbf{g}}{4\pi G} + \frac{\mathbf{g}^2}{8\pi G c^2} = \frac{1}{4\pi G}\left\{-A\nabla \cdot [f(r)\mathbf{r}_u] + \frac{A^2 f^2(r)}{2c^2}\right\}, \quad (19\text{-}3.2)$$

which upon differentiation and simplification gives

$$\frac{d}{dr}f(r) + \frac{2}{r}f(r) - \frac{A}{2c^2}f^2(r) = 0. \quad (19\text{-}3.3)$$

The general solution of this equation is

$$f(r) = \frac{2c^2}{Ar + 2Bc^2r^2}, \quad (19\text{-}3.4)$$

where $B$ is an arbitrary constant.

Thus, by Eqs. (19-3.1) and (19-3.4), the most general expression for a spherically symmetric gravitational field in the region where $\rho=0$ is

$$\mathbf{g} = \frac{2Ac^2}{Ar + 2Bc^2r^2}\mathbf{r}_u, \quad (19\text{-}3.5)$$

where $A$ and $B$ are to be determined from the boundary conditions [Condition (d) of Section 19-2].

For this field to be repulsive ($g>0$) outside some "critical" radius $r_c$, and attractive ($g<0$) within $r_c$, we must have $\mathbf{g}=0$ at $r=r_c$, or

$$A + 2Bc^2r_c = \infty, \quad (19\text{-}3.6)$$

which is impossible for a finite $r_c$. Hence there can be no spherically symmetric antigravitational field outside a mass distribution if the field within the distribution is everywhere attractive. Consequently, a spherical antigravitational body must be held together by some nongravitational forces in addition to the gravitational ones.

Several other important conclusions concerning gravitational fields in mass-free space can be made from Eq. (19-3.5).

First, let us note that for $r \rightarrow \infty$, Eq. (19-3.5) reduces to

$$\mathbf{g} = \frac{A}{Br^2}\mathbf{r}_u = \pm G\frac{M}{r^2}\mathbf{r}_u, \tag{19-3.7}$$

where we have set $A/B = \pm GM$. Therefore in the limit $r \rightarrow \infty$, the field of a spherical mass $m_0$ is just a point-mass field of an "effective" positive or negative mass $M$ ($M$ must be determined from boundary conditions at the surface of $m_0$).

Next, let us consider the possible values of the arbitrary constants in Eq. (19-3.5). To do so, we shall rewrite Eq. (19-3.5) as

$$\mathbf{g} = \frac{2c^2}{r + 2B'c^2r^2}\mathbf{r}_u, \tag{19-3.8}$$

where we have set $B' = B/A$. Let us now assume that the gravitational field represented by Eq. (19-3.8) is created by a spherical mass of radius $r_0$, and that the field at the surface of the mass is $g_0$. Substituting $r_0$ and $g_0$ into Eq. (19-3.8) and solving it for $B'$, we obtain

$$B' = \frac{2c^2 - g_0 r_0}{2g_0 c^2 r_0^2}. \tag{19-3.9}$$

Assuming that $B'$ in Eq. (19-3.8) is arbitrary, we can have $B' = 0$, $B' < 0$, or $B' > 0$. Let us consider these cases in some detail.

$B' = 0$. Substituting Eq. (19-3.8) with $B' = 0$ into Eq. (7-3.37) (the energy equation) and integrating over all space external to $r_0$, we obtain $U = -\infty$, in violation of Condition (a) of Section 19-2. Thus $B' = 0$ is impossible, unless the range of validity of Eq. (19-3.8) is limited to a finite region of space, such as a spherical cavity within a spherical mass distribution.

$B' < 0$. This is the condition for the normal (attractive) Newtonian gravitational field. However, there may exist a "critical" distance $r_c = -1/2B'c^2$ for which Eq. (19-3.8) gives

$g = \pm\infty$ ($g > 0$ for $r < r_c$ and $g < 0$ for $r > r_c$). In this case, the field violates Condition (a) as well as Condition (d). Therefore $B' < 0$ with $r_0 < r_c$ is also impossible, except, of course, when the region under consideration is a spherical cavity whose radius is smaller than $r_c$ (the field in the cavity is then antigravitational). If $r_0 = r_c$, we have the case of a sphere representing a "black hole"[6] of the general relativity theory. However, the resulting field violates Condition (a) and, which is even more important, if $\mathbf{g}_0 = -\infty \mathbf{r}_u$ is substituted in Eq. (19-1.6), and Eq. (19-1.6) is integrated over the volume of the sphere (radius $r_0$), one obtains for the true mass of the sphere $m_0 > \infty$, which cannot be. Thus, according to our theory, black holes (and therefore "gravitational collapse"[7]) are impossible (at least for spherically symmetric mass distributions).

$B' > 0$. This is the condition for a purely antigravitational field. For this field, Eq. (19-3.9) imposes an important condition on $g_0$ and $r_0$:

$$g_0 r_0 < 2c^2. \tag{19-3.10}$$

The significance of this condition will be apparent from the example that follows.

▼

**Example 19-3.1** Construct an antigravitational mass distribution by combining mass distributions given by Eqs. (19-2.3) and (19-2.18).

According to Examples 19-2.1 and 19-2.3, the fields associated with the two mass distributions are

$$\mathbf{g} = -G\frac{m}{a^3}\mathbf{r} \quad \text{for} \quad r \le a \tag{19-2.1}$$

and

$$\mathbf{g} = -G\frac{m}{a^2}(2e^{1-r/a} - 1)\mathbf{r}_u \quad \text{for} \quad a \le r \le 2a. \tag{19-2.17}$$

The combined field is continuous at $r=a$ and becomes antigravitational when $r>(\ln 2+1)a$.

For the field to be antigravitational everywhere outside $r=2a$, the condition given by Eq. (19-3.10) must be satisfied. This can be achieved by imposing an appropriate restriction on $m$. If we substitute $g$ given by Eq. (19-2.17) for $g_0$ in Eq. (19-3.10) and set $r=r_0=2a$, we find that the restriction is

$$m < \frac{c^2 a}{G(1-2/e)}. \tag{19-3.11}$$

Let us make

$$m = \frac{3c^2 a}{4G(1-2/e)}. \tag{19-3.12}$$

The field outside of the mass distribution is given by Eq. (19-3.8). To determine the value for $B'$ appearing in this equation, we use Eqs. (19-3.9) and (19-2.17) with $r=r_0=2a$. After elementary calculation we find that

$$B' = \frac{1}{12c^2 a}. \tag{19-3.13}$$

Substituting this $B'$ into Eq. (19-3.8) and eliminating $c^2$ by means of Eq. (19-3.12), we finally obtain for our antigravitational field

$$\mathbf{g} = G\frac{m}{a^2}\frac{16(1-2/e)}{6r/a+(r/a)^2}\mathbf{r}_u \quad \text{for} \quad r \geq 2a. \tag{19-3.14}$$

A graphical representation of this field and of the corresponding mass distribution (true mass) is given in Fig. 19.2. Starting at infinity and proceeding toward the origin, we find that from $r=\infty$ to $r=1.69a$ the field is antigravitational (repulsive) with a maximum at $r=2a$. At $r=1.69a$ the field becomes zero. From there on the field is an ordinary gravitational (attractive) field with a minimum at $r=a$ and diminishing to zero at $r=0$.

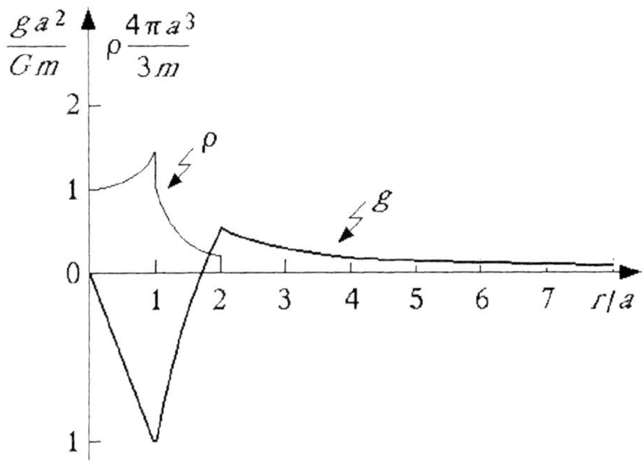


*Fig. 19.2 An example of an antigravitational field and of the corresponding mass distribution. (The scale for the field is twice as large as the scale for the mass density.)*

Of course, the mass distribution shown in Fig. 19.2 cannot be maintained by gravitational forces alone because, unless the distribution is kept together by some other forces, all the mass located at $r > 1.69a$ would be ejected by antigravitational repulsion, until the radius of the distribution decreases to $r = 1.69a$. The external gravitational (or antigravitational) field of the remaining distribution would then completely disappear, and the distribution would become a "hidden mass" that neither exerts nor experiences any gravitational forces on or from the surrounding bodies.

Observe that $m$ appearing in Eqs. (19-2.1), (19-2.17), and (19-3.14) is not the true mass of the spherical distribution under consideration; it is a quantity associated with the true mass of the central part ($r \leq a$) of the distribution through Eqs. (19-2.5) and (19-2.6). By Eqs. (19-2.5) and (19-3.14), the true mass of the central part is $m_0 \approx 1.3m$.

As can be seen from Fig. 19.2, the maximum density of the true mass occurs at $r=a$. Since, by Eq. (19-3.12), $m=2.48c^2a/G$, the radius of the central part of the mass distribution is $a=Gm/2.48c^2$. By Eq. (19-2.3), we then have for the maximum mass density of the distribution

$$\rho_{\max} \approx \frac{5c^6}{G^3m^2}. \qquad (19\text{-}3.15)$$

Equation (19-3.15) shows that for a very large mass, the density of an antigravitational mass distribution can be very small. This may be an important factor for the stability of the galaxies in the Universe.

▲

## 19-4. Discussion

For almost a century now, Newton's theory of gravitation has been abandoned in theoretical physics (but not in practical space exploration) in favor of Einstein's general relativity theory. Some authors even insist that the general relativity theory is the definitive theory of gravitation. However, the generalized Newtonian theory of gravitation outlined in this book points out a path for an unquestionably viable new inquiry into the nature and properties of gravitational fields and interactions. The generalized Newtonian theory is based to a large extent on the idea that the gravitational-cogravitational field is a seat of momentum and energy. One of the consequences of this idea is the supposition, discussed in this chapter, that gravitation is caused not only by a true mass but also by the equivalent mass of the gravitational field energy. Plausible as it is, this supposition is contrary to the general relativity theory. Moreover, even the existence of gravitational field energy is contrary to the general relativity theory. It is important therefore to clarify the reasons why general relativity theory denies the existence of gravitational field energy and it is important to examine the validity of these reasons.

The basic gravitational equation of the general relativity theory is Einstein's gravitational field equation

$$R_{ik} - \frac{1}{2}Rg_{ik} = - G\frac{8\pi}{c^4}T_{ik}. \tag{19-4.1}$$

The sources of gravitation appear in this equation in the form of the energy-momentum tensor $T_{ik}$. This tensor includes all types of mass densities and all types of energy densities (electric, magnetic, thermal, etc.) except for the energy density of the gravitational field itself. The determining reason for this is quite simple: in spite of many efforts, no energy-momentum tensor has been found for the gravitational energy (only a "pseudotensor" has been obtained). Various plausibility arguments have therefore been suggested to justify the absence of gravitational energy as a source of gravitation in Einstein's field equation.[8] Since it would be difficult (if not impossible) to accept the existence of gravitational field energy without accepting this energy as a source of gravitation, these arguments are also the arguments against the presence of gravitational energy in the gravitational field.

The two strongest plausibility arguments for excluding gravitational energy as a source of gravitation are:

(1) Predictions of the general relativity theory obtained with the aid of Einstein's field equation without gravitational energy as a source of gravitation have been found to agree with observations.

(2) Einstein's "equivalence principle" forbids gravitational energy to be a source of gravitation.

However, a careful examination of these arguments shows that neither of them is truly convincing or compelling.

The first argument is easily refuted by the fact that all presently verifiable predictions of the general relativity theory are in the domain of weak fields, where, as follows from the material presented in this chapter, the effects of gravitational energy are hardly prominent.[9]

The second argument appears to be much stronger than the first. What it means is that since, according to Einstein, a gravitational field is equivalent to a certain accelerated frame of reference, and since there apparently is no special energy in the space defined by the accelerated frame of reference, no energy should be present in the space containing the gravitational field (this is known as the "nonlocalizability" of gravitational energy).[8,10] An analysis of this argument shows, however, that it is based on an unprovable premise and that it can be refuted by reversing it. Indeed, let us suppose that a gravitational field is a seat of gravitational energy. The equivalence principle demands then that a certain energy density would appear in the space defined by the equivalent reference frame. But how will this energy manifest itself? The only presently known way in which it could be detected is by its gravitational effects. However, since the equivalent reference frame is flat and boundless, the "equivalent" energy density, as seen in this frame, must be uniform and must occupy all space. But, as is well known, a uniformly distributed mass (energy) occupying all space produces no gravitational effects [see Eq. (19-1.5); if $\nabla\rho=0$ or $\nabla\rho_g=0$ everywhere, $\mathbf{g}=0$, too]. Hence the "equivalent" energy is not detectable, or, as an observer in the equivalent reference frame would say, is "absent."

Thus the absence of space energy in an accelerated reference frame does not prove the nonexistence or nonlocalizability of gravitational field energy, and hence the equivalence principle does not forbid its appearance as a source term in Einstein's gravitational field equation. Therefore the exclusion of the gravitational energy as a source of gravitation in the general relativity theory is merely a matter of practical necessity (since no tensor has been found for it). Hence all presently known results of the general relativity theory based on Einstein's field equation cannot be considered as reliable when these results involve gravitational fields whose gravitational-energy mass is comparable with the true mass of the system (see the next chapter for a further discussion of questionable aspects and ambiguities of the general relativity theory). And therefore the fact that the results obtained in this chapter are in conflict with the general relativity theory does in no way indicate that these results are wrong. The conflict cannot be resolved by plausibility arguments. Only reliable observational data can truly resolve it .[11]

Let us now summarize what our theory of nonlinear gravitational fields has indicated:

1. The gravitational force acting on a body in a gravitational field is determined not only by the mass of the field-producing body, but also by the gravitational field energy of the field-producing body.

2. Antigravitational bodies can exist in the Universe.

3. The mass of the Universe, of a galaxy, or of a stellar object can be much larger than the present astrophysical measurements indicate, since there can exist objects of negative or of zero apparent mass. The latter objects would constitute "hidden" masses insofar as they do not produce or experience gravitational effects.

4. Spherical "black holes" cannot exist, and "gravitational collapse" is impossible. Indeed, according to the general relativity theory, a sphere creates an "unescapable" gravitational field and becomes a "black hole" after its radius becomes smaller than the "gravitational radius" [12]

$$r_g = G\frac{2m}{c^2}. \tag{19-4.2}$$

But the radius of the central mass of the mass distribution shown in Fig. 19.2 is *smaller* than the gravitational radius, yet the field at this radius is zero rather than immensely strong, as is required for black holes.

5. Since "gravitational collapse" is impossible, and since antigravitational mass formations are possible, the normal state of

the Universe appears to be an alternating expansion and contraction.

6. Since a "hidden" mass is an object whose overall rest mass is zero, such a mass could conceivably move with a velocity equal to (or even larger than) the velocity of light.

These are fascinating and intriguing conclusions. Are they true or are they false? Only time will tell.

## References and Remarks for Chapter 19

**1.** This chapter is based on Chapter 8 of Oleg D. Jefimenko, *Causality, Electromagnetic Induction and Gravitation,* 2nd Ed., (Electret Scientific, Star City, 2000) pp. 140-160 (same pages in the first edition of the book in 1992).

**2.** A similar equation including the equivalent mass of the gravitational field energy as well as of the cogravitational field energy [Eq. (6-2.38)] was suggested by F. Hund, "Zugänge zum Verständnis der allgemeinen Relativitätstheorie," Z. Physik, **124**, 742-756 (1947). We do not include the cogravitational energy in our discussion, although it may make a significant contribution in the case of rapidly moving or rapidly rotating bodies.

**3.** A somewhat similar equation (for fields *external* to mass distributions) was suggested by L. Brillouin and R. Lucas, "Le Relation Masse-Énergie en Gravitation," J. Phys. Radium **25**, 229-232 (1966). See also M. Mannheimer, "L'Énergie au Champ de Gravitation," Ann. Phys. (Paris) **1**, 189-194 (1966) and **2**, 57-60 (1967); Léon Brillouin, *Relativity Reexamined* (Academic Press, New York, 1970) pp. 87-95. A related publication is P. C. Peters, "Where is the energy stored in a gravitational field?" Am. J. Phys. **49**, 546-569 (1981).

**4.** Equation (19-1.5) is the time independent version of Eq. (3-1.1).

**5.** See Oleg D. Jefimenko, *Electricity and Magnetism*, 2nd ed., (Electret Scientific, Star City, 1989) p. 60, Problem 2.22.

**6.** See, for example, H. C. Ohanian and R. Ruffini, *Gravitation* *and Spacetime*, 2nd ed., (W. W. Norton, New York, 1994) pp. 437-438.

**7.** See, for example, Ref. 6, pp. 447-448, 489-496.

**8.** See, for example, C. W. Misner, K. S. Thorne, and J. A. Wheeler, *Gravitation* (W. H. Freeman, San Francisco, 1973) pp. 466-468.

**9.** See, however, Sections 20-1 and 20-2 for the effect of gravitational energy on Mercury's perihelion precession.

**10.** Nevertheless, in a glaring contradiction, gravitational waves in the general relativity theory are assumed to carry gravitational energy with them! See, for example, Ref. 6, pp. 241-301.

**11.** As we know from Chapter 9, the generalized Newtonian theory of gravitation is compatible with the special relativity theory. However, the nonlinear gravitational equations discussed in this chapter are incompatible with relativistic transformations equations. Although there is a widespread opinion that all correct physical theories and equations must be compatible with the special relativity theory, the incompatibility of the nonlinear gravitational equations with this theory does not mean that the equations are wrong. In this connection it may be noted that there are other examples of perfectly viable equations which are incompatible with the special relativity theory. Maxwell's electromagnetic equations *in their vector form* present the most prominent example of such incompatibility [see Oleg D. Jefimenko, *Electromagnetic Retardation and Theory of Relativity*, 2nd ed., (Electret Scientific, Star City, 2004) pp. 158-165]. Furthermore, in the real world, the special relativity theory is itself only approximately correct. This is because this theory is only applicable to inertial systems, but true inertial systems do not really exist. In the end, the only reliable criterion of the correctness (or erroneousness) of our nonlinear gravitational equations is the agreement (or disagreement) of these equations with the experimental data within the range of applicability of the equations.

**12.** Ref. 6, pp. 438-439.

# 20

# MERCURY'S PERIHELION PRECESSION AND ANALYSIS OF RELATED CALCULATIONS

One of the most important consequences of the generalized theory of gravitation is that in time-dependent systems gravitational interactions involve not only the usual Newtonian attraction but also additional forces associated with the motion of interacting bodies. This effect is particularly significant because of its relevance for explaining certain discrepancies between observed and calculated properties of planetary motion. The best known such discrepancy is in the precession of the perihelion of Mercury. In this chapter we shall see how the generalized theory of gravitation resolves this discrepancy and shall analyze the resolution suggested by the general relativity theory.

## 20-1. Mercury's Anomaly

In the middle of the 19th century, Urbain Le Verrier found that Newton's gravitational law was incapable of explaining certain discrepancies between the observed and calculated parameters of planetary motion. In particular, he computed the secular perturbations of the motion of Mercury under the action of other planets and found that there was an inexplicable "residual" precession of Mercury's perihelion. According to the presently accepted data, the precession of Mercury's perihelion as observed relative to the Earth's equinox reference line is 5600 arc seconds of arc per century of which 5025 arc seconds are attributed to the precession of the equinoxes. In the heliocentral system the precession of Mercury's perihelion is therefore approximately 575 seconds of arc per century, of which 532 seconds can be attributed to Newtonian attraction between Mercury and other planets, while about 43 seconds cannot be explained on the basis of Newton's gravitational law.

For more than 150 years now, physicists, mathematicians and astronomers have been attempting to explain the residual precession of Mercury's perihelion, and it is now generally accepted that no theory of gravitation can be regarded as correct or complete if it cannot provide a convincing explanation for Mercury's residual precession. How is Mercury's residual precession explained by the generalized theory of gravitation?

First of all, according to the generalized theory of gravitation, we do not really know the exact value of Mercury's residual precession. Clearly, the value of 43 seconds of arc per century obtained exclusively on the basis of Newton's gravitational law is highly questionable. Newton's gravitational law is truly accurate for interactions between stationary bodies only. As has been shown throughout this book, gravitational interactions between moving bodies involve not only the usual Newtonian attraction but also many additional forces associated with the motion of the interacting bodies. Therefore, according to the generalized theory of gravitation, the calculation of the *main* precession of Mercury's perihelion based only on Newtonian attraction cannot be accurate, and hence the residual precession obtained on the basis of such calculations cannot be accurate either. Here is a short (but by no means complete) list of omissions resulting in errors in calculations based on Newtonian attraction only:

(1) Retardation in the propagation of gravitation is not taken into account; in particular (and this is extremely important), the

fact that an outer planet, according to Section 15-1, exerts not only a radial force on the inner planets, but also a force in the direction of the velocity of the outer planet.

(2) Cogravitational fields are not taken into account. In this connection it should be noticed that there are several sources of cogravitational fields affecting the motion of Mercury including: (a) the orbital motion of outer planets (Section 15-1), (b) the rotation of outer planets [Section 15-2 and Eq. (13-2.46)], (c) the rotation of the central body (Section 15-2), and (d) the rotation of the Galaxy (galactic cogravitational field) (Section 15-2).

(3) Gravikinetic fields are not taken into account (Chapters 11, 12).

(4) Nonlinearity of the gravitational field is not taken into account (Section 19-2).

(5) Possible errors in the determination of the masses of the Sun and planets because of the effect of the gravitational field energy of the Sun (negative "shielding" mass of the field energy) is not taken into account (Section 19-2).

(6) The fact that large masses may accelerate faster than small masses is not taken into account (Example 13-2.8).

(7) The fact that, since the residual precession is a second order effect, calculation of forces between outer planets and Mercury must be done in accordance with the special relativity theory (observe that the mass of an *orbiting* planet is the *transverse* mass of relativistic mechanics, and that a mass of a planet subjected to a force in the direction of the motion of the planet is the *longitudinal* mass).[1]

Clearly, not all omissions listed above may lead to serious errors in the calculation of the main precession of Mercury's perihelion, and some of them are obviously insignificant. But the fact remains that unless all the possible forces and interactions that can potentially affect the main perihelion precession of Mercury are taken into account, the true value of the residual perihelion precession of Mercury (if it exists at all) remains unknown.

## 20-2. Mercury's Residual Precession According to the Generalized Theory of Gravitation

Let us assume that even after all the effects mentioned in the preceding section are taken into account there still remains some residual precession of Mercury's perihelion. What, then, may be the cause of the residual precession?

From the viewpoint of the generalized theory of gravitation there are two most likely causes: the existence of the negative mass-energy (mass of the gravitational field energy) near the Sun and the failure to use the force equation of the special relativity theory while calculating the effect of the Sun on the motion of Mercury. The need to use the negative gravitational mass-energy follows from Chapter 19. The need to use the relativistic force equation follows from the fact that, as was shown by various authors in the 19th century, in order to explain the residual precession of Mercury, terms involving $1/c^2$ need to be taken into account.[2] Let us therefore calculate the contribution of these effects to the precession of Mercury's perihelion.[3]

According to the special theory of relativity, the acceleration **a** of a body subjected to the action of a force **F** is determined by the equation[1]

$$\mathbf{a} = \frac{\mathbf{F} - (\mathbf{F}\cdot\mathbf{v})\mathbf{v}/c^2}{m/(1-v^2/c^2)^{1/2}} = \left[\frac{\mathbf{F}}{m} - \frac{(\mathbf{F}\cdot\mathbf{v})\mathbf{v}/c^2}{m}\right](1-v^2/c^2)^{1/2}, \quad (20\text{-}2.1)$$

where $\mathbf{v}$ is the velocity of the body and $\gamma = (1 - v^2/c^2)^{1/2}$. With terms smaller than $v^2/c^2$ omitted, Eq. (20-2.1) becomes

$$\mathbf{a} = \frac{\mathbf{F}}{m}(1 - v^2/2c^2) - \frac{(\mathbf{F}\cdot\mathbf{v})\mathbf{v}/c^2}{m}. \quad (20\text{-}2.2)$$

If **F** is in a radial direction, Eq. (20-2.2) can be written as

$$\mathbf{a} = \frac{F}{rm}[\mathbf{r}(1 - v^2/2c^2) - (\mathbf{r}\cdot\mathbf{v})\mathbf{v}/c^2]. \quad (20\text{-}2.3)$$

Using in Eq. (19-2.16) $M$ for the mass of the Sun and $m$ for the mass of Mercury and introducing $F$ from Eq. (19-2.16) into Eq. (20-2.3), we have

$$\mathbf{a} = -\frac{GM}{r^3}[\mathbf{r}(1 - v^2/2c^2) - (\mathbf{r}\cdot\mathbf{v})\mathbf{v}/c^2](1 + v^2/2c^2), \quad (20\text{-}2.4)$$

and once again omitting terms smaller than $v^2/c^2$,

$$\mathbf{a} = -\frac{GM}{r^3}\mathbf{r} + \frac{GM}{r^3c^2}(\mathbf{r}\cdot\mathbf{v})\mathbf{v}. \quad (20\text{-}2.5)$$

The first term on the right side of this equation represents the acceleration of Mercury due to the usual Newtonian attraction, the second term is an additional acceleration due to the negative gravitational mass-energy together with the relativistic correction of the Newtonian force. Observe that this term is in the direction of the orbital velocity of Mercury. Let us write it as

$$\mathbf{a}' = \frac{GM}{r^2c^2}v^2\cos\alpha\,\mathbf{v}_u, \quad (20\text{-}2.6)$$

where $\alpha$ is the angle between the radius vector of Mercury in its orbit and Mercury's orbital velocity vector $\mathbf{v}$, and $\mathbf{v}_u$ is a unit vector along $\mathbf{v}$. It is the acceleration $\mathbf{a}'$ that should be responsible for the residual precession.

We can calculate the effect of $\mathbf{a}'$ on Mercury's orbital motion by using Lagrange's perturbed orbit equation[4] for the argument of perihelion $\omega$ of a planet

$$\frac{\partial\omega}{\partial\varphi} = -\frac{\cos\varphi}{GMe}r^2R + \frac{(2 + e\cos\varphi)\sin\varphi}{GMa(1 - e^2)e}r^2S, \quad (20\text{-}2.7)$$

where $\varphi$ is the polar angle ("true anomaly") of the planet, $e$ is the eccentricity of the orbit, $r$ is the distance between the Sun and the planet, $R$ is the radial component of the acceleration of the planet (due to perturbing force), $a$ is the semimajor axis of the orbit, and $S$ is the component of the planet's acceleration perpendicular to the radial component (due to perturbing force).

The component of $\mathbf{a}'$ along the radius vector is

$$R = \frac{GMv^2}{r^2c^2}\cos^2\alpha \tag{20-2.8}$$

and the component in the direction perpendicular to the radius vector is

$$S = \frac{GMv^2}{r^2c^2}\sin\alpha\cos\alpha \tag{20-2.9}$$

For an elliptic orbit, $\sin\alpha$ and $\cos\alpha$ are given by

$$\sin\alpha = \frac{1 + e\cos\varphi}{(1 + e^2 + 2e\cos\varphi)^{1/2}} \tag{20-2.10}$$

$$\cos\alpha = \frac{e\sin\varphi}{(1 + e^2 + 2e\cos\varphi)^{1/2}}. \tag{20-2.11}$$

Since $\mathbf{a}'$ is small compared with $\mathbf{a}$, we can assume that the orbital velocity of Mercury is not affected by $\mathbf{a}'$ and can be found from the usual equation for unperturbed Keplerian motion:

$$v^2 = \frac{GM(1 + e^2 + 2e\cos\varphi)}{a(1 - e^2)}. \tag{20-2.12}$$

Combining Eqs. (20-2.8)-(20-2.12) and simplifying, we obtain

$$r^2R = \frac{(GM)^2e^2}{c^2a(1 - e^2)}\sin^2\varphi \tag{20-2.13}$$

and

$$r^2S = \frac{(GM)^2e}{c^2}\sin\varphi. \tag{20-2.14}$$

Substituting Eqs. (20-2.13) and (20-2.14) into Eq. (20-2.7), we have

$$\frac{\partial\omega}{\partial\varphi} = -\frac{GMe}{c^2a(1 - e^2)}\sin^2\varphi\cos\varphi + \frac{GM(2 + e\cos\varphi)}{c^2a(1 - e^2)}\sin^2\varphi \tag{20-2.15}$$

or

$$\frac{\partial \omega}{\partial \varphi} = \frac{2GM}{c^2 a(1 - e^2)} \sin^2 \varphi . \qquad (20\text{-}2.16)$$

Integrating over $\varphi$ from 0 to $2\pi$, we finally obtain for the perihelion advance

$$\Delta\omega = \int_0^{2\pi} \frac{2GM}{c^2 a(1 - e^2)} \sin^2 \varphi \, d\varphi = \frac{2\pi GM}{c^2 a(1 - e^2)} . \qquad (20\text{-}2.17)$$

Substituting the values $G = 6.67 \times 10^{-11} \mathrm{m^3 kg^{-1} s^{-2}}$, $M = 1.99 \times 10^{30} \mathrm{kg}$, $c = 2.99 \times 10^8 \mathrm{ms^{-1}}$, $a = 5.77 \times 10^{10}$ m, and $e = 0.206$ in Eq. (20-2.17), we find for the residual precession of Mercury's perihelion

$$\Delta\omega = 1.69 \times 10^{-7} \text{ rad per revolution,}$$

or, taking into account that Mercury revolves 415 times per century,

$$\Delta\omega = 14 \text{ arc sec per century.}$$

Qualitatively this result is in accord with observations, but quantitatively it is about 1/3 of the presently accepted value of 43 arc seconds per century. However, according to the generalized theory of gravitation, this apparent discrepancy actually speaks in favor of the theory, because, as has been explained in the preceding section, the residual precession of 43 seconds per century was obtained by taking into account only the Newtonian attraction between Mercury and outer planets. In reality, according to the generalized theory of gravitation, Mercury is subjected to a very complex system of forces, so that it is most unlikely that the calculations based on Newtonian attraction alone (and without using the relativistic force equation at that) could yield correct results for the residual precession.

## 20-3. Einstein's Formula for Mercury's Residual Precession

By the end of the 19th century it was well know that Gerber's formula[5]

$$\Delta\omega = \frac{6\pi GM}{c^2 a(1 - e^2)}, \tag{20-3.1}$$

where all the symbols are as in Eq. (20-2.17), yielded the 43 arc seconds residual precession of Mercury. It was the greatest triumph of Einstein's general relativity theory when in 1916, on the basis of this theory, Einstein obtained the same formula. In fact, to this day most of the credibility of the general relativity theory is directly attributable to Einstein's derivation of Eq. (20-3.1) and to the amazing accuracy with which the general relativity theory has explained Mercury's residual perihelion precession. And yet, from the viewpoint of the generalized theory of gravitation presented in this book, and according to Section 20-1 in particular, it is the very accuracy of Einstein's result that makes the validity of his explanation of Mercury's residual precession highly questionable. Let us therefore discuss in some detail Einstein's explanation of Mercury's precession.

We now know from Einstein's correspondence with his friends and collaborators that in developing the general relativity theory he hoped from the very start to provide an explanation of Mercury's anomaly.[6] The inescapable conclusion is that, to the extent possible, he constructed the theory so that it would yield Eq. (20-3.1).

Einstein himself, while first presenting Eq. (20-3.1),[7] pointed out that this equation did not follow uniquely from his theory. And, what is especially important as far as his explanation of Mercury's residual precession is concerned, we now know that the theory does not yield an unambiguous expression for the velocity of gravitation and for the relation between the velocity of gravitation and the velocity of light.

The problem with the velocity of light starts at the very beginning of the mathematical formulation of the general relativity theory, when the velocity of light is expressed in terms of the so-called "geometrized units," according to which $c = G = 1$ (that is, the velocity of light and the universal constant of gravitation are assumed to be pure numbers equal to "one").[8] Setting $c = 1$ actually excludes the velocity of light from the formulation of the theory and makes the velocity of light irrelevant for the theory.

As a physical quantity, the velocity of light $c$ is first introduced into the general relativity theory when the theory, in its limiting case, is made compatible with Poisson's equation of the Newtonian theory of gravitation, at which time Einstein's gravitational field equation (the basic equation of the general relativity theory)

$$R_{ik} - \frac{1}{2} R g_{ik} = - G \frac{8\pi}{c^4} T_{ik} \tag{19-4.1}$$

is finally obtained. However, since the velocity of light as a physical quantity is ignored in the main process of the formulation of the theory, $c$ in Eq. (19-4.1) is basically undetermined. Moreover, it is not at all clear whether $c$ in Eq. (19-4.1) and therefore in Einstein's version of Eq. (20-3.1) stands for the velocity of light, the velocity of gravitation or for some other velocity.

In this connection let us refer to Einstein's explanation of "Mach's principle." Explaining Mach's principle on the basis of the general relativity theory, Einstein gave a quantitative formulation of this principle in the form of an equation closely resembling the electromagnetic equation for the force acting on a moving charged particle of magnitude "one" in the presence of an electric and a magnetic field:[9]

$$\frac{d}{dt}[(1 + \bar{\sigma})\mathbf{u}] = c^2 \nabla \bar{\sigma} + \frac{\partial \mathbf{A}}{\partial t} - [\mathbf{u} \times (\nabla \times \mathbf{A})] \tag{20-3.2}$$

with

$$\overline{\sigma} = \frac{G}{c^2}\int\frac{\rho}{r}dV' \tag{20-3.3}$$

and

$$\mathbf{A} = \frac{4G}{c^2}\int\frac{\rho\mathbf{v}}{r}dV', \tag{20-3.4}$$

where **u** and **v** are the velocities, respectively, of the force-experiencing and force-producing mass, $c$ is the velocity of light, and where the remaining symbols are as elsewhere in this book.

Clearly, the right side of Eq. (20-3.2) is similar to the right side of our gravitational force equation, Eq. (2-2.6), combined with Eqs. (3-3.4), (3-3.1), (3-3.7), and (3-3.6), except for two major differences: First, $c$ in our Eq. (3-3.6) [which corresponds to Einstein's Eq. (20-3.4)] is very clearly the velocity of gravitation rather than the velocity of light. Second, Einstein's Eq. (20-3.4) contains the factor "4" which is absent in our Eq. (3-3.6). Einstein's factor is counterintuitive and very puzzling, since it implies that the mass current produced by a moving mass distribution of density $\rho$ is $\mathbf{J} = 4\rho\mathbf{v}$ rather than $\mathbf{J} = \rho\mathbf{v}$ as follows from our derivations in Section 3-1 and as would be expected on the basis of general considerations of the mass-current concept.

The problem with the velocity of light and with Einstein's factor "4" becomes fully apparent in connection with the determination of the velocity of gravitation on the basis of the general relativity theory. As is known, the weak-field linearized form of general relativity theory splits gravitation into a gravitational field proper and a second field analogous to our cogravitational field. From Einstein's Eq. (19-4.1) the following equations for **g** and **K** (using our notation) are then obtained[10]

$$\nabla\cdot\mathbf{g} = -4\pi G\rho, \tag{20-3.5}$$

$$\nabla\cdot\mathbf{K} = 0, \tag{20-3.6}$$

$$\nabla\times\mathbf{g} = -\frac{\partial\mathbf{K}}{\partial t}, \tag{20-3.7}$$

and

$$\nabla\times\mathbf{K} = -4\frac{4\pi G}{c^2}\mathbf{J} + 4\frac{1}{c^2}\frac{\partial\mathbf{g}}{\partial t}. \tag{20-3.8}$$

These equations are similar to our Eqs. (7-1.1)-(7-1.4) except for the extra factor "4" on the right of Eq. (20-3.8). One can easily see by repeating the calculations presented in Section 18-1 and using Eq. (20-3.8) instead of Eq. (7-1.4) that the factor "4" in Eq. (20.3.8) results in the velocity of gravitation equal to $c/2$.

A slightly different linearization of Einstein's Eq. (19-4.1) yields equations similar to Eqs. (20-3.5)-(20-3.8), except that instead of Eq. (20-3.7) the following equation is obtained[11]

$$2\nabla\times\mathbf{g} = -\frac{\partial\mathbf{K}}{\partial t}, \tag{20-3.9}$$

resulting in the velocity of gravitation equal to $c/\sqrt{2}$.

Finally, according to yet another calculation based on linearization of Einstein's Eq. (19-4.1) in combination with Lorentz transformation equations for time and space,[12,13] the following equations are obtained (again, we write these equation in our notation)

$$\nabla\cdot\mathbf{g} = 4\pi G\rho, \tag{20-3.10}$$

$$\nabla\cdot\mathbf{K} = 0, \tag{20-3.11}$$

$$\nabla\times\mathbf{g} = -\frac{\varepsilon\,\partial\mathbf{K}}{c\,\partial t}, \tag{20-3.12}$$

and

$$\nabla\times\mathbf{K} = \frac{4\pi G\varepsilon}{c}\mathbf{J} + \frac{\varepsilon}{c}\frac{\partial\mathbf{g}}{\partial t}, \tag{20-3.13}$$

where $\varepsilon$ is an undetermined parameter. According to Eqs. (20-3.10)-(20-3.13), the velocity of gravitation is now $c/\varepsilon$.

In view of all these conflicting results we are forced to conclude that there is no real evidence that $c$ in the general relativity theory represents either the velocity of light (as generally believed) or the velocity of gravitation. We are also forced to conclude that if $c$ represents the velocity of light, then, according to Eqs. (20-3.5)-(20-3.13), the velocity of gravitation, contrary to the general perception, is not equal to the velocity of light.

As far as the Mercury's anomaly is concerned, the uncertainty of the meaning of $c$ in Einstein's version of Eq. (20-3.1) combined with the non-uniqueness of Einstein's derivation of this equation makes the significance of the numerical value obtained by Einstein for Mercury's residual precession questionable.

We shall close our discussion of Mercury's residual precession by mentioning yet another problem with the general-relativistic explanation of the precession. As has been pointed out in the literature on various occasions,[14] an important inconsistency of the general-relativistic treatment of Mercury's perihelion precession is that only the residual precession is explained, and it is explained under the assumption that Mercury's *main* perihelion precession has been correctly calculated on the basis of Newton's theory of gravitation. The question arises: why not compute Mercury's *entire* precession on the basis of the general relativity theory and see whether or not the general relativity theory can correctly account for the entire precession? The answer is very simple: the general relativity theory provides no methods for calculating gravitational effects of bodies in translational or orbital motion and for calculating gravitational effects of *moving* outer planets in particular. In fact, in contrast to the generalized Newtonian theory of gravitation presented in this book, the general relativity theory is only pertinent to gravitational fields created by stationary masses.

## References and Remarks for Chapter 20

**1.** See, for example, Oleg D. Jefimenko, *Electromagnetic Retardation and Theory of Relativity*, 2nd ed., (Electret Scientific, Star City, 2004) pp. 193-196.
**2.** See, for example, N. T. Roseveare, *Mercury's perihelion from Le Verrier to Einstein* (Clarendon Press, Oxford, 1982).
**3.** The calculations presented below require some knowledge of celestial mechanics. Assuming that readers either possess such knowledge or will consult appropriate textbooks, we present these calculations with a minimum of explanations.
**4.** See, for example, L. G. Taff, *Celestial Mechanics*, (Wiley, New York, 1985), p. 315. We are using a modified version of the Gaussian form of the equation obtained by replacing $d\omega/dt$ with $d\omega/d\varphi$; for details see, for example, A. F. Bogorodsky, *Universal Gravitation*, (Naukova Dumka, Kiev, 1971) pp. 23-28.
**5.** This formula first appeared in P. Gerber, "Die raümlich und zeitliche Ausbreitung der Gravitation," Z. Math. Phys. **43**, 93-104 (1898).
**6.** Einstein, in a 1907 letter to C. Habicht, wrote: "Now I am busy on a relativistic theory of the gravitational law with which I hope to account for the still unexplained secular changes of the perihelion movement of Mercury." See C. Seelig, *Albert Einstein, A Documentary Biography*," (Staples, London, 1956) p. 76.
**7.** A. Einstein, "Erklärung der Perihelbewegung des Merkur aus der allgemeinen Relativitätstheorie," in Königlich Preussische Akademie der Wissenschaften, Sitzungsberichte, Berlin, 1915, pp. 831-839.
**8.** See, for example, C. W. Misner, K. S. Thorne, and J. A. Wheeler, *Gravitation* (W. H. Freeman, San Francisco, 1973) p. 36. The theory so constructed violates the requirement of dimensional homogeneity of physical equations and therefore is a strictly mathematical theory where all the operations are strictly mathematical operations having no physical significance. Moreover, by excluding dimensional constants, the theory ignores the

corresponding physical laws. See Oleg D. Jefimenko, *Electricity and Magnetism,* 2nd ed., (Electret Scientific, Star City, 1989) pp. 1-14.

**9.** A. Einstein, *The Meaning of Relativity*, 5th ed., (Princeton U. P., Princeton, NJ, 1956) p. 100. Mach's principle played an important role in Einstein's search for a new theory of gravitation and in his formulation of the general relativity theory.

**10.** V. B. Braginsky, C. M. Caves, and K. S. Thorne, "Laboratory experiments to test relativistic gravity," Phys. Rev. D. **15**, 2047-2068 (1977). In this paper there is an extra term in the equation similar to our Eq. (20-3.5), which is irrelevant for the present discussion.

**11.** F. G. Harris, "Analogy between general relativity and electromagnetism for slowly moving particles in weak gravitational fields," Am. J. Phys. **59**, 421-425 (1991). In the second part of this paper the right side of Eq. (20-3.9) is made equal to zero, which does not make much sense for time-variable fields.

**12.** S. M. Kopeikin, "Speed of gravity and gravitomagnetism," Int. J. Mod. Phys. D **13**: 2345-2350 (2006).

**13.** Some authors assume that Einstein's gravitational equations are compatible with Lorentz transformation equations for time and space. However, no proof of such compatibility has ever been presented.

**14.** See J. Earman and M. Janssen, "Einstein's Explanation of the Motion of Mercury's Perihelion" in J. Earman, M. Janssen and J. Norton (Editors), *The Attraction of Gravitation: New Studies in the History of General Relativity*, (Birkhäuser, Boston, 1993) pp. 129-172 (pp. 161, 162 in particular); see also Oleg D. Jefimenko, *Causality, Electromagnetic Induction and Gravitation,* 2nd Ed., (Electret Scientific, Star City, 2000) pp. 137, 138.

# APPENDIX

# VECTOR IDENTITIES

In the vector identities listed below $\varphi$ and $U$ are scalar point functions; **A**, **B**, and **C** are vector point functions; X is a scalar or vector point function of primed coordinates and incorporates an appropriate multiplication sign (dot or cross for vectors).

*Box product*

(V-1) $$\mathbf{A}\cdot(\mathbf{B}\times\mathbf{C}) = \mathbf{B}\cdot(\mathbf{C}\times\mathbf{A}) = \mathbf{C}\cdot(\mathbf{A}\times\mathbf{B})$$

(V-2) $$\mathbf{A}\cdot(\mathbf{B}\times\mathbf{C}) = (\mathbf{A}\times\mathbf{B})\cdot\mathbf{C} = -(\mathbf{B}\times\mathbf{A})\cdot\mathbf{C}$$

*"BAC CAB" expansion*

(V-3) $$\mathbf{A}\times(\mathbf{B}\times\mathbf{C}) = \mathbf{B}(\mathbf{A}\cdot\mathbf{C}) - \mathbf{C}(\mathbf{A}\cdot\mathbf{B})$$

*"Do nothing" identity*

(V-4) $$(\mathbf{A}\cdot\nabla)\mathbf{r} = -(\mathbf{A}\cdot\nabla)\mathbf{r}' = \mathbf{A}$$

*Identities for the calculation of gradient*

(V-5) $$\nabla(\varphi U) = \varphi\nabla U + U\nabla\varphi$$

(V-6) $$\nabla(\mathbf{A}\cdot\mathbf{B}) = (\mathbf{A}\cdot\nabla)\mathbf{B} + \mathbf{A}\times(\nabla\times\mathbf{B}) + (\mathbf{B}\cdot\nabla)\mathbf{A} + \mathbf{B}\times(\nabla\times\mathbf{A})$$

(V-7) $$\nabla\varphi(U_1\cdots U_n) = \sum_{i=1}^{n}\frac{\partial\varphi}{\partial U_i}\nabla U_i$$

*Identities for the calculation of divergence*

(V-8) $$\nabla\cdot(\varphi\mathbf{A}) = \varphi\nabla\cdot\mathbf{A} + \mathbf{A}\cdot\nabla\varphi$$

(V-9) $$\nabla\cdot(\mathbf{A}\times\mathbf{B}) = \mathbf{B}\cdot\nabla\times\mathbf{A} - \mathbf{A}\cdot\nabla\times\mathbf{B}$$

(V-10) $$\nabla\cdot\mathbf{A}(U_1\cdots U_n) = \sum_{i=1}^{n}\nabla U_i\cdot\frac{\partial\mathbf{A}}{\partial U_i}$$

*Identities for the calculation of curl*

(V-11) $$\nabla \times (\varphi \mathbf{A}) = \varphi \nabla \times \mathbf{A} + \nabla \varphi \times \mathbf{A}$$

(V-12) $$\nabla \times (\mathbf{A} \times \mathbf{B}) = (\mathbf{B} \cdot \nabla)\mathbf{A} + \mathbf{A}(\nabla \cdot \mathbf{B}) - (\mathbf{A} \cdot \nabla)\mathbf{B} - \mathbf{B}(\nabla \cdot \mathbf{A})$$

(V-13) $$\nabla \times \mathbf{A}(U_1 \cdots U_n) = \sum_{i=1}^{n} \nabla U_i \times \frac{\partial \mathbf{A}}{\partial U_i}$$

*Repeated application of* $\nabla$

(V-14) $$\nabla \cdot (\nabla \times \mathbf{A}) = 0$$

(V-15) $$\nabla \times \nabla U = 0$$

(V-16) $$\nabla \times (\nabla \times \mathbf{A}) = \nabla(\nabla \cdot \mathbf{A}) - \nabla^2 \mathbf{A}$$

*Identities for the calculation of line and surface integrals*

(V-17) $$\oint \mathbf{A} \cdot d\mathbf{l} = \int \nabla \times \mathbf{A} \cdot d\mathbf{S} \qquad \text{(Stokes's theorem)}$$

(V-18) $$\oint U d\mathbf{l} = \int d\mathbf{S} \times \nabla U$$

*Identities for the calculation of surface and volume integrals*

(V-19) $$\oint \mathbf{A} \cdot d\mathbf{S} = \int \nabla \cdot \mathbf{A} dV \qquad \text{(Gauss's theorem)}$$

(V-20) $$\oint U d\mathbf{S} = \int \nabla U dV$$

(V-21) $$\oint \mathbf{A} \times d\mathbf{S} = - \int \nabla \times \mathbf{A} dV$$

(V-22) $$\oint (\mathbf{A} \cdot \mathbf{B}) d\mathbf{S} - \oint \mathbf{B}(\mathbf{A} \cdot d\mathbf{S}) - \oint \mathbf{A}(\mathbf{B} \cdot d\mathbf{S})$$
$$= \int [\mathbf{A} \times (\nabla \times \mathbf{B}) + \mathbf{B} \times (\nabla \times \mathbf{A}) - \mathbf{A}(\nabla \cdot \mathbf{B}) - \mathbf{B}(\nabla \cdot \mathbf{A})] dV$$

(V-23) $$\oint \mathbf{A}(\mathbf{B} \cdot d\mathbf{S}) = \int [(\nabla \cdot \mathbf{B})\mathbf{A} + (\mathbf{B} \cdot \nabla)\mathbf{A}] dV$$

*Helmholtz's (Poisson's) theorem*

(V-24) $$\mathbf{V} = -\frac{1}{4\pi}\int_{\text{All space}} \frac{\boldsymbol{\nabla}'(\boldsymbol{\nabla}'\cdot\mathbf{V}) - \boldsymbol{\nabla}'\times(\boldsymbol{\nabla}'\times\mathbf{V})}{r}\,dV'$$

*Operations with $\boldsymbol{\nabla}$ in Helmholtz's (Poisson's) integrals*

(V-25) $$\boldsymbol{\nabla}'\frac{(\mathrm{X})}{r} = \frac{\boldsymbol{\nabla}'(\mathrm{X})}{r} + \mathbf{r}_u\frac{(\mathrm{X})}{r^2}$$

(V-26) $$\boldsymbol{\nabla}\frac{(\mathrm{X})}{r} = -\mathbf{r}_u\frac{(\mathrm{X})}{r^2}$$

(V-27) $$\frac{\boldsymbol{\nabla}'(\mathrm{X})}{r} = \boldsymbol{\nabla}\frac{(\mathrm{X})}{r} + \boldsymbol{\nabla}'\frac{(\mathrm{X})}{r}$$

*Retarded (causal) integrals*

(V-28) $$\mathbf{V} = -\frac{1}{4\pi}\int_{\text{All space}} \frac{\left[\boldsymbol{\nabla}'(\boldsymbol{\nabla}'\cdot\mathbf{V}) - \boldsymbol{\nabla}'\times(\boldsymbol{\nabla}'\times\mathbf{V}) - \dfrac{1}{c^2}\dfrac{\partial^2\mathbf{V}}{\partial t^2}\right]}{r}\,dV'$$

(V-29) $$\mathbf{V} = -\frac{1}{4\pi}\int_{\text{All space}} \frac{\left[\boldsymbol{\nabla}'^2\mathbf{V} - \dfrac{1}{c^2}\dfrac{\partial^2\mathbf{V}}{\partial t^2}\right]}{r}\,dV'$$

(V-30) $$U = -\frac{1}{4\pi}\int_{\text{All space}} \frac{\left[\boldsymbol{\nabla}'\cdot(\boldsymbol{\nabla}'U) - \dfrac{1}{c^2}\dfrac{\partial^2 U}{\partial t^2}\right]}{r}\,dV'$$

*Operations with $\boldsymbol{\nabla}$ in retarded (causal) integrals*

(V-31) $$\boldsymbol{\nabla}'[\mathrm{X}] = [\boldsymbol{\nabla}'\mathrm{X}] + \frac{\mathbf{r}_u}{c}\frac{\partial[\mathrm{X}]}{\partial t}$$

(V-32) $$\boldsymbol{\nabla}[\mathrm{X}] = -\frac{\mathbf{r}_u}{c}\frac{\partial[\mathrm{X}]}{\partial t}$$

(V-33) $$[\boldsymbol{\nabla}'\mathrm{X}] = \boldsymbol{\nabla}[\mathrm{X}] + \boldsymbol{\nabla}'[\mathrm{X}]$$

(V-34) $$\frac{[\nabla' \mathrm{X}]}{r} = \nabla \frac{[\mathrm{X}]}{r} + \nabla' \frac{[\mathrm{X}]}{r}$$

(V-35) $$\nabla \frac{[\mathrm{X}]}{r} = -\frac{\mathbf{r}_u[\mathrm{X}]}{r^2} - \frac{\mathbf{r}_u}{rc}\left[\frac{\partial \mathrm{X}}{\partial t}\right]$$

(V-36) $$\left[\frac{\partial \mathrm{X}}{\partial t}\right] = \frac{[\partial \mathrm{X}]}{\partial t}$$

## DIMENSIONS OF GRAVITATIONAL AND COGRAVITATIONAL QUANTITIES

According to a convention, to indicate that only the dimensions of a quantity are being considered, the symbol designating the quantity is placed between square brackets.

[length] = [L]
[mass] = [M]
[time] = [T]

$[\mathrm{G}] = [\mathrm{L}^3\mathrm{M}^{-1}\mathrm{T}^{-2}]$
$[\mathrm{g}] = [\mathrm{LT}^{-2}]$
$[\mathrm{K}] = [\mathrm{T}^{-1}]$
$[\mathrm{I}] = [\mathrm{MT}^{-1}]$
$[\mathrm{J}] = [\mathrm{ML}^{-2}\mathrm{T}^{-1}]$
$[\mathrm{A}] = [\mathrm{LT}^{-1}]$
$[\varphi] = [\mathrm{L}^2\mathrm{T}^{-2}]$
$[\mathrm{A}_g] = [\mathrm{L}^2\mathrm{T}^{-2}]$
$[\varphi_c] = [\mathrm{LT}^{-1}]$
$[\mathrm{P}] = [\mathrm{MT}^{-3}]$
$[\mathrm{G}_f] = [\mathrm{LMT}^{-1}]$
$[\mathrm{d}] = [\mathrm{L}^3\mathrm{T}^{-1}\}$

# INDEX

# INDEX